SOCIAL GEOGRAPHY

An Introduction to Contemporary Issues

John Cater

Co-ordinating Head of Department
Edge Hill College of Higher Education
and

Trevor Jones

Senior Lecturer in Geography
Liverpool Polytechnic

Edward Arnold
A division of Hodder & Stoughton
LONDON NEW YORK MELBOURNE AUCKLAND

© 1989 John Cater and Trevor Jones

First published in Great Britain 1989

Distributed in the USA by Routledge, Chapman and Hall, Inc.
29 West 35th Street, New York, NY 10001

British Library Cataloguing in Publication Data

Cater, John
Social geography: an introduction to
contemporary issues
1. Great Britain. Human geographical features
I. Title   II. Jones, Trevor
304.2′0941

ISBN 0–7131–6474–3

Library of Congress Cataloging-in-Publication Data

Cater, John, 1953–
    Social geography: an introduction to contemporary issues/John
    Cater and Trevor Jones.
        p.   cm.
        Bibliography: p.
        Includes index.
        ISBN 0–7131–6474–3
        1. Sociology. 2. Anthropo-geography. I. Jones, Trevor, 1940–
        II. Title.
    HM51.C336   1989   89–66
    304.2—dc19   CIP

Typeset in 10/11 Garamond by Colset Private Limited, Singapore
Printed and bound in Great Britain for Edward Arnold, the educa-
tional, academic and medical publishing division of Hodder and
Stoughton Limited, 41 Bedford Square, London WC1B 3DQ by
Bookcraft, Avon

Contents

4 Gender

5 Racial and Ethnic Minorities

6 The Urban Neighbourhood

7 Contemporary Rural Society

List of Tables

List of Figures

Acknowledgements

The publisher would like to thank the following for permission to reproduce copyright material:

B T Batsford for figure 6.1 from *The Sociology of Planning: A Study of Social Activity on New Housing Estates* by Carey and Mapes; Cambridge University Press for figure 7.2 from 'An index of rurality for England and Wales' by P J Clake published in *Regional Studies* 11 (1977); Croom Helm for figure 7.1 from *An Atlas of Rural Protest in Britain 1448–1900* by A Charlesworth; Her Majesty's Stationery Office for figure 3.1 from Home Office 1984: *Criminal Statistics, England and Wales* reproduced with permission of the Controller of Her Majesty's Stationery Office; The Institute of British Geographers for figure 3.2 from 'The study of delinquency areas: a social geographical approach' by D T Herbert published in *Transaction* NS1 (1976) and for figure 5.1 from 'The distribution and diffusion of the coloured population in England and Wales 1961–1971' by P Jones published in *Transactions* NS3 (1978); Methuen & Co for figure 7.3 from *Accessibility: The Rural Challenge* by M J Moseley; The University of Chicago Press for figure 2.1 from *The City* by Park, Burgess and McKenzie.

Issues in Social Geography

Introduction

Social Geography's Identity Crisis

Social geography is one of the products of the transformation in geographical thought which occurred during the 1960s. Its emergence as a recognized field of enquiry was a logical consequence of the new items appearing on the geographer's agenda in that decade: the search for theories of spatial and social order, for links with the mainstream of social science and for 'real world' relevance. Charting the progress of the new discipline in the early 1960s, a contemporary observer noted a shift away from the human geographer's traditional preoccupation with humanity's interdependence with the physical environment towards a recognition of the relative autonomy of human culture and social institutions. Hence the rise of a new branch of the discipline, social geography, concerned with 'the theoretical location of social groups and social characteristics, often within an urban setting' (Pahl, 1965a, 82).

While Pahl's statement is an entirely faithful record of an intellectual watershed, it nevertheless leads to a definition of social geography at first sight so loose as to be virtually meaningless, one which fails to distinguish any body of subject matter exclusive to the discipline. If we accept literally that social geography is concerned with the location and spatial organization of all social phenomena, we portray it as a ragbag which replicates the entire field of human geography and plagiarizes large chunks of sociology, demography and history for good measure.

In fairness to Pahl, we must of course recognize the embryonic state of the discipline at the time of writing. As Jones and Eyles (1977, 1) remind us, any new discipline must undergo a formative stage 'before there is general agreement about substantive issues, about methodology and about the conceptual framework'. No clear identity could have been expected in the fluid experimental climate of the early / mid 1960s. And yet the subsequent passage of more than 20 years seems to have brought social geography no nearer to a mature identity. If the maturity of a discipline consists of establishing internal cohesion and external boundaries ׀ with other disciplines, then social geography is conspicuously retarded. Whether this lack of internal cohesion and external delineation really matters is, of course, a moot point. According to Knox (1987, 1), social geography 'encompasses an eclectic mixture of ideas, theories and empirical research . . . [and] several different approaches to knowledge and understanding'. As with any other branch of geography, there are vital inputs from 'sister' disciplines: 'the obvious academic connection is between geography and sociology

. . . but there are other necessary ties with, for example, planning, history, demography and economics' (Jones and Eyles, 1977, 5). Indeed it may be that the very mark of the maturity of a discipline is the confidence to dispense with boundaries for all but pragmatic purposes.

Thus, far from exhibiting anxiety symptoms, many social geographers seem unconcerned by the apparent boundlessness of their field. For Knox (1987), its sheer breadth of vision and its ability to incorporate numerous social sciences within a single interdisciplinary perspective are causes for rejoicing. That single perspective is, of course, the *spatial* perspective, the unifying element around which all branches of modern human geography are organized. Social geography's core theme might be said to be spatial pattern as an expression of and an influence on social process. This theme, whose origins are to be found far back in the work of Park (1926), would seem to provide a clear and valid identity for the discipline.

Not all practitioners would share this confidence. Jones (1980, 258) expresses concern for a discipline which is 'still trying to establish an identity. . . . Only most diffidently can we begin to express a certain agreement on topics and aims and even more hesitantly on methodology.' Despite these sentiments, Jones's chief source of anxiety is not so much the ill-defined topic areas or the overlap with other disciplines but the endlessly shifting, conflicting interpretations to which social geographers are prone. Since abandoning the traditional Man : Environment theme, the discipline has taken on board a succession of fresh philosophies and perspectives – Ecology, Behaviourism, Managerialism, Marxism/Structuralism, Humanism – each with its zealous disciples intent on destroying established 'truth' and imposing a new version of social and political reality. Notwithstanding the threat to scholarly continuity (which might be synonymous with complacency), our own view is that this ideological ferment is evidence of social geography's rude health. Indeed, one of the principal aims of the present book is to reflect the range of options open to the social geographer striving to make an honest appraisal of social injustices and inequalities.

Social Awareness and Public Issues

For the purposes of the present book, with its proclaimed devotion to 'issues', the vital strand in social geography is its potential relevance to contemporary questions of public concern and political debate. Part of the development of the discipline has been a dawning consciousness of this potential, as evinced by the growing output of problem-oriented works (e.g. Coates, Johnston, and Knox 1977; Smith, 1977; Herbert and Smith, 1979) by writers not content to simply record the existence of poverty, ghettos and slums as 'facts of life' but eager to acknowledge that deprivation carries with it social, political and even moral implications.

It is clear that these and many other geographers have narrowed their field of interest to specialize in the study of wealth allocation and *consumption*. This is somewhat obscured by variations in terminology and emphasis according to writer and context. For some the key word is 'welfare', for others it is replaced by 'resources', 'social well-being' or 'quality of life', but essentially all these terms indicate a preoccupation with consumption and especially with the mechanisms which ensure unequal access to the material (and, sometimes, psychic) rewards conferred by society as a whole. Above all, the concern is with the *victims* of the process, those who fail to qualify for what might be deemed a legitimate share of any given form of wealth. Evidently there is a sense in which social geography may be seen as the geography of *non*-consumption, given its interest in those excluded from resources and opportunities.

We would not venture to suggest that this study of how resources are allocated for consumption represents the whole of social geography. In passing, however, it might be suggested that such an orientation might give the discipline the unifying core of substantive matter which some of its practitioners believe it needs to establish a true identity. Irrespective of whether or not this approach is ever destined to find universal favour, it is nevertheless the guiding principle around which the present book is shaped.

In so defining our subject matter, we are among those who have responded to Smith's (1974, 289) plea that social geography should revolve around the inter-related set of questions, 'Who gets what, where and how?' We would join others in adding to this agenda the essential question 'why?', but we have still failed to stake out a special field of interest shared with no other discipline. What social science is *not* centrally interested in resource allocation? Economics, for example, is traditionally defined as the 'science' of precisely that. Even so, the social geographer occupies a vital strand within this seamless web by virtue of his stress on the spatial dimension, the 'where' component in Smith's sequence of questions. The role of the spatial (territorial, physical, locational) dimension is competently summarized by Knox (1987, 2):

> Geography's traditional concern with interrelationships between people and their physical and social environments provides a basis for study which . . . is broader than most. Moreover, space itself should not simply be regarded as a medium in which social, economic and political processes are expressed. It is of importance in its own right in contributing both to the pattern of urban development and to the nature of the relationships between different social groups within the city. While not necessarily the dominant factor in shaping patterns of social interaction, distance is undeniably important as a determinant of social networks. . . . [It] also emerges as a significant determinant of the quality of life in different parts of the city because of variations in physical accessibility to opportunities and amenities. . . . Because the benefits conferred by proximity to these amenities contribute so much to other people's welfare, locational issues also often form the focus of inter-class conflict within the city, thus giving the spatial perspective a key role in the analysis of urban politics.

As the early Chicago ecologists perceived, social phenomena exist, not as abstractions, but as concrete manifestations within a physical setting (Peach, 1975a). Social Geography addresses itself both to the spatial expression of institutions as witnessed by the incidence of inequality, poverty, racism and exploitation; and to the active (though by no means determinant) role of space in influencing these social structures (Harvey, 1973). This is the first principle around which the present book is organized.

There is also a second principle, that of dissent. At all stages we have attempted to reflect the cardinal political and ideological differences surrounding the issue of resource allocation. According to Newman (1973), social theorists can be split into two opposed camps divided by the relative emphasis placed on conflict or harmony. Harmony theories are those which portray human society as fundamentally tending towards convergence, order and consensus and as a mechanism for distributing benefits to all its members. Social disorder tends to be explained as a temporary malfunction, as is illustrated by the ecologist's view of the inner city as a dynamic zone-in-transition. The struggle for limited resources is conceptualized as 'competition': a contest presided over by neutral umpires such as the State, with all entrants agreed on the rules and reasonably satisfied with their prizes. It is not possible to locate the major spatial paradigms neatly within the harmony/conflict dichotomy but we might certainly regard the purer forms of behaviourism as subscribing to a predominantly harmonious view of the world.

By contrast, conflict theorists, including those labelling themselves radical geographers, would present resource allocation as a contest in which certain dominant interest groups make up their own rules and own the bat and ball! Springing to prominence within geography in the 1970s, the managerialist version of conflict theory highlights the 'gatekeeping' institutions, whose function it is to ration scarce resources, a process which almost inevitably leads to the disqualification of the more vulnerable social groups. The (neo-) Marxist variant of conflict theory poses much more fundamental questions, being more concerned with why resources should be scarce in the first instance (Gray, 1977), and presenting such social disorders as poverty and racism as preconditions for the necessary sources of cheap labour upon which the expansion of capitalist economies depends.

For two decades social geography has provided an arena within which these and other world-views have struggled for intellectual supremacy. In general this has been a rejuvenating rather than a destructive influence, a means by which the discipline has gained a broader perspective and a growing sophistication. It has also, given the deep-rooted positivist tradition in geography (Guelke, 1978; Hay, 1985), afforded an insurance against degenerating into pure empiricism. No realistic text on social geography can now be written which does not incorporate these shifts and counter-shifts in interpretation.

Organization of the Book

While any division of a wide range of social scientific subject matter into discrete chapters is in a sense artificial, it is nevertheless an essential convention without which the material would be unmanageable. The organization of this book follows from our conception of social geography as an enquiry into resource allocation and, true to the eclectic character of the discipline, we have adopted variable criteria for our chapter divisions. Chapters 1 and 2 discuss certain key resources, work and housing, and the means by which they are distributed. Particularly important here is the part played by spatial properties such as distance and location in governing access to social goods, although spatial factors are always considered in relation to the operation of aspatial processes. The third chapter, which considers crime and social disorder, lies on a different plane from those that precede it, being concerned with a 'negative resource'. It is, however, a logical progression since no discussion of criminality should be divorced from the questions of social exclusion and allocative failure considered in the first two chapters.

Chapters 4 and 5 view the struggle for resources from the angle of the participants themselves. In line with most socio-geographic analyses, this section focuses primarily on victims, the female population, racial minorities; those who are excluded from their due as members of an affluent society. In the final two chapters the areal dimension becomes the explicit focus around which material is organized, placing the allocation of resources and the distribution of recipients in a spatial context. Two case-study areas are used in this attempt to integrate society and space, the urban neighbourhood community and rural society. To what extent are social bonds formed within localized spaces? What is the relationship between such spaces and the external world? Where does the local community fit within the wider system of resource allocation?

Scope of the Book

Having outlined our substantive matter, we must define its geographical boundaries in the literal sense. This book does not set out to make sense of global society at all possible spatial scales. Its overwhelming concern is with the advanced capitalist realm and, in practice, most of our theory and empirical cases are drawn from the English-speaking world, with a bias towards the United Kingdom and, to a lesser extent, the United States. Within this broad realm, the basic unit of investigation is the city system. This is not to ignore the rural sphere, which is considered in full detail in Chapter 7 and is also treated in several other chapters, but simply to lend due weight to the dominant and determinant position of urbanism in mature capitalist societies. It is acknowledged that, within these societies, urban and rural are now often inseparable as analytical entities (Pahl, 1970; Hall 1973 et al; Saunders, 1981).

Finally let us note that, although this book is essentially a critical review of literature in the field, it is not intended to replicate existing standard texts. In its organization of material, its pluralist approach to interpretation and its method and style of exploration, it can lay considerable claim to individuality. It is to be hoped that this approach will prove helpful, comprehensible and, above all, stimulating.

1

A Social Geography of Work

1.1 Introduction

This book opens with a chapter on work for the very good reason that work is the key to who gets what where, the question which we have posed as definitive of social geography. If our central concern is with people as consumers, in inequalities in access to consumption resources by place, group and class, then logic dictates that the workplace be placed first in the sequence of the various mechanisms by which wealth is distributed. Quite simply consumption cannot occur without production. To be enjoyed, wealth must first be created.

This truism does not mean, however, that the geography of consumption is no more than the geography of production, that the places which produce most wealth are also those which enjoy the highest standards of well-being. There is, of course, a mass of empirical evidence suggesting a considerable degree of spatial covariance between production and productivity on the one hand and income and consumption on the other (see for example, the Central Statistical Office's Annual *Abstract of Statistics* for the United Kingdom or the Bureau of Census's *Statistical Abstract* for the United States). This is particularly marked at the regional scale (Central Statistics Office, 1988a). Yet, as *social* geographers we cannot conduct our analysis entirely from what Massey (1984a, 194) calls 'the vantage point of production'. We cannot rest easily with a one-dimensional equation which ignores the means by which the product is distributed, both between those who actually create it (employers, managers and workers of various grades) and those who play no *direct* role (children, pensioners and other non-active dependants). Furthermore, an individual worker's real income may not be solely determined by a wage or salary earned in the workplace. In advanced industrial society, government fiscal intervention plays a highly significant part in redistributing income and in the case of certain key resources such as educational and health provision state intervention may actually outweigh the role of the workplace in allocating life chances. An obvious inference to be drawn here is that the division of the spoils both inside and outside the sphere of production is as much a *political* process as a technical one. We cannot simply assume a harmony of interest between all parties to the production process. On the contrary who gets what where may be the outcome of a struggle between various opposed parties, an outcome decided by the relative power of the groups involved.

Whatever the importance of allocative systems outside the workplace, it remains true that the sphere of production must be the starting point for any examination of social and hence spatial inequality. Work is the principal passport to the enjoyment of consumption resources and the individual's place in the division of labour the critical determinant of his/her life chances. Note here that the division of labour, that ancient concept first codified by the classical economists of the last century, is currently reappearing in the geographical literature in the new guise of the *spatial division of labour* (Massey, 1979, 1984a). The basic proposition here is that the modern production process is not only minutely broken down into a host of narrow tasks each performed by a different section of the workforce: it is also spatially differentiated in that many of these specialized tasks are highly localized, over-represented here, under-represented there, absent elsewhere. In the very broadest sense this has always been recognized by regional and economic geographers. Indeed the existence of regional specialization might be said to have constituted the very *raison d'être* of these disciplines from their inception. But specialization has rarely been explicitly acknowledged as part of a division of labour and in the absence of that concept geography has lacked a theoretical tool which could link spatially uneven patterns of production with socio-spatial inequalities in consumption. The cardinal importance of the division of labour is simply that the differentiation of producers into separate tasks is the basic mechanism by which social inequality is created. In advanced industrial society, the workplace is a graded hierarchy of functions, each carrying its own level of reward, and as such it is the ultimate source of *class divisions* in society. Hence, when we examine the question of spatial inequality (whether it be 'rich' versus 'poor' regions, urban deprivation or whatever the spatial scale), what we are actually studying is the geography of class.

As we shall see later, not all forms of inequality can be reduced to class as such. Nevertheless, the intermeshing of space and class must offer an extremely promising avenue of exploration. As a means of theorizing geographical inequality it is to say the least a new departure, since it is only within the present decade that social scientists have made any serious progress towards this fusion of the social and the spatial. Confident statements such as 'social processes are constructed over space' (Gregory and Urry, 1985, 6), and 'the theory of class repeatedly bumps into the hard reality of space' (Walker, 1985, 165) are of distinctly recent vintage. Even so there is a long road to travel before we can begin to elaborate on the work of this new space and class school. It would be a grave error to dismiss all that has gone before in the fields of industrial geography and regional analysis. Each of these traditions can provide valuable empirical material and conceptual insight. Even though the geography of production is not in itself the geography of employment or of social well-being, it is nevertheless a necessary starting point, enabling 'a bridge to be built between "industrial" and "social" geography.' (Massey, 1984a, 195). We need initially to understand the processes which create a geography of production before we can appreciate the impact of the latter on regional class structure. Accordingly the next section will deal with evolving theories of location and regional development.

At this point we should note that the space–class approach can also help interpret what currently appears to be the greatest of all the many social inequalities generated by the production process – the yawning gap between those who have a job and the army of the workless. With total unemployment in the OECD nations standing in excess of 24 million at the time of writing (Department of Employment, 1988), the temptation to retitle this chapter 'Geography of the Workless' is almost irresistible. Quite clearly the role of some individuals in the division of labour is paradoxically to be without labour altogether, with certain occupational classes bearing a disproportionate burden of technological and organizational changes in the production process. It goes without saying that specific regions and localities likewise suffer disproportionately and, of all the various facets of the 'regional question', it is the politically charged issue of regional unemployment which has received the greatest share of attention from academics and policy-makers. Once again, as we shall see, the rubric of space and class opens up particularly promising pathways to explanation.

1.2 Uneven Development and the Geography of Production

Ever since the January 1933 nadir of the inter-war depression, when the British unemployment rate exceeded 23 per cent of the insured population and reached spectacular levels in local blackspots such as Jarrow (77 per cent) and Merthyr Tydfil (68 per cent) (McCrone, 1969), spatially uneven development has been recognized as a major issue of social policy. Far from being confined to Britain, the 'regional problem' has become a preoccupation for other advanced West European states, most of whom have adopted interventionist policies designed to cushion or reverse the tendency. Spatially biased growth is now widely regarded as a socially regressive process, concentrating the rewards of industrial capitalism in a few favoured locations and its penalties elsewhere.

Quite naturally geographers have played a prominent role in the proliferating literature concerned with diagnosing, explaining and prescribing for the regional problem. Quite naturally, too, their view has been that the qualities of space itself – or more precisely those of relative economic space – hold one of the keys to explaining uneven development. Much geographical analysis of the question has hung upon what Rich (1980) calls the 'locational hypothesis', which argues in effect that all regions in any given economic system are competing for a share of the total economic activity generated by that system; but that, by virtue of their location, some possess relative advantages for production and are, therefore, able to attract an over-large share of producers at the expense of other regions. As Massey (1979, 234) phrases it, 'At any point in time . . . . there is a given uneven geographical distribution of the conditions necessary for profitable, and competitive, production'. All this is implicitly predicated upon the original locational principles of Alfred Weber (1909), who argued that firms will locate at the point where the cost of assembling power, materials, labour and other inputs is at a minimum, hence the magnetic pull of coalfields for the great nineteenth-century heavy industries seeking to minimize the huge transport costs on their most expensive input (Wrigley, 1962).

Later writers have argued persuasively that, throughout modern history, the determinant 'location factor' has actually been *population* rather than natural resources. Particularly influential here is Clark (1967), who portrayed the great nineteenth-century expansion of European manufacturing industry as gravitating overwhelmingly towards pre-existing clusters of high population density. The logic of this is that access to potential customers and workers represents the most vital of all industrial resources. In order to measure this form of locational advantage, Clark *et al.* (1969) have applied the concept of *population potential*, an index of the centrality of a given region in relation to the population distribution of the entire economic system. More recently Keeble *et al.* (1982) have used the derived index *economic potential* (which substitutes

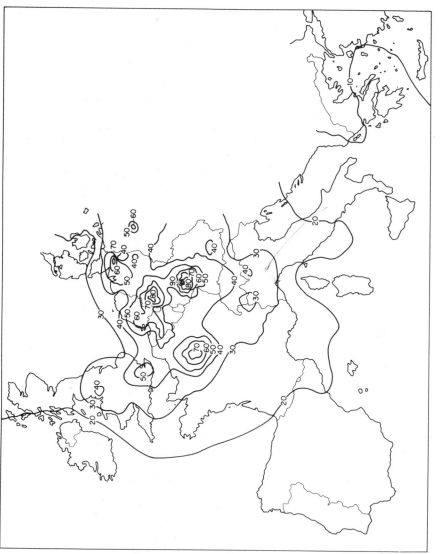

Figure 1.1 Regional economic potentials, European Community twelve, 1977.
Source: Keeble, Owens and Thompson 1982b, 429.
Note: Isopleth values are percentages of the maximum regional economic potential within the expanded European Community.

GDP for population: See footnote to Fig. I.1. for formula and clarification) to plot the centrality of regions within the EEC. From this they derive a threefold classification of EEC regions into central, intermediate and peripheral. In doing so, they show the space economy of the EEC as currently centred upon a gigantic supra-national core region (Seers, 1979; King, 1982) conforming (if only approximately) to the 'Golden Triangle' beloved of popular journalese.

Although the authors caution us against the inference that economic potential (centrality) can be equated with economic performance, they stress at the same time that proximity to customers and suppliers is a supreme asset for producers: 'For manufacturing and higher-order service industry, the most obvious advantage of centrality is *accessibility to markets* for products, whether these are intermediate components supplied to other manufacturers, final demand capital or consumer goods, or financial, business or other services (Keeble *et al.*, 1982, 61-2, authors' italics). *Ceteris paribus*, then, we might expect the large-scale firm seeking Europe-wide outlets and contacts to favour a location within the Euro core, where opportunities for commercial interaction are maximized.

One highly significant implication of this reasoning is that regional growth is cumulative and self-perpetuating. If large central agglomerations are attractive to firms and people, then by attracting those firms and people they become even larger and still more attractive. This self-expanding 'snowball effect' is one of the classic principles of regional development theory and has been noted under the various labels of 'agglomeration economies' (Weber, 1909), 'cumulative causation' (Myrdal, 1957) and 'regional multiplier' (Isard, 1956). No less significant is the fact that, almost inevitably, cumulative growth in some locations occurs at the expense of others. Agglomeration in some regions produces deglomeration in others. Myrdal's early contribution to this analysis (Gaile, 1980) was to identify the basic mechanism ('backwash effects' (Myrdal, 1957)) by which development in some regions brings about underdevelopment in others. In effect, core regions suck in resources from the periphery, thereby enhancing their own development and destroying that of the periphery (Table 1.1.). This fundamentally exploitative process has been analysed at the world scale under the label 'unequal exchange' (Amin, 1976).

Given that growth creates the conditions for more growth while decline breeds further decline, we would expect a continual widening of the regional gap, a built-in tendency towards divergence between core and periphery. And indeed historical data suggest that the EEC core has been increasing its share of the European population for a century or more. Yet since the 1960s there is evidence of a reversal. In Western Europe both the supra-national core itself and the various national industrial core regions have been performing less well on various economic indicators than many of the peripheral regions. In Britain, Fothergill and Gudgin's (1982) exercise demonstrates that ever since the 1950s the fastest employment growth (especially in manufacturing) has taken place away from the traditional industrial agglomerations; and that it is regions with a large rural component such as East Anglia, Wales and South West England which have performed best in this respect. These findings are parallelled by Bentham's (1985) study of earnings trends, which shows that some of the most rapid rises in wages have occurred in some of the most remote areas, such as the Grampian region of Scotland. Focusing on the EEC as a whole, Wabe (1986) highlights the extent to which the regions of the EEC core have suffered *deindustrialization* (i.e. loss of manufacturing jobs) since the 1973 oil crisis, with old-established agglomera-

Table 1.1. Core–periphery relations: Myrdal's 'backwash' effects

Process	Impact on core regions	Impact on peripheral regions
Inter-regional trade and competition	Lower cost production; capture of national markets (including the periphery); stimulus to output; rising employment and income	Higher cost production, undercut by core industry; loss of markets; deindustrialization; rising unemployment; falling real incomes
Selective migration	Labour deficit, high wage-rate regions; inflow of able-bodied young adult labour, often the more skilled and ambitious workers	Loss of skills and labour power; ageing population with or increased proportion of dependants
Capital flows	Influx of capital seeking a higher rate of return; boost to investible surplus	Lower returns on saving and investment; loss of investible surplus

Note: Myrdal, a non-geographer, lays great stress on *spatial interaction*, the exchange of resources (goods, money, people) across terrestrial space, and the way that a systematic pattern of regional advantage/disadvantage appears to be a built-in feature of competitive unregulated markets.

tions such as North West England, the West Midlands and the French Lorraine bearing a particularly heavy burden. Such job growth as has occurred has been predominantly located outside the Eurocore. Quite startlingly, the only large contiguous block of regions to generate rapid expansion since 1973 are those of the Italian Mezzogiorno, long regarded as the most problematic section of the EEC periphery and ranked at the bottom of Keeble *et al.*'s (1982) table of economic potential scores. Why has locational advantage apparently ceased to count?

Some light might be shed here by shifting the enquiry away from locational advantage towards *structural change*. As Rich (1980) points out, structural change has always vied with locational advantage as a means of explaining uneven development. In fact, ever since the Barlow Report (Royal Commission on the Distribution of Industrial Population, 1940), official British thinking on the rise and decline of regions has been heavily permeated by the structural change factor. According to Barlow, regional concentrations of job loss in pre-war Britain had occurred where regional economies were over-specialized in antiquated activities using outdated methods to create uncompetitive products. And indeed this is a very plausible account. It goes almost without saying that structural change has been one of the definitive features of the modern economy ever since the Industrial Revolution (in itself a most profound set of structural changes). In its constant search for profitability, through increasing efficiency and cost cutting, modern industry has generated an unceasing flow of new marketable commodities and of new means by which to make and distribute them. In parallel with this product innovation and method innovation, the industrial firm has also been transformed by various innovations in organization, ownership and control, most notably a remarkable increase in scale and a concentration of ownership in fewer and fewer hands (Prais, 1976).

Singly and in combination, these tendencies have exerted a profound effect on the spatial form of the economy. Perhaps the most profound of all in Britain was the reorganization of space accompanying the first phase of the industrial revolution, when the spatial shift from dispersed rural to concentrated urban settlement was the expression of a structural shift in the national economy from an agricultural to an industrial dominance. During this phase, regions like South-West England, East Anglia and the Scottish Highlands became economically marginalized by their inability to attract a share of the new economy and their consequent reliance on a traditional sector (agriculture) which lagged behind in terms of productivity, income-earning and job-creating capacities. The subsequent shifting patterns of industrial geography might also be explained by the death of old generations of technology and activity and the birth of new ones. A case in point would be the gradual but marked shift of the British economy from World War I onwards, away from the former industrial leaders in Northern England, South Wales and Central Scotland towards the Midlands and the South-East. On the continent, equivalent cases would be the more recent decline of established agglomerations such as the Ruhr, Wallonia and the Nord together with the rise of modern growth leaders in South Germany and Flanders. New waves of industrial development do not necessarily favour the same locations as former waves: 'the new technology embodied in a new investment may enable and require a changed location . . .' (Massey, 1979, 240). For all its persuasiveness the structural change thesis gives a somewhat limited interpretation. As Rich (1980) reminds us, it is of little value if, as is so often the case, it is treated as separate from and alternative to the locational thesis, since structure and space should be recognized as interdependent at the economic level. Structural change does not cause regional change independently of locational factors: rather it acts to change the relative patterns of locational advantage and disadvantage (Rich, 1980). For example, coalfields held many of the locational aces at a particular stage in the development of the British economy but subsequent changes in the structure of that economy – particularly the shift towards light consumer-orientated fabrication industries – rendered that advantage obsolete and substituted new 'location factors'.

Taken in isolation, the structural factor is a poor predictor of regional economic performance, particularly with reference to net employment change. In principle we should be able to explain growth or decline in any region by reference to the structure of its economy. Lagging regional economies composed mainly of senescent industries which are nationally in decline should suffer rates of decline roughly proportional to the national rate of contraction in those industrial sectors. In practice, however, a great deal of growth and decline cannot be accounted for by this procedure (Fothergill and Gudgin, 1982; Keeble et al. 1982), since a given branch of industry tends to perform differently in different regions. Fothergill and Gudgin's (1982) work is especially revealing here, since it finds that regional job gains or losses are related only very weakly to industrial composition but very strongly to *spatial* composition. Geography does indeed matter (Massey and Allen, 1984). More specifically these authors find that the more heavily urbanized a region, the greater the likelihood of its experiencing manufacturing job loss. 'The more a region is dominated by large urban agglomerations, the more it is likely to decline' (Fothergill and Gudgin, 1982, 69). As we show in Table 1.2. this relationship can be even more simply expressed by using the elementary measure of population density as an indicator of the presence or absence of large clusters of people and activity. Even at the

Table 1.2. English regions: employment performance and sparsity of population

| | Rankings | |
	Employment performance[1]	Sparsity of population[2]
East Anglia	1	1
South West	2	2
North	3	3
East Midlands	4	4
Yorks and Humberside	5	5
North West	6	8
West Midlands	7	6
South East	8	7

Notes: 1. According to Fothergill and Gudgin, 1982.
         2. Density of population rankings (Eurostat) arranged in reverse order.

notoriously coarse scale of British Standard Regions, the correspondence between crude density and economic performance rankings is a striking one. There are many possible explanations for this, the most obvious being that, beyond a certain size of industrial and demographic concentration, the agglomeration tendency is replaced by one of deglomeration, as economies of scale give way to diseconomies. To use Myrdal's (1957) terminology, backwash effects are replaced by spread effects and the evidence is that the modern space economy is now dominated by centrifugal rather than centripetal forces as of old. This is clear from Fothergill and Gudgin's interpretation: ' . . . manufacturing is in decline in the cities because a higher proportion of firms in cities are in ''constrained locations'', restricted by old-fashioned premises, hemmed in by existing urban development and with no room for expansion' (1982, 68). In other words, long-established agglomerations eventually become saturated. Their very success in the past has jeopardized their present and future prospects. Pressure on space not only creates problems for physical expansion but also raises the cost of land and of transport; and these costs fall upon the firm not only in a direct sense but also indirectly in pressures for rising wages from a labour force whose own real income is threatened by the increasing cost of housing and of travel to work.

   In summary then the changing economics of the free market has wrought fundamental spatial change, the most recent twist being a marked tendency towards industrial movement from core to periphery. But we should note also that these trends have been reinforced, shaped and in some cases amplified by Government intervention. The importance of Government policy is amply recognized by the modern regional literature, much of which is taken up with analysis of the aims, methods and results of state policy, in particular regional planning, which is explicitly directed at relocating industrial investment from core to peripheral regions. Although definitions of the problem and methods of attacking it vary considerably from country to country, it is fair to say that all Western European states operate in some form a system of regional subsidies and penalties aimed at inducing private capital to invest in designated problem regions, usually areas of high unemployment and/or low incomes. Such policies are generally justified both on social grounds (regional planning as a kind of geographical arm of the welfare state); and on economic grounds (under-utilized regional resources, especially of labour, are a brake on national economic performance).

While all this is fairly uncontroversial, there is ample room for debate when it comes to evaluating the impact of regional strategy. Here there is an acute problem of disentangling the effects of government policy from those of private sector decision-making. Many writers have argued that the kind of employment decentralization that has occurred during the post-war period would have inevitably taken place as employers sought less congested locations and an experienced and plentiful labour force; others have seen the state as a key-motivator of industrial movement.

In Britain this issue has been a controversial one; geographers and economists have frequently voiced strong criticisms of regional planning policy (Chisholm, 1974). By the 1970s, however, opinion had begun to soften, some authors moving towards the view that the core–periphery gap was now narrowing and that government incentives and controls had played a critical role in this. Keeble (1977, 5) writes confidently of ' . . . a substantial regional planning impact upon inter-regional economic disparities . . . '. Far from being confined to rural and semi-rural localities like East Anglia, South West England and the outer South-East, employment dispersal had also benefited remoter regions of previous industrial decline, regions specifically designated as Development Areas qualifying for government regional aid. It has been estimated that between 1960 and 1976 government financial assistance was responsible for creating some 540,000 new jobs in the Development Areas. (Moore *et al.*, 1977). The inter-regional impact of this can be partially gauged by the *relative* fall in Development Area unemployment rates. In Scotland relative unemployment fell from twice to one and a quarter times the UK average between 1965 and 1977, a trend broadly echoed by Wales and the North (Keeble, 1984, 41). A general move towards convergence is confirmed by numerous other studies using a range of indicators (Moore and Rhodes, 1973; Keeble 1976; 1977). Quite naturally, then, the 1970s was a decade of cautious optimism on the part of many influential opinion-formers, a period in which 'erstwhile sceptics' were obliged to admit that ' . . . the major single influence . . . [for regional convergence] . . . has been regional economic policy, working through shifts in manufacturing location and especially industrial movement' (Keeble 1977, 5). It was now possible to take a sanguine view even of areas like Merseyside, long regarded as a particularly intractable concentration of hard-core unemployment, economic decay and industrial blight. From its inception as a Development Area in 1949 until 1975, Merseyside received no less than 82,600 new manufacturing jobs created by firms who had been induced to relocate (Cornfoot, 1982). Among these new entrants the three giant car plants (Ford, Vauxhall, Standard Triumph) were seen as the heralds of a new economic dawn and one hard-headed commentator was moved to observed: ' . . . forces were set in motion which . . . should forge a new more successful economic base for the region . . . an industrial structure predisposed to decline is being converted into one predisposed to growth' (Lloyd, 1969, 408).

Unhappily, however, subsequent developments (or more properly 'under-developments') in the Development Areas have conspired to confound the Keeble–Lloyd scenario. On Merseyside itself, the continuing decline of the old staple maritime activities has now been joined by an unpredicted collapse of many of the industries of the 'new more successful economic base'. This decline

Table 1.3. Major job losses on Merseyside, 1975–9 and 1980–5

Name of firm	Number of redundancies	Year
a. 1975–9		
British Leyland	3,750	1976–8
Dunlop	2,600	1979
Plessey Telecommunications	2,400	1975–9
Lucas Victor	2,400	1978–9
Mersey Docks and Harbour Board	2,280	1975–9
Cammell Lairds (British Shipbuilders)	1,900	1978–9
Courtaulds	1,600	1976–8
Thorn EMI	1,600	1976
Spillers	1,227	1978
BICC	1,180	1977
GEC	1,000	1978
KME	740	1979
Western Ship Repairers	675	1978
Tate and Lyle	600	1976
Bird's Eye Foods	450	1978
Meccano	425	1978
Other notified losses of 100–399 jobs	2,795	—
Sub-total (1975–9)	27,422	
b. 1980–5		
United Biscuits	4,000	1982–3
Ford Motor Company	3,500	1980–5
Liverpool City Council	3,000	1981
Plessey Telecommunications	2,830	1980–4
Tate and Lyle	1,700	1981
Pilkingtons	1,700	1981
Mersey Docks and Harbour Board	1,520	1980–1
Lucas Aerospace	1,177	—
Associated Biscuits	1,164	—
BL	1,100	1981
Pressed Steel Fisher (BL)	1,100	1981
Courtaulds	900	1981
GEC	700	—
Barker and Dobson	647	—
Northgate Group	600	—
Cadbury Schweppes	500	—
CBS Engineering Co.	487	—
Metal Box	470	—
Lyons Maid	450	—
Cousins' Bakeries	450	—
S. Reece and Sons	420	—
Tillotsons	411	—
Other losses of fewer than 400 jobs*	25,348	—
Sub-total (1980–5)	54,174	
Total (1975–85)	81,596	

Note: *—includes losses of over 400 jobs not publicized (especially losses by contraction over a period of time)

Sources: Merseyside Socialist Research Group (1980); Townsend (1983); personal communication, Dr. Martyn Nightingale, MSRG (1986).

has been extensively chronicled by the Merseyside Socialist Research Group (1980) in a document whose bitterness captures precisely the experiences of the local workforce (see Table I.3 and also Gould and Hodgkiss, 1982).

Faced with similar events elsewhere on the periphery (Townsend, 1983), past optimists have been forced to concede that the ' . . . broad picture of economic convergence . . . should . . . be qualified . . . since about 1978 convergence appears to have been replaced by divergence . . . ' (Keeble, 1984, 42–3).

Yet this is not the only 'qualification'. Much earlier it was evident to some observers (notably Hudson, 1978) that, at a time of spectacularly rapid national deindustrialization, the *relative advance* achieved by the periphery was actually an *absolute loss* of jobs. Hence Merseyside in 1988, with a registered unemployment rate of 18.4 per cent (unthinkable a decade and a half earlier) is 'better off' in relation to the UK average than it was in the 1950s, when its unemployment rate never exceeded 4.2 per cent but rarely fell much below twice the national average! Similarly the great relative employment 'gains' made by Scotland conceal a loss of over one in ten manufacturing jobs between 1965 and 1975 (Hudson, 1978).

The deindustrialization of the British (indeed virtually the entire EEC) economy since 1970 has been analysed *ad nauseam*, although with several pertinent commentaries by geographers on its regional consequences (Massey, 1979; Massey and Meegan, 1982, Fothergill and Gudgin, 1982). In coarse outline what has happened is that, at a time of global recession, British industry has suffered from a shrinking share of a stagnant world market. Its response to the falling rate of profit has been to speed up a number of technological and organizational innovations, which have had the effect of reducing costs and also the number of workers employed. The process is described by Massey (1979, 236–7) as follows:

> Certain characteristics and requirements of production . . . , in combination with particular spatial conditions, form the basis of the development of a new division of labour. Such characteristics and requirements include the increasing size of individual firms, and of individual plants, the separation and hierarchization of technical, control and management functions, and the division, even within production, into separately functioning stages. . . The growing intensity of competition in recent years has led to increased pressure to cut labour costs and increase productivity, and this in turn has produced an apparent acceleration of the processes of standardization of the commodities produced (thus reducing both the number of workers for any given level of output, and the level of skills required of them), (and) of automation.

In all fairness none of this constitutes an outright dismissal of the Keeble interpretation. In a period of overall national employment decline, the economic performance of Development Areas can hardly be judged a failure on the basis of their inability to expand job opportunities: this would be to condemn the weakest for failing to swim against a tide powerful enough to sweep away even the strongest. Given the pace of national job loss, the only fair yardstick is whether regional decline is faster or slower than the UK average. Even when we find that Scotland and the North are now declining faster than the UK (Keeble, 1984), this could logically be laid at the door of the 1979 Conservative Administration, one of whose first acts was to dismantle much of the existing regional aid machinery and to drastically reduce the territorial scope of regional incentives (Townsend, 1980; Fig. 1.2). This is to argue that regional policy *is* an

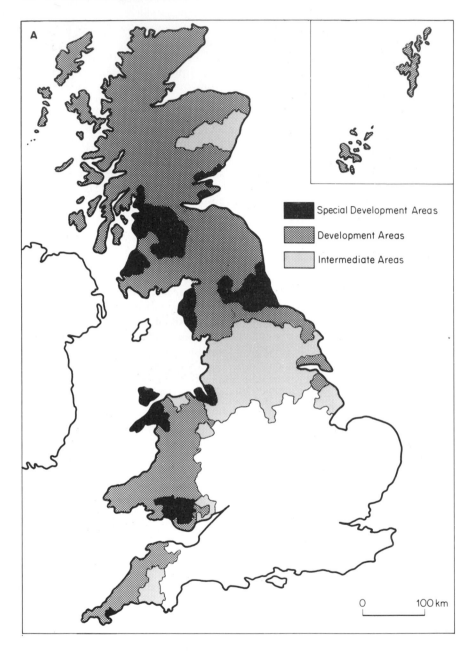

Figure 1.2A Areas in receipt of Regional Aid, 1979.
B Areas in receipt of Regional Aid 1984.

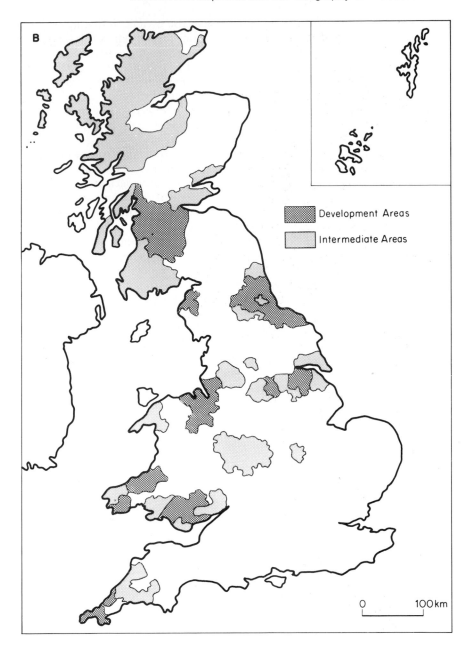

B

Development Areas
Intermediate Areas

0   100 km

effective medicine and that the recent relapse of the patients is due to a drastic reduction in dosage at the very time when most needed. Seen in this light, the regional achievements of the 1966–79 governments assume a more meritorious aspect: the remotest peripheral regions suffered less than elsewhere from the overall negative employment trend, a benefit due in no small measure to positive discrimination in their favour on the part of the state. Projecting this reasoning into the future, we can expect regional resurgence to occur only in the event of renewed government commitment to regional employment creation – hopefully in the context of a global economic recovery.

There is, however, one basic flaw in the above reasoning, one which had already been exposed by critics of regional policy in the 1970s. This is its bias towards the sheer *quantity* of employment and its relative lack of stress on the problem of job *quality*. Essentially this echoes one of the dominant thought-strands within the planning profession itself, particularly during the formative years of regional planning. While it would be unjust to overlook the early planners' concern for job enrichment, diversity of opportunities and skill training in the Development Areas, a perusal of the first generation of 'Regional Plans' reveals an overwhelming preoccupation with employment creation *per se*.

Reflecting their political masters' determination to avoid a repetition of the horrors of 1930s-style mass unemployment, planners were concerned to create work at all costs. In effect work was treated as a virtue in itself, a rather uncritical acceptance of a central Victorian value and not a view which would be shared by many workers in low paid and/or repetitive arduous menial tasks. It is an obvious (but often forgotten) feature of industrial capitalism that many of its essential work tasks are anything but virtuous for those who must perform them in unpleasant conditions for low earnings. As Brown (1978) reminds us, insofar as paid employment involves subordination to authority and the expenditure of time and energy for someone else's profit, it contains 'an inherent coercive element' (Brown, 1978, 85). For many employees, work is the ultimate unfreedom: there is only one thing worse than work and this is its arbitrary removal (Seabrook, 1982).

Having established this, however, we must note the equally salient fact that subordination is a matter of degree. In the advanced capitalist economy, a significant and growing number of employees are autonomous or semi-autonomous in respect of the work process, while others enjoy a share of authority itself. In other cases a high degree of subordination may be compensated by high earnings. In short employment is a highly complex rather than a monolithic condition. Workers are differentiated along numerous dimensions, including material rewards, skill, creativity, job security, autonomy, responsibility and authority. It is this division of labour which is now held by many geographers to be crucial in the interpretation of the regional question.

1.3 The Spatial Division of Labour

How does the spatial division of labour in the workplace act to create spatial inequality? After all, there is no reason in abstract principle why every region and locality should not experience an identical pattern of work differentiation. Stripped to its bare essentials, the thesis of the spatial division of labour argues that, on the contrary, the specialized work functions of the modern economy are unevenly distributed over space, with an especially heavy and growing pre-

Table 1.4 Occupational grouping of head of household, United Kingdom regions, 1985

Region	Managerial & professional	Other non-manual	All non-manual	Skilled manual	Other manual	Self-employed	Unemployed or econ. inactive
North	11.4	4.8	16.2	11.9	13.2	7.3	49.5
Yorks and Humber	14.5	5.4	19.9	12.3	14.1	8.3	45.0
East Midlands	17.1	5.3	22.4	13.4	13.2	8.2	42.5
East Anglia	16.7	8.0	24.7	9.4	14.7	11.1	39.9
South East	22.6	7.7	30.3	8.7	11.9	10.3	38.6
South West	16.9	7.5	24.4	9.2	12.6	11.4	42.1
West Midlands	15.5	5.4	20.9	12.6	14.0	7.5	44.6
North West	15.0	5.9	20.9	10.7	13.6	7.8	46.4
England	18.0	6.5	24.5	10.2	13.2	9.2	42.4
Wales	12.6	4.9	17.5	9.8	13.1	9.0	50.2
Scotland	14.2	6.3	20.5	11.1	15.5	6.5	46.1
Northern Ireland	12.7	7.1	19.8	9.0	13.4	9.8	47.6
United Kingdom	17.3	6.5	23.8	10.3	13.4	9.0	43.2

Source: based on *Regional Trends* 22, Table 8.8, p. 104. London: Central Statistical Office 1987.

Note the concentration of workers in the managerial and professional classes in the South East 'core', where 36.5% of the economically active heads of household fit into this category; this compares with 22.6% in the Northern region and 24.2% in Northern Ireland. Areas such as the South West, South East and East Anglia also contain a large proportion of self-employed workers. Those unemployed or otherwise economically inactive are over-represented in the 'periphery', notably Wales, Northern Ireland and Scotland, the North and North West.

ponderance of high-level activities – control, management, research, innovation, high-skill production, high-order services – in a small proportion of the
national space, in core regions, usually the centres of finance and / or government
administration. Outside these 'command centres' workers in other areas
(peripheral regions) are increasingly required to perform routine production
tasks, which are not only less well rewarded in material terms but are also subject
to the authority of core regions. Hence questions of power, autonomy,
dominance, subordination and dependence now become definitive of the
core–periphery relationship, which can no longer be seen simply as a dualism
between 'rich' and 'poor' regions. Relative affluence is of course still of the
highest importance but is now seen as a symptom of an economic power relationship between core and periphery.

Some evidence relating to regional occupational specialization in the United
Kingdom is presented in Table I.4. Metaphorically speaking, the core regions are
the brains and nerve-centres of the modern space economy, while the periphery
performs the role of limbs and muscles. The historical processes through which
this system has evolved are reviewed in later sections. Before proceeding to this,
however, we must recognize that even in its aspatial guise the division of labour
concept presents many intellectual problems, not least its value-laden nature.
This is no politically neutral construct and its very use by geographers denotes a
move away from the anodyne apolitical approach which has traditionally
dominated the analysis of the geography of work.

1.3.1 Social Theory and the Division of Labour

By and large the dominant perception of the division of labour – i.e. the
fragmentation of the work process both within and between individual workplaces and between industries; and the consequent allocation of each worker to a
narrow and specialized range of tasks – is one of a socially and economically
beneficial institution. Historically an ever more complex division of labour has
been one of the technical keys to increased productivity and an enhancement of
the wealth available to the population as a whole. An obvious objection to this is
that not all participants in the productive system benefit equally, since the
division of labour is also a major generator of inequality: between those who own
and control it and those who do not; between workers in various branches of
industry; and between various occupations within the same plant and branch
(Brown, 1978). In short, the division of labour is the basic source of class divisions
(though this point needs qualification: see Rattansi, 1982; MacKenzie, 1982).

Until fairly recently this objection lay muted. Theoretically, inequalities in
pay, status, work satisfaction and so on could be rationalized by reference to
Durkheim's (1947) notions of functional *inter*dependence, the argument that
all workers of whatever grade are mutually essential to one another, that the
whole benefits from the parts and *vice versa* (Salaman, 1981). Stratification is
'. . . inescapable, a functional necessity for any complex industrial economy', as
Westergaard and Resler (1975, 14) ironically observe. In any case, during the
post-war boom years there was a virtual consensus that income and other class differentials were narrowing and that a progressive expansionary economy was
opening up all manner of opportunities for social mobility – workers were no
longer *trapped* in low-paid unsatisfying work.

Certainly over the longer term there has been a far-reaching transformation in the British division of labour and most of the changes which have occurred since the last century appear entirely benevolent. Brown (1978) notes the following trends:

- A shift from low-productivity industrial sectors to higher-value-producing activities: e.g. agriculture's share of total employment fell from 22 per cent in 1851 to less than 2 per cent today
- rising occupational status: '. . . an enormous absolute and relative increase in the number of non-manual workers' (Brown, 1978, 69), with professional employees alone expanding from 141,000 (1861) to 3.6 million (1984) in the tertiary (services) sector, denoting a shift from the factory to the ostensibly much more pleasant and rewarding conditions of the office.

Not only does the labour market itself appear far less oppressive and exploitative but its worst remaining effects have been progressively mitigated by the rise of the 'welfare state', the public provision of 'free' benefits for the worst casualties of low pay and unemployment. Note here that this kind of provision has an unintended but nevertheless vital geographical component, in that it automatically discriminates in favour of those localities where the casualties of the labour market are concentrated, it is a means of core–periphery redistribution and a principal means of narrowing regional income disparities.

By the end of the 1960s, however, this bland model was being actively demolished by researchers engaged in 'rediscovering' poverty and in producing a mounting volume of evidence on the chronic persistence of permanent deprivation, often spatially concentrated (Townsend, 1979 Townsend *et al.*, 1988). One of the most telling summaries of this issue is contained in Westergaard and Resler (1975), who completely dismember the remaining myths of mobility, income convergence, the effectiveness of welfarism and redistribution and the alleged death of class. Such empirical findings demanded theoretical reformulation, especially when coupled with evidence of rising dissatisfaction on the part of the victims of inequality:

> . . . domestic tensions within Western societies came into focus once more in the 1960s . . . a recurrence of sporadic 'direct action' – such as 'squatting' by homeless families, rent strikes, worker occupation of factories – outside the limits of conventional . . . opposition (Westergaard and Resler, 1975, 18).

Evidently the consensus was at an end and theories of social conflict were once more on the agenda. Not surprisingly this period saw a reawakening of interest in Marxian social theories, not least Marx on the division of labour (Giddens and MacKenzie, 1982). Among recent contributors to a Marxist interpretation is Braverman (1974) who, in a work subtitled 'The Degradation of Work', argues that the capitalist labour process is now far more narrowly subdivided than formerly; that this fragmentation continues and is likely to do so for the foreseeable future; and that one of its consequences is the gradual *deskilling* of labour, with the work process substituting automation and task fragmentation for human aptitudes. Both Braverman (1974) and Marglin (1976) are concerned with the worker's increasing loss of control over work, itself a major consequence of deskilling. 'The issue is one of *class* rather than "technical" or "neutral" efficiency . . . divided labour is conquered labour' (MacKenzie, 1982, 78 and 77). One of the major contributions of recent radical industrial geography has been to show, firstly, how the brunt of this 'work degradation' is borne by specific regions and localities and, second, how space is now used by modern capital as an extra dimension in the logic of 'divide and conquer'. Before unravelling these spatial themes, however, we should remind ourselves that Marxist analysis of the social relations of production starts from the more fundamental proposition that it is the ownership or non-ownership of capital which determines class membership in capitalist society. Production takes place through a combination of *capital* (its personnel being labelled the bourgeoisie), the minority who own the means of production, and *labour* (the proletariat), the propertyless majority owning nothing but its own labour power. By definition the relationship between these two elements is quintessentially exploitative because capital, by virtue of its monopoly over productive property, is able to expropriate the *surplus value* of labour – surplus value being the whole of the worker's product other than the *means of subsistence*, the bare minimum consumption needs of the worker and his/her dependants. Apart from this means of subsistence, without which the *reproduction of labour power* could not occur, all value is expropriated by the property-owning class in the form of capital accumulation. Reduced like this to its simplest abstract fundamentals, Marx's class model can be no more than ' . . . a guideline for thought which cannot bear the weight of much historical specificity' (Walker, 1985, 169). Even if the original formulation captured the essence of class domination at the time of writing, subsequent developments – the rise of the corporation and the separation of ownership and control, the rise of a managerial 'class' and of bureaucratic hierarchies in the workplace, the expansion of state intervention, both in the ownership of production and in the subsidization of workers' real wages (the welfare state) – have combined to blur, distort and complicate its naked lines. Among the many apparent gaps between theory and reality is the proliferation of 'intermediate classes' or 'middle strata', who appear to belong wholly to neither capital nor labour. Marx's own attitude to these groupings was that they were transitional: 'In the longer term . . . intermediate groupings will become absorbed into one or the other of the two great classes' (MacKenzie, 1982, 64). Yet in reality they seem to have multiplied and it is this stubborn persistence of social layers such as the 'petty bourgeoisie', 'the aristocracy of labour' and the 'white-collar intelligentsia' which has occupied the minds of subsequent social theorists Marxist, non-Marxist and anti-Marxist alike. Table 1.5 presents a summarized version of various attempts to refute or revise the original bi-polar model. Apart from their tendency to write the great mass of the population out of the working class (see Poulantzas, 1975; MacKenzie, 1982 and Walker 1985 for critical comments), these alternative approaches suffer from a static preoccupation with class boundaries, an obsession – entirely foreign to the spirit of Marx himself – with rigid 'class boxes'. The point is that capital and labour were never presented as historically frozen categories, locked into some unchanging equilibrium of dominance–dependence uninfluenced by human agency. On the contrary, the relationship is a dynamic one, with the dominant class continually subject to challenge and resistance and therefore continually obliged to take positive action to reassert itself. As Walker writes, 'Class relations must be repeatedly formed and reformed under changing circumstances . . . ' (1985, 168). And, centrally for our argument, the changing organization of space has been a key element in capital's struggle to maintain both its rate of accumulation and its political domination of labour. For this reason the Marxist concept of class struggle underpins (either overtly or implicitly) much of the work to be reviewed in the following sections.

Table 1.5 Class positions and boundaries

a. Class positions according to Weber
  i.   Dominant propertied and entrepreneurial class
  ii. Propertyless white collar intelligensia
  iii. Petty bourgeoisie
  iv. Manual working class: a. skilled
                             b. semi-skilled
                             c. unskilled

b. Class positions according to E.O. Wright[a]

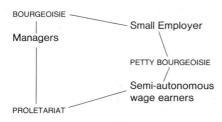

c. Class positions according to Poulantzas

BOURGEOISIE
Owners and Managers

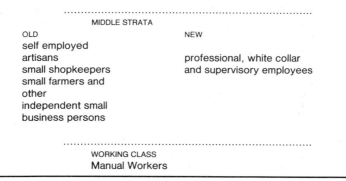

....
MIDDLE STRATA

OLD   NEW
self employed
artisans   professional, white collar
small shopkeepers   and supervisory employees
small farmers and
other
independent small
business persons

....
WORKING CLASS
Manual Workers

Source: MacKenzie (1982)
Notes: [a] – classes in upper case letters, contradictory class positions in lower case letters

1.3.2 Dominant and Dependent Regions

As originally formulated, dependency theory was addressed to core–periphery relationships at the global ('North–South') level but more recently various writers have applied its logic to regional and international disparities within the West European space-economy (Seers, 1979; Seers *et al.*, 1979; King, 1982). Like Keeble and colleagues, these authors identify a major supranational core (the 'megacore' – King, 1982) at the heart of the subcontinent but in this case the definition of 'coreness' is *economic power*, rather than geographical centrality, economic potential or wealth *per se*. In essence the relationship between core and periphery is presented as one of *unequal exchange* (Amin, 1976), in which the net benefits of trade, migration, capital flows and other transactions accrue to the core at the expense of the periphery.

Superficially this approach might be dismissed as no more than a modernized version of regional divergence theory *à la* Myrdal, merely one of a number of 'classless theories of exploitation between core and periphery' (Walker, 1985, 165): apolitical and lacking awareness of the true nature of spatial conflict. Yet even though no theory of class relations is explicitly spelled out, there can be no doubt that the entire thrust of dependency logic rests on an assumption of class domination across space. As in the global context, so in the intra-European context the essence of regional interaction is the exploitation of peripheral *labour* by core *capital*, notably the giant multinational corporations whose headquarters are overwhelmingly centred in the EEC core (and of course in various extra-European portions of the world core such as the USA and Japan). Hence, a dominant theme in the volume edited by Seers *et al.* (1979) is the way in which core capital generates abnormally high returns ('superprofits' to use Lenin's (1964) terminology) at the expense of peripheral labour, either through its use of migrant labour (see also Castles and Kosack, 1973; Castles, 1984) or increasingly through the establishment of branch activities in low-wage nations and regions. In this sense Gunder Frank's early diagnosis of class expropriation across space –

> . . . the metropolis expropriates economic surplus from its satellites and appro-priates it for its own development. The satellites remain under-developed for lack of access to their own surplus . . . This . . . . relationship runs through the entire world capitalist system in chainlike fashion from its uppermost world metropolitan centre through each of the various national, regional, local and enterprise centres. (Frank, 1969, 9, 10)

is broadly applicable at the intra-continental (and indeed the intra-national) scale.

Thus for all its failure to confront class conflict as an explicit theme (and this in the final analysis proves to be a crippling weakness), the dependency perspective as applied by Seers and followers has nevertheless provided radically new insights into the geography of work. In particular the rapid growth of new satellite indus-trial offshoots and of tourism in peripheral regions as far apart as Ireland, Iberia and Southern Italy now appears less an unalloyed blessing for peripheral labour and more a means of boosting capital accumulation at the expense of that labour. In Europe the growth of the peripheral branch plant economy has inspired a large literature (see for example Firn, 1975; Townroe, 1975; Dicken, 1976; 1986; Watts, 1981), which notes the following disadvantages for peripheral workers and their localities.

- increasing subordination to external ownership and control
- repatriation of profits, dividends and other returns to capital
- earnings less than equivalent workers in the core
- subjection to factory work discipline
- squeezing out of local entrepreneurs
- confinement within low skill/low reward activities
- lack of linkages and of the multiplier effect
- despoliation of landscape and environment

In this context King (1982) argues that the peripheral regions of Europe should be regarded as actually *under-developed*, distinct from the 'Third World' in degree rather than kind.

It may seem strange to liken southern Europe with the Third World . . . but the fact is that countries like Portugal, Spain and Turkey do have much in common, in terms of the structure of their economies and societies and their external relationships, with some of the more prosperous countries of the less developed world such as Mexico, Venezuela and Chile (King, 1982, 221).

It would be by no means far-fetched to extend this reasoning to many regions and portions of nations within the so-called megacore. Certainly the post-1960 period in Britain has witnessed the creation of increasingly dependent branch-plant economies in localities such as the North-East, Central Scotland and Merseyside, which were once characterized by substantial local ownership of industrial/commercial capital. A process which was once hailed as 'redevelopment' can now be seen as underdevelopment through satellization (MSRG, 1980). Such regions have been demoted from their historic status as cores whose prosperity (itself selectively distributed) derived from the expropriation of the economic surplus of other places, in particular the colonies of the old British Empire.

1.3.3 Space and Class

For the social geographer, the real interest of the foregoing lies in its (usually implicit) conception of space as a crucial factor in class conflict, in essence the use of space by capital in its constant struggle to maximize its expropriation of labour. Seen from this angle, spatially uneven development no longer appears as an unfortunate by-product of capitalist production to be regretted by high-minded citizens of all classes and persuasions: it now resembles an institution which is *directly functional* for capital accumulation and directly inimical to the interests of workers. For Massey (1979, 242) ' . . . spatial inequality may be positively useful for unplanned private production for profit'. Though radical geographers have been mostly concerned with the 'hypermobility' of capital and its most recent spatial strategies, we should note that throughout the entire period from the first Industrial Revolution onwards the ruling class has continually been obliged to wrestle with the problems posed by space. Capital has constantly altered its patterns of spatial organization in order to maintain its extraction of surplus value and its political dominion over its workforce. In short, space has always acted as a critical element in the continued reproduction of capitalist class relations.

This theme occupies a central place in the writings of Harvey (especially 1975, 1981, 1982), who points out that over a century ago Marx and Engels themselves were well aware of the problem of territorial organization in capitalist class relations. Indeed they saw the new geographical configurations of the nineteenth century – industrial production highly concentrated in space, together with the parallel concentration of the workforce in large urban clusters – as one of the basic contradictions of the capitalist system. On the one hand concentration provided economies of scale necessary to maximum capital accumulation. On the other it posed a political threat in that it 'proletarianized' (Carchedi, 1977) a formerly docile and deferential labour force. In pre-capitalist rural society, despite feudal oppression and exploitation, social order was maintained through the dispersal of the population in small isolated parochial communities. Under these conditions rural workers were not conscious of themselves as a class with common grievances (Bell and Newby, 1971). But now that industrialization had herded vast numbers on to factory floors and urbanization had herded them into a common living space, such a consciousness would blossom and would lead to resistance, political organization and the ultimate revolutionary overthrow of capital. Spatial proximity is thus a necessary (though not sufficient) precondition for socialist revolution.

Why has this revolution been a non-event in most advanced industrial societies? Quite clearly this cannot be explained entirely in geographical terms. Historically, the continued reproduction of capitalist domination has not depended wholly upon spatial strategies or in many instances on conscious ruling-class strategies of any kind. The counter-revolutionary process begins in the largely aspatial world of the workplace itself, where various authors (notably Lockwood, 1982; Salaman, 1982 and Morgan and Sayer, 1985) have noted that widespread and continuing acceptance of authority is the 'normal' state of affairs: and that not withstanding intermittent strikes and other forms of breakdown, this acquiescence stems from *pragmatism* (work is unavoidable) and *fatalism* (I have no control over my own destiny). Coupled with conscious strategies of control (Salaman, 1982; Benwell CDP, 1978), this accommodation by workers serves to maintain a certain equilibrium, though not one which can be taken for granted. Beyond this, however, in the world which forms the context of the workplace, further counter-revolutionary processes are at work, with capital utilizing the spatial properties of location, distance, proximity and mobility to defeat working-class aspirations. How precisely does it achieve this? From the earliest days of industrialization, capital has been ' . . . forced to seek external relief . . . [to] . . . resolve its internal problems through spatial expansion' (Harvey, 1982, 414). These 'internal problems' are precisely those introduced above: the contradictions posed by the necessary agglomeration of production and labour force in confined space, leading to rapid unplanned urbanization, poor housing, high disease mortality and all the unspeakable symptoms of urban crisis denounced by the nineteenth-century social commentators (Engels, 1845; Booth, 1902). 'The accumulation of capital and misery go hand in hand, concentrated in space' (Harvey, 1982, 418), and the 'creation of a penurious rabble' (Harvey, 1982, 414) spelt potential danger for the capitalist order. Yet the existence of undeveloped external space (both domestic and colonial) acted as a safety valve, enabling capital to counter any threat to its political domination or its rate of accumulation. The emerging power of shop-floor organization and the consequent pressure for wage rises was answered by importing migrant labour from this external rural space, 'preproletarian' labour, politically unorganized and lacking class consciousness. In cases where such workers had migrated across ethnic and religious divides, their cultural estrangement from indigenous labour was such as to drive a wedge through class unity and to impede proletarian organization, whether for reform or revolution. In England the role of Irish migration in nineteenth-century class struggle was a highly salient one (Miles, 1982). Such migrants (like countless others before and since) constituted a 'reserve army' of surplus labour which, when tapped during boom periods of high labour demand, enabled wages to be pressed back down to subsistence level.

At this point we should emphasize that, while this account has recognized *space* as a contributory factor in class struggle, it has so far ignored the impact of *place*: i.e. variations in class configuration from one locality to another. It is a major contention of the space and class school that class theory, as propounded by sociologists, depicts the processes of class formation as occurring in a spaceless world, or at least one in which there is no bounded social space smaller than the nation-state (Massey, 1984a; Walker, 1985). Yet the realization is now dawning that the model of a 'national' working class is to say the least misleading. In reality, the nature of the working class – its consciousness of itself, its industrial and political responses to capital, the outcomes of these actions – is extremely variable from one locality to another, sometimes critically so. We should regard 'class formation as a geographic process' embodying 'social structures differentiated over space' (Gregory and Urry, 1985, 5 and 6). Class practices are a product of a mass of geographically varying conditions and processes, among them the degree of isolation from other regions, the presence or absence of pre-capitalist cultural institutions, demographic composition of the work-force and its religious, ethnic and linguistic affiliations – to name but a random few. (See Cooke 1985 for a systematic summary of the 'multiple influences upon socio-spatial diversity' (1985, 222)).

On this issue, Massey's (1984a) case study of capitalist development and proletarian response in South Wales provides a number of valuable insights into the impact of the geography of production on local work patterns and local class relations. Clearly nineteenth-century South Wales was in many ways socially representative of coal mining areas elsewhere in Britain, created by the regional division of labour typical of the period. Indeed we might regard it as an extreme version of its type, an unusually fertile breeding ground for the labour movement – 'virtually synonymous with strong labourist culture' (Morgan and Sayer, 1985, 383) – where at times during the last century the power of organized labour was sufficient to wrest abnormally high returns from capital. Williams (1985, 185–6) writes of 'some of the most remarkable working class communities in Britain. . . Wages, though they fluctuated wildly, were often six times those of farm workers and the skilled men were among the worker aristocrats, most of them lived better than small farmers.' (See also Williams, 1980.)

This industrial muscle was closely related to the unusual solidarity, cohesiveness and self-consciousness of labour which, according to Massey (1984a), were produced by a combination of factors including:

- the narrow range of activities in an economy overwhelmingly skewed towards mining and iron-steel production. Worker resistance is greater in coherent communities dominated by one factory or one industry (Massey, 1984a), and in South Wales mining and metals together accounted for three-quarters of the insured population as late as 1930 (Morgan and Sayer, 1985, 385). The experience of heavy, dangerous, dirty work and the perception of an oppressive boss-class was shared by the vast majority of local workers (Cooke, 1985, 223).
- the relative absence of hierarchical control in the workplace, where the pace of production was set not so much by management or supervisors as by key workers such as puddlers, whose monopoly of vital skills gave them unusual bargaining strength and autonomy
- the relative absence of intermediate class strata. The bulk of the workforce was unambiguously proletarian.
- neighbourhood and communal solidarity outside the workplace.

(For a comparison with another coalfield locality see Benwell CDP, 1978.) Reinforcing all this we might identify an 'ethnic' dimension. By and large capital in South Wales was English-owned, notably in iron and steel, while labour power was mostly Welsh: ' . . . the land's mineral wealth was expropriated by other luckier people and harnessed to human skills, usually native . . . ' (Williams, 1985, 4). In a very real sense modern class exploitation reproduced the age-old colonial exploitation and was all the more perceptible for that.

South Wales, then, epitomized the class resistance which capital encounters when it agglomerates the dispossessed in close mutual proximity. The foregoing account is of course drastically over-simplified. Historical reality is invariably clouded by contradictory forces acting to reverse, nullify, reduce or divert the dominant tendencies and even in South Wales labour solidarity was threatened by various factors: a growing influx of non-Welsh migrants; the strength of the nonconformist chapel; the stubborn persistence of the traditional *gwerin* ideology, the Welsh perception of themselves as a non-class nation, a notion originating from an earlier rural society (Williams, 1985, 237–8). Nevertheless the salient image is one of a muscular male-dominated proletariat in direct confrontation with an equally identifiable oppressor class. Once again this bears heavily upon the space and class thesis that the 'national working class' is a dangerous concept. We cannot project the South Wales model on to nineteenth and early twentieth-century Britain as a whole (Cooke, 1985) since the specializations of the spatial division of labour produced vast regional and local contrasts in class composition and other necessary conditions for class action: Table 1.6. provides one example of such a contrast. Historically, there were two conditions – *casualism* and *ethnicity* – which distinguished Merseyside from South Wales. Although, as Table 1.6 confirms, the two localities shared a marked degree of industrial specialization, the gulf between them could hardly have been greater in respect of occupational structure and class practices. In contrast to the solid, highly organized labour of the Welsh coal and steel industries, Merseyside's dominance by maritime activities was inevitably associated with casualism, the consequences of which were graphically (if somewhat moralistically) described by Ramsay Muir in 1907:

> . . . a large proportion of the men employed have no permanent work, but must submit to periods of idleness alternating with periods of sudden heavy labour, extending over long hours and inevitably followed by fatigue, reaction and the over-easy consolation of alcohol (Muir, 1970 reprint, 306).

This situation had obtained throughout much of the nineteenth century and had not substantially altered by 1921, when over one-quarter of Liverpool's male

Table 1.6 The industrial division of labour in South Wales and Merseyside, 1931–2. (Percentage of total workforce engaged in selected key sectors)

	Glamorgan & Monmouthshire (1)	Merseyside (2)	Gt Britain
Mining and Quarrying	29.0	0.1	5.7
Metal Manufacture	7.1	1.0	2.1
Shipbuilding and Shipping	n/a	15.9	3.1
Transport and Distribution	24.8	31.2	22.6
Food, Drink and Tobacco	2.4	9.9	3.4

(1) Taken from Lee (1979)   (2) Adapted from Caradog Jones (1934)
Note Whilst ideally an earlier (preferably late nineteenth century) year might have been selected for this comparison, it is fair to say that the basic anatomy of employment created by the Victorian period was still broadly intact as late as the 1930s and perhaps even more recently.

labour force were still employed as labourers, seamen, messengers and porters (Caradog Jones, 1934, 22). These workers constituted a semi-permanent labour reserve, surplus to the requirements of capital for much of the time and bearing on their own shoulders the costs of the daily fluctuations in port activity.

The consequences of this for working-class resistance are not difficult to divine. The presence of a large permanent 'lumpen' stratum, subject to constant unpredictable shifts in employment and morally isolated from the 'respectable' working class in skilled or stable occupations is clearly not conducive to unified working-class consciousness. It would be an overstatement to claim that the 'rootless and volatile' character of the reserve army bred a culture of hedonistic escapism rather than one of organized class resistance: nevertheless it is true that the normal expressions of labour resistance – unionism and the Labour Party – were for long retarded in this region. Note here that the Labour Party did not win a Liverpool Parliamentary seat until 1925, nor majority control of the local council until 1955 (MSRG, 1980, 76).

In addition to casualism, a further obstacle to class action was the deep and often bitter ethnic-religious cleavages, the legacy of heavy Irish immigration, which set Protestant against Catholic and native against immigrant. These internal divides ensured that sectarian politics often swamped class politics, with the 'loyalist' vote being delivered by arrangement with the Orange Lodges to the Conservative Party and the Catholics maintaining the Irish Nationalist Party as the second largest on the City Council as late as the 1920s (MSRG, 1980, 76). Given these historical legacies of casualism and ethnic boss politics, it is hardly surprising if Liverpool local politics are somewhat unorthodox today.

Apart from illuminating some of the processes of regional and local class formation, these case studies also hint at the mythical nature of much that still passes for wisdom on the regional question. In particular the notion of spatial conflict as rich regions versus poor now appears as a smokescreen obliterating the basic conflict between the two primary classes. Until the inter-war depression each of our case-study areas possessed all the hallmarks of the prosperous core region, sucking in resources from a wide orbit, both domestic and colonial, and using them to fuel vigorous industrial-commercial growth. Merseyside is especially noteworthy in this respect. With Liverpool mercantile and finance capital at its heart it functioned as a metropolis in its own right (surpassed only by London), a regional, national and international nerve centre whose fabulous wealth was hymned both by contemporary commentators and later historians. The stock image of this and of other mature industrial centres is one of faded grandeur, once prosperous but now fallen upon hard times.

Yet this 'Vanished Golden Age' scenario contains one obvious but incessantly repeated fallacy: very little of the gold actually rubbed off on those who sweated to produce it. As Lawton (1982, 1) notes, the Victorian–Edwardian era in Liverpool created ' . . . not only its waterfront and fine commercial and civic buildings but also . . . persistent social problems rooted in marked class differences and pervasive poverty.' Muir, an Edwardian himself, is even blunter on the subject of class expropriation: ' . . . the city's prosperity depends upon casual labour, the most degrading as well as the most insecure form of employment'. (1970, 306). There was no Golden Age for the tens of thousands of Irish cellar-dwellers of Scotland Road.

Thus to speak of 'rich' regions, as if the entire populace were composed of a single uniformly affluent class is profoundly misleading, even mischievous, whether applied to past or present. Under capitalism, each region is class-divided and indeed the richer and more powerful the 'region' the deeper the polarization. This is (or should be) a central theme in the analysis of modern metropolitan core regions where, even more than Victorian Liverpool, the concentration of a large bourgeoisie together with its high-level financial, commercial and decision-making functions generates not only a 'middle class' of privileged workers but also an even larger 'service' stratum of superexploited labour. Of notable interest here is Stedman Jones's (1971) account of nineteenth-century London and Simmie's (1985) review of recent trends in the capital's labour market: a decline in the number of regular jobs for skilled industrial manual labour and a rise in 'low-level services such as pubs, restaurants and clubs' (Simmie, 1985, 566), which now account for about one in three jobs.

Contained within all this is a basic spatial contradiction for labour. Class expropriation over space is a double-barrelled process, with core-based capital simultaneously expropriating both its own local labour and that of a multitude of peripheral regions. But the victims of this process are not united in resistance. On a national and international scale, the emergence of labour as a unified self-conscious class organized to combat capital in all places and at all times is seriously impeded. Distance acts to fragment labour into a host of mutually antagonistic groups and class membership is transcended by divisions of nationality, language, culture and place. In summary, regionalism/localism is both an enabling and a limiting condition for labour. On the one hand, as we have seen, localized concentrations of workers can develop formidable cultures of resistance. Yet at the same time such place-bound loyalties inhibit the formation of sympathetic ties with workers elsewhere and hence of *universal* political and industrial organization. Given capital's increasing mobility in the present century, its ability to by-pass the static working-class communities and to move into external space to capture new supplies of virgin labour, this lack of universality has proved to be an increasingly crippling weakness.

1.4 The New Spatial Division of Labour

Whatever the spatial variations, it is a general contention of Massey and like-minded writers that capitalist industrialism created place-bounded working-class cultures (it is tempting to see these as urban-industrial versions of the Vidalian *pays*), which were a source of strength for labour in its resistance to capital accumulation. Evidently the most fundamental twentieth-century change in the geography of work is that this 'geographical solidarity of labour' (Massey, 1984a, 57) has lost its potency. The erosion of these proletarian fortresses began in the inter-war years, slackened during the post-war boom and reasserted itself even more vigorously from the mid-1960s. This seven-decade period has seen a widespread dismantling of old regional economies, with industrial capital deserting its original centres of accumulation (and class confrontation) in search of new external space. Capital's twentieth-century eagerness to retreat is seemingly in direct proportion to its eagerness to invest and some of the worst sufferers are the very regions which were the most vigorous innovators of the earlier era. Of North-West England it has been observed that 'This region . . . which from 1978 to 1982 lost over a quarter of a million jobs, was the birthplace of the industrial revolution' (Massey and Meegan, 1985, 13). Still more poignantly, the town which acted as midwife to the steam locomotive had been reduced to industrial rubble even earlier: the 1939 unemployment figure in Merthyr Tydfil stood at 70 per cent ' . . . and the planning organization Political and Economic Planning (PEP) proposed that the whole population be shipped out to the river Usk' (Williams, 1985, 252).

Though the underdevelopment of the first industrial areas is conventionally rationalized as resulting from 'world recession', 'technological change', 'monetarism' or even simply 'progress', radical geographers would regard these as verbal cloaks for continuing class conflict. But class conflict has now assumed a new spatial form. No longer is the industrialist confined to a single plant or a single labour market as of old. Whereas, for example, a mid nineteenth century iron-master such as Crawshay had his entire capacity located in Merthyr Tydfil and therefore was obliged to stand or fall with the workers of that town, the late twentieth-century company is not so constrained. In the late twentieth century 'One of capital's crucial advantages over labour is its great and increasing geographical mobility' (Massey, 1984, 57): employers are now able to by-pass organized labour by shifting ' . . . beyond the geographical bounds of the workers with whom they are negotiating' (Massey, 1984a, 42).

Above all else, this new mobility of capital is related to the rise of the giant corporation. Ever since capitalism's birth, its development has been characterized by a constant rise in the size of firm and a concentration of ownership in fewer hands, a tendency which was noted by the classical social theorists of the last century (see Marx and Engels, 1848; 1969 reprint) and has continued ever since to the point where advanced economies are dominated by a handful of enterprises of unprecedented size (Table 1.7). They have aptly been dubbed 'prime movers' (Lloyd and Shutt, 1983), because of the inordinate degree to which workers, consumers and smaller firms are determined by them. 'The institution which most changes our lives . . . is the modern corporation. Week by week, month by month, year by year, it exercises a greater influence on our livelihood and the way we live than unions, universities, politicians, the government' (Galbraith, 1977, 257).

The power of the corporation is now legendary. It has been subject to a torrent of analysis, popular and academic, which it is not our intention to reproduce here (see Dicken, 1986). Selectively, from a geography of work perspective, it is the following characteristics of the large corporate firm which are most germane:

– its capacity to shed labour while maintaining or increasing production, a cost saving achieved through the application of improved technology to both production and work organization.
– its multi-plant structure, which is almost by definition multi-locational also, often involving production activities in several different nations or regions; and the separation (spatial as well as organizational) of routine production from the higher-level scientific, management and control functions. The corporation embodies an occupational regional division of labour in itself.

It is this relatively recent capability of maintaining coherence over a wide (sometimes global) span which gives the corporate firm its spatial liberation. Apart from its nerve-centre location, none of its regional centres of production is permanently indispensable to its operations (Damette, 1980) and the company enjoys considerable freedom to close down plant deemed to be sub-optimal and to open up replacements or expansions in new locations. In the post-1973 climate

Table 1.7 The Twenty Largest Industrial Firms in Western Europe, 1982 (ranked by value of output)

Rank	Company	Hadquarters	Number of Employees	Activities
1	Royal Dutch Shell	The Hague	161,000	Oil and petroleum
2	British Petroleum	London	118,200	Oil and petroleum
3	Unilever	London	300,000	Detergents and food products
4	Cie. Francais des Petroles	Paris	48,115	Oil and petroleum
5	Veba	Dusseldorf	83,936	Electricity, lignite, petroleum
6	B.A.T. Industries	London	177,000	Tobacco, retailing, paper
7	Electricity Council	London	156,850	Electricity production/distrib.
8	Renault	Paris	231,700	Vehicle manufacture
9	Philips	Eindhoven	372,600	Electrical goods
10	Volkswagen Audi Group	Wolfsburg	257,900	Vehicle manufacture
11	Ste. Nationale Elf Aquitaine	Paris	37,000	Petrol, oil, gas, sulphur
12	Siemens	Munich	339,000	Electrical/general engineering
13	Daimler-Benz	Stuttgart	183,392	Motor vehicles and engines
14	Peugeot	Paris	245,000	Vehicle manufacture
15	Electricite de France	Paris	108,150	Electricity production/distrib.
16	BASF	Ludwigshafen	116,518	Chemicals and plastics
17	Hoescht	Frankfurt	186,850	Chemicals, dyes and plastics
18	Bayer	Leverkusen	182,000	Chemicals
19	Thyssen	Duisberg	152,089	Steel, manufactured goods
20	Nestle	Vevey (Switz.)	152,653	Chocolate, milk, food products

of recession and falling profits, emphasis has been on rationalization, the reduction of total capacity and its concentration in the most efficient plant and locations.

In the opinion of several recent writers, access to readily exploitable labour has now become far and away the most important consideration in the firm's locational strategy. Hypermobility allows the manufacturing corporation to tap into 'labour reserve regions' – spatial concentrations of surplus labour – by relocating parts of its low-skill assembly operations: ' . . . the advantages to capital of decentralizing the most labour-intensive parts of the production process to cheap labour regions can be considerable' (Carney *et al.*, 1980, 22). Certainly within the EEC, the existence of labour reserves on the periphery has during the past two decades been a vital weapon against rising labour costs (Lapple and van Hoogerstrand, 1980; Damette, 1980). For Hudson (1983, 223–4), these labour-reserve regions come in two guises:

> Conventional regional typologies identify both agricultural and mature coal-mining and industrial regions as important categories of 'problem region'. The existence of these types of 'problem region' can be reinterpreted in terms of the processes involved in the production and reproduction of regional labour reserves and the dynamics of the process of capital accumulation.

In the case of the *agricultural periphery* we see an outward diffusion of the corporate branch plant economy to take advantage of labour surplus to the requirements of the pre-existing local economy, a clear-cut and direct example of capital's use of external space. The 'cheapness' of this labour is a combined function of its over-supply, its low expectations and its lack of industrial or political organization. Hudson points out that the oversupply of labour in the EEC periphery has been made manifest by mechanization, farm amalgamation and other agricultural developments which have had the effect of flushing out a hidden surplus which had previously reproduced itself at substandard levels on sub-marginal plots or in part-time, seasonal or ill-paid farm employment. Much of this has now been displaced, with the EEC agricultural labour force falling from 22 millions to 10 millions between 1950 and 1970 (Hudson, 1983). In effect, then, agricultural displacement has released a new proletariat ripe for harvesting by multinational industrial capital. It is no coincidence that the fastest industrial growth rates have occurred in regions of the Italian Mezzogiorno, Greece, Iberia and Ireland, where the creation of this potential proletariat has been most substantial (Hudson, 1983). Moreover, in Southern Italy, there is an ample hidden reservoir still to be tapped. Its magnitude can be judged from the information that, if regional agricultural labour productivity were raised to the EEC average, over 400,000 producers would be surplus to requirements.

In the matter of the *industrial 'problem regions'*, their recent economic history represents not so much the opening up of virgin territory as the obliteration of the past, the decomposition of an established proletariat and the recomposition of the population as reserve labour. Returning again to Wales, we find the erstwhile industrial staples reduced to a shadow of their former selves and replaced – though not so thoroughly as to eliminate acute unemployment – by a new branch plant economy with prominent Japanese involvement (Massey, 1984a). Inevitably this has broken the old bastions of labour resistance. Williams (1985, 298) laments the malaise of 'The Wales TUC, weakened and losing both numbers and funds . . . '. Compliant new categories of labour have been absorbed into production. Especially significant here is the shift from a heavily male-dominated economy to one where over 40 per cent of the jobs are now performed by women (Massey, 1984a; Williams, 1985).

None of this is peculiar to South Wales. Females have always constituted a major portion of the industrial reserve army and their incorporation into new branch plant economies has been a common feature of depressed industrial regions from Scotland to Lorraine. Other fairly common features have been the low uptake of redundant males displaced from the traditional highly unionized sectors of the economy; the problems of gaining union recognition from many new firms; and the flimsy impermanence of many recently arrived factories (MSRG, 1980).

This last is hardly unexpected, since one of the definitive requirements of reserve labour is *expendability*, to be turned off like a tap during periods such as the post-1973 and post-1979 recessions. One special version of the 'decomposed' industrial area is the *inner city*, a term usually employed to denote the oldest core areas of the largest urban agglomerations. For geographers and planners the progressive post-war collapse of the inner urban economy has forced a rethinking of spatial categories, demonstrating that marginalized spaces are not always created at the regional scale but are frequently very localized. This clearly obliges us to add a further refinement to the argument of this chapter. So far we have inferred that the spatial division of labour produces a polarity between metropolitan core regions – dominant concentrations of capital together with privileged upper strata of labour – and peripheral regional labour reserves. While substantially correct, it is also true that the metropolitan region itself contains the largest absolute numbers of reserve labour and that these workers are intensely concentrated into a small number of local neighbourhoods, primarily in the inner zone or in decanted public-sector housing estates. Hence in recent years a vigorous debate about the merits of regional policy versus urban policy and the emergence of the view ' . . . that the most significant social if not economic problems demanding government attention in Britain today are those specific to the inner-city areas of the country's major conurbations in *whichever* region these are located, rather than in the broadly defined development area regions as a whole.' (Keeble, 1977, 6).

Here we should stress that the spatial division of labour concept does not propose a crude dichotomy between 'bourgeois' and 'proletarian' regions: indeed the English myth of a 'cloth-cap North' and a 'rolled-umbrella South' is among the sillier of a number of silly national self-images. Class divisions do not coincide precisely with space divisions and never have, especially at the regional scale. More recently, however, increasing polarization at the intra-urban scale has brought such a coincidence somewhat closer at a local level. A geographical concentration of the less privileged strata in the inner city is one of the more important recent consequences of the spatial division of labour process.

The political-economic perspective advanced so far in this section presents an alternative to the traditional explanations for the ills of the inner city. While explanations invoking congestion, obsolete infrastructure, technological change and planning blight are all technically correct, they ignore the way in which the inner city functions in the interests of capital accumulation as a reservoir of surplus labour. This is not to say that it is structurally identical to the old industrial periphery. Other more specific forces have been operative to create surplus labour, one of the most crucial being the immigration of a black sub-proletariat, whose reserve status is confirmed by racism (Chapter 5). Largely as a result of this, the inner city presents conditions even less favourable for class solidarity and resistance than the industrial periphery.

1.5 The Role of the State

So far we have highlighted the interaction between classes with little mention of the state, the institution which is commonly depicted as interposing between them. The great mass of the literature on government spatial policy implicitly discusses the state as a kind of neutral umpire reconciling clashes of interest between various factions within society. Since the last century, social and economic life has become increasingly complex, hence an ever-growing need for this kind of intervention.

The specifically geographical aspect of the problem is, in basic terms, one of a collision between the immediate locational needs of firms and the social costs thereby generated. It is deemed to be in the public interest for firms to be profitable (a sacred maxim of classical economics) but equally, if the pursuit of profitability demands locational strategies which entail spatial inequality, regional hardship and area deprivation, then the public interest is actually threatened. It is the role of government regional policy to reconcile this paradox by enabling firms to relocate in a socially beneficial manner without losing efficiency or incurring unacceptable costs.

Since the 1960s this *pluralist* model of the modern state has received mounting criticism. A model which tacitly accepts all the institutions of liberal democracy at face value, it depicts the public as sovereign (through the ballot-box, etc.) and presents the officers of the state (including planners) as servants of the public. Yet, as Dunleavy (1980) and others (Pickvance, 1977; Saunders, 1979, Taylor and Johnston, 1984) have shown, this is only one of a number of alternative presentations. More recent formulations such as *elite theory* and *managerialism* take a more critical view (Chapter 2 contains an extended discussion of managerialism). For the purpose of the present chapter, we focus on a *Marxist* approach to spatial policy, for the very good reason that this perspective forms a natural extension of the class approach of the preceding sections.

1.5.1 Marxism, the State and Spatial Policy

Despite certain qualifications, modern Marxist theories remain firmly wedded to the classical proposition that the state is subordinate to capital. Neutral it is not (even though it must to a degree bend to interest groups), neither does it constitute a ruling class, or a section thereof, in its own right. The economically dominant class also controls the organs of state power. According to Pickvance (1981, 231), ' . . . the State has a built-in link to the capitalist class', a view affirmed by Taylor and Johnston (1984, 24) who see the role of the British state, irrespective of the party in government, as ' . . . supporting British capital and, in particular, enabling the accumulation of capital to operate as smoothly and profitably as possible'. While in general planners and other state officers are certainly innocent of any conscious or conspiratorial design, it is nonetheless the case that the actions of the capitalist state do in practice favour the dominant class over and above other interests. A number of recent writers have attempted to demonstrate that this principle of 'public policy subordination to private interests' (Martinelli, 1985, 52) applies to government management of the space-economy as much as to any other aspect of state policy. (See for example Carney, 1980; Pickvance, 1981 on Britain; Secchi, 1977 on Italy; Martinelli, 1985; Preteceille, 1981; Zukin, 1985 on France.) Distilling from these and other sources, we arrive at a reinterpretation which portrays the state as an active agent in the creation of new regional labour reserves and their incorporation by highly mobile industrial capital. For Damette (1980), state intervention is ' . . . essential to the establishment and reproduction of basic labour reserves' and certainly in post-war Western Europe it is government action which has been instrumental in the rapid displacement of labour from agriculture and from older industries, thus creating the two types of 'problem' region. In the agricultural case, Western European governments have in varying degrees subsidized and managed the process of modernization which has been responsible for flushing out the great rural labour surplus discussed previously. It is also the state which has subsidized the great rundown and 'streamlining' of the traditional coalfield industries, often through the contraction of its own nationalized coal and steel enterprises (Hudson, 1983). Moreover, it is the state which, as provider of unemployment and social security benefits, enables such marginalized labour to be maintained in reserve. (On the role of the state in the reproduction of labour power see Taylor and Johnston, 1984, 33 ff and chapter 2 of this book.)

It is at this point that the critique can be extended to cover regional policy and other areally-defined activities such as urban aid, which are after all the most self-consciously *spatial* forms of state intervention. In a sense spatial policy follows one step behind the policies discussed above. If the above policies are largely instrumental in the *creation* of labour surplus regions, then regional policy is largely instrumental in *releasing them for use* by private firms. Conventionally, regional policy is judged by its ostensible aim of reducing geographical inequalities and it is by this yardstick that it is frequently condemned as a failure. By contrast, Marxist method focuses not on the benefits which policy should confer on the public but on the service which it performs for capital. Here the relevant authors are unequivocal that the primary purpose of spatial policy is one of enabling private firms to readily 'plug into' regional/local labour reserves, by providing the necessary physical infrastructure, subsidizing initial investment outlays and generally defraying the costs of relocation in innumerable ways. Thus it is claimed that ' . . . state regional policy has become . . . a means by which accumulation by industrial capitalists is bolstered in a period of falling profitability' (Pickvance, 1981, 241).

Notwithstanding its complexities and apparent self-contradictions, the Marxist critique represents an important attempt to break away from the pluralist notion of spatial policy as 'redistributive justice'. Reduced to its core, the critique exposes spatial policy as principally concerned with matching the geography of capitalist production to that of labour power – at minimal cost and evident benefit to capital. Yet for all its forcefulness, this interpretation begins to look increasingly flawed with the passage of time. Over the last decade or more, its rationale has been undermined by two major events: firstly the growing irrelevance of the reserve army of labour in capital's locational strategy; secondly a drastic reduction in the degree and type of state intervention and, concomitantly, a retreat from the time-honoured conventions of regional policy.

On the question of reserve or surplus labour, Ginatempo (1985, 100) writes of a population which has grown and continues to grow beyond the needs of capital. This excess population is a structural phenomenon which requires a new theoretical definition . . . this 'overpopulation' cannot be absorbed in a new economic expansion because, even if the latter were possible, it would be based on labour-saving investment . . . this kind of overpopulation is not functional for capitalist development.

Although this author is dealing specifically with the Italian Mezzogiorno, with all its local peculiarities, there is no reason why his central argument should not apply to similar geographical concentrations of marginalized labour elsewhere in the advanced capitalist states. In particular the structural position of many workers in the inner cities of North-West Europe, notably blacks and immigrants, exhibits many parallels. As many writers (notably Pahl, 1985; Blackaby, 1979) have emphasized, since the 1960s advanced manufacturing technology and labour-saving investment have drastically reduced the modern firm's dependence on labour. Consequently, any future rise in output resulting from an upturn in the global economy is likely to substantially exceed the rise in demand for labour. As Pahl (1985) points out, it seemed for some time as if the expansion of the tertiary (services) sector would compensate for de-industrialization but since the mid-1970s this inter-sectoral labour transfer has ceased to be effective.

This progressive 'abolition' of employment in the formal sense of the word is a vital consideration for any future geography of work and the role of the state therein. In the event of a future manufacturing boom firms will certainly be interested in 'plugging in' to local and regional labour reserves but never to the extent of absorbing all those currently unemployed, underemployed, unregistered and otherwise marginalized. Hence the rationale for state intervention on behalf of capital is critically reduced. And indeed at the policy level a significant diminution of state activity has already occurred. Post-1979 Conservative Britain is the most vivid example of this, though similar trends have to some degree been followed throughout almost all the advanced capitalist states. In Britain the post-1979 Administration's regional aid cuts are consistent with its wider aim of reducing public expenditure and state involvement across the entire spectrum of activities formerly considered 'fair game' for government intervention.

This is not to say that the government has been idle and has failed to substitute other policy measures directed at employment creation. In the next (and final) section of the chapter we shall evaluate these, together with certain potential socialist alternatives.

1.6 Radical Alternatives to the Jobs Crisis

Any truly radical response to the employment crisis of the 1980s must start from the naked premise that full employment as traditionally defined is a thing of the past. Employment in the modern economy is now dominated by a *corporate level* of production and exchange ('Monopoly Capital' in Baran and Sweezy's (1966) vocabulary), consisting of very large-scale private firms both in manufacturing and in services, together with a variety of equally large enterprises and organizations owned by central and local government. As we have seen, the now prevailing trend within these organizations, more especially those privately owned, is for a given quantity of work to be performed by a shrinking workforce, with capital progressively substituted for labour. No foreseeable factor will reverse this trend and the corporate level can no longer be relied on to provide jobs for all. Consequently – in the interests of either common humanity or political stability, depending on one's value-bias – other levels of the economy must be stimulated to perform that task.

1.6.1 Small Enterprise

Quite predictably in view of its fundamentalist beliefs in private enterprise, the British Government's response has been to promote both the philosophy and the practice of small business, in effect to transfer the political spotlight from monopoly capital to petty capital. Philosophically the small entrepreneur embodies all the 'Victorian values' of self-help, thrift, industriousness, self-sacrifice and pioneering innovative spirit which hold a key place in the belief-system of radical Toryism. In practical terms small enterprise is thought to be capable of generating the jobs necessary to compensate for the failures of the corporate level and has even been hailed as a potential regenerator of an entire 'ailing' economy (Birch, 1979). To this end self-employment has been stimulated by an unprecedented battery of incentives and support systems. (In Britain these include the Enterprise Allowance Scheme, the Small Firms Service, the Loan Guarantee Scheme, and the Business Expansion Scheme (Department of Employment, 1986b).)

In addition to these, the most clearly 'locational' in intent is the *Enterprise Zone*, a dramatic departure from the old-style industrial estate, which in the words of the planner often credited with its invention, is a hotbed of ' . . . fairly shameless free enterprise . . . outside industrial and other regulations . . . we would aim to recreate the Hong Kong of the 1960s in inner Liverpool and Glasgow' (Hall, 1982, 417). Despite emanating from a 'socialist or (perhaps more accurately) social democrat' (Hall, 1982, 416), these words were music to the ears of the political right who on their accession to power proceeded to designate eleven such enterprise zones, whose numbers have subsequently expanded to 25 (see Fig. 1.3).

Both the zones themselves and the role of small business in general have received scathing criticism. The enterprise zones have been criticized for failing to provide jobs of sufficient quantity to make a significant dent in unemployment (for example the 25 Enterprise Zones had by 1984 created only 8065 new jobs in total, many of which were simply transfers in from the immediate locality (Hall, 1984; Lloyd, 1984)): and more bitterly still for the type of work which they create (O'Dowd and Rolston, 1985). Harrison (1982, 424) warns that the zones are likely to attract low-wage employers who 'follow especially authoritarian or arbitrary personnel practices', while Massey, concerned with the 'Silicon Valley' style imagery employed to sell Enterprise Zones as seedbeds of innovation remarks: 'Many small firms are of the ''scrap metal dealer'' variety; their relationship to high technology could not be more distant' (Massey, 1982, 430). With their exemptions from local rates, together with other concessions, the zones continue to offer a state subsidy to capital in much the manner of old-style regional policy. Furthermore, their exemption from many forms of factory legislation gives firms an even greater opportunity than previously to exploit marginalized labour. Their only fundamental distinction is one of style – it is

Figure 1.3 Enterprise Zones in Great Britain 1984.

now small and medium-sized rather than large capital which stands to reap these benefits at the expense of both the taxpayer and the worker.

At the broader level of small business as a whole, much scepticism has been expressed about the level and direction of benefits which can be generated by an expansion of petty capitalist production and distribution. Crucially at issue here is the question of *ownership*. In principal, the leading virtue of small business expansion is its creation of *self*-employment in addition to employment *per se*. In its switch from large to small scale it represents expanded opportunities for individuals to own and control the means of production. Immediately this opens up visions of an almost utopian classless future economy operated by independent producers no longer subject to imposed authority and discipline, no longer expropriated, no longer alienated from their own product. Unhappily for this pleasant scenario, small business has been shown to be a poor vehicle for economic progress – it tends to confer meagre rewards (low profits and incomes for long, arduous work (see Bechofer *et al.*, 1971; Scase and Goffee, 1980, Aldrich *et al.*, 1983) and to operate at a chronic disadvantage in markets dominated and manipulated by large capital (Jones and McEvoy, 1986). We must naturally qualify this by recording that the term 'small business' covers a wide and heterogeneous range of concerns (Scase and Goffee, 1980; Massey, 1984a) and that many of them escape the generalizations made here. Even so, an uncomfortably large proportion fall into what Wright Mills (1956) called the '*lumpenbourgeoisie*', a term implying that ownership in itself is no guarantee of material rewards or status: and that there is a category of capital owners whose structural position is parallel to that of the lumpenproletariat. These business owners are lumpen in that their product or service is surplus to rational market requirements (i.e. uncompetitive) and hence they represent a form of concealed underemployment, surviving only by dint of extremely hard work for uneconomic returns. Consequently the expansion of the small business sector often merely switches people from one form of 'lumpenness' to another.

Despite Cooke's (1985) claim that the small business bandwagon is 'discredited', the radical left attack on the entrepreneurial lobby has actually been somewhat fragmented and lacking theoretical coherence. For Marxists in particular it is hard to come to terms with the continued survival, indeed the current proliferation, of petty capital, a stratum which according to the Communist Manifesto should by now have 'sunk into the proletariat' (see Scase, 1982). All over Western Europe the numbers of self-employed have risen since 1980, reversing a trend of many decades (Boissevain, 1984) and in North America it has been claimed that petty capitalism has been responsible for no less than two-thirds of new jobs created in the 1970s (Birch, 1979).

Growth, however, is not to be confused with development, nor proliferation with economic health (Ward and Reeves, 1980). A modern Marxist response to the small business phenomenon must start with the axiom that petty capital survives almost by default. As a subordinate and dependent form, it is highly functional for monopoly capital, operating in markets where the yield is too low for rational profit-seeking capital to exploit (Jones and McEvoy, 1986). For example, small retailing bears the costs of selling (realizing surplus value on the products of industrial capital) in marginal markets like the inner city: small workshops supply large firms with cheap materials (see the literature on 'diffuse' industrialization: e.g. Hudson, 1983; Ginatempo, 1985; Martinelli, 1985).

How does Geography affect UK society?
Gender.

More women in part time work than men - are more likely to be in areas such as retail.

Some areas are seen as "out of bounds", e.g. unlit streets, certain pubs unless in a group of women.

At some Universities, one gender dominates a subject, although this is being redressed.

Some areas are more family orientated - Universities such as Lancaster house predominantly single students with houses as well.

Traditionally cities have been seen as more "equal" but have also been seen as "unsafe" for lone women.

There is effort being made to make professions "ungendered" – eg, plumbing, engineering

However, gender differences can still be seen in the portrayal of some sports – unequal winnings at Wimbledon

Gender divide not seen in educati

The UK is moving away from the idea of where somebody's "place" should be,

Often it is the sweated labour of proprietors and workers in this lower level of the economy which allows these costs to be borne.

We would also note that petty capital serves various political and ideological purposes for the capitalist state. It mops up a portion of the unemployed, thereby averting potential social unrest; it incorporates former workers into a property-owning conservative mentality (See Bechhofer *et al.*, 1974 on the values of petty property owners and Bechhofer *et al.*, 1974 and Aldrich *et al.*, 1986 on the right-wing voting behaviour of the shopkeeping fraternity); it seeks to replace proletarian cultures of resistance with a new 'entrepreneurial' culture.

We wind up this argument by noting that at the regional level there is a strong correlation between spatial concentrations of petty capitalists and various indicators of poverty and labour surplus. This is no mere accident but a striking confirmation of the underlying role of small 'enterprise' as a survival strategy, a means by which part of the surplus population reproduces itself at minimal cost to the state. A Marxist critique must certainly be aware of small business policy as one means of managing the chronic problem of unnecessary reserve labour power.

Recommended Reading

Most research by geographers in this field has concentrated on the economic geography of industry rather than the social geography of work. One of the best attempts to marry the two together has been made by Doreen Massey, and her *Spatial Divisions of Labour: Social Structures and the Geography of Production* (1984) underpins much of the debate in this chapter. Of the more traditional work on industrial location, Keeble (1976) is now rather dated but provides a competent review. The geographer's obsession with regional policy and regional economic problems is given thorough and conventional coverage by Law (1981), among many others. The appropriate sections of Lloyd and Dicken's *Modern Western Society* (1981) succeed in presenting a straightforward but more wide-ranging socio-economic review, and Peter Dicken (1986) updates this picture of industrial change on an international scale in parts 1 and 3 of *Global Shift*.

Returning to Britain, the edited collection from Ron Martin and Bob Rowthorn (1986), *The Geography of De-Industrialisation* includes chapters by many of the more significant commentators on Britain's recent industrial experience, reviewing the legacy of the United Kingdom's past role in the world economy, regional and urban dimensions of post-war decline, the response of the state, and the effects on differing social and economic groups. More detailed work on the nature of recent economic change is provided by Alan Townsend's (1983) painstaking study, *The Impact of Recession*. The same author has, with Jim Lewis, edited a new collection entitled *The North-South Divide* (1989). Finally the uninspiringly titled *The United Kingdom* by Ray Hudson and Allan Williams (1986), one of a series of economic and social studies of Western Europe, provides a clear and up-to-date review of contemporary Britain, moving from the international and national to the regional and urban scales.

2
Housing

2.1 The Functions of Housing

Unlike many of the other consumer items to be considered in this book, shelter answers an absolute need, being vital to human survival and reproduction. It must be included among those necessities which Laski (1968) describes as 'the irreducible minimum of human wants'. In advanced urban society, however, housing is generally required to satisfy a very wide range of demands over and above mere protection from the elements (Dickens *et al.*, 1985, 193). For convenience, we shall define three levels of satisfaction which housing is expected to confer upon its occupants: they are material, symbolic and external.

2.1.1 Material Utility

In considering the material benefits derived from housing, we must make an immediate distinction between 'needs' and 'wants'. In all societies at all times and places housing satisfies a need in the sense that bodily well-being and survival are dependant upon shelter. 'Wants' on the other hand are *culturally* (not biologically) determined. The typical European or North American household makes unprecedentedly high demands upon its dwelling place. More than a place to eat, sleep and procreate, the modern home is required to provide privacy, security, warmth and comfort together with ample facilities for leisure, family life, storage of possessions, entertainment of visitors and maintenance of hygiene – the list could be almost endless (Berry, 1974, 2–3).

Housing then is judged against cultural norms. When, for example, we use the term 'slum', we do not necessarily refer to dwellings which place their inhabitants' lives at risk (though sometimes this may indeed be the case): rather we mean dwellings which society deems to be unsatisfactory for continued occupation (Kirby, 1979). Criteria for judging houses will tend to be based on the conviction that every person is entitled to a life in decent conditions consistent with human dignity (Drewnowski, 1974). For the social commentator, the insuperable problem is that cultural yardsticks are almost impossible to apply in practice, since there can never be absolute agreement among all members of society on what precisely constitutes an 'acceptable' or 'normal' standard. The convention among social scientists is to fall back upon the somewhat arbitrary official standards laid down by Government. (See Kirby, 1979 on the official yardsticks adopted by various advanced nations.) One person per room is officially designated as 'overcrowded', a bureaucratic conception of the minimum living space consistent with human dignity, while those lacking or sharing an

Table 2.1 Presence of Housing Amenities, England and Wales 1971–1985

	1971	1975	1979	1985
a. Bath or Shower				
Sole use	88	92	95	98
Shared use	3	3	3	1
Lack	9	5	3	1
b. Inside W.C.				
Sole use	87	92	94	98
Shared use	2	3	2	1
lack	10	5	3	1
c. Central Heating				
Full central heating	27	38	47	63
Night storage heaters	8	10	7	6
No central heating	65	53	45	31
d. Persons per Room				
Less than 0.50	37	39	41	45
0.50 to 0.99	49	48	49	47
1.00 to 1.49	13	11	9	7
1.50 and greater	1	1	*	*

Source: Table 5.2 and 5.3, *General Household Survey 1985* London: HMSO, 1987.
Notes: all figures are percentages, and may not total 100 because of roundings
*   * – less than 0.5 per cent

inside toilet, a bathroom, a kitchen or hot and cold running water are deprived of a 'basic amenity'. According to these official standards, Britain's housing problem is a thing of the past. As Table 2.1 shows, only 2 per cent of households lack sole use of a bath or shower and a similar proportion lack an inside toilet.

Important physical improvements have been achieved. In practice, however, this analysis demonstrates the inadequacy of official statistics, which fail to reflect the reasonable expectations of householders in the late twentieth century. One in three households, many of them elderly, lack central heating; countless thousands more lack the resources to use it. Eight per cent of households live at least four storeys up in unpopular flatted accommodation, four million people are deprived of gardens and safe playspace for children. In 1981 1.2 million dwellings were officially regarded as unfit for human habitation and 1.1 million required repairs of £7,000 or more at contemporary prices. Even the small percentage of dwellings lacking one or more basic amenities totalled one million in absolute terms. Although the three categories are not mutually exclusive, 2.1 million dwellings in England and Wales, 11 per cent of a total stock of 19.1 million, were deficient in at least one respect.

In some cases we even fail to deliver the basic amenity of shelter. Despite the 1977 Housing (Homeless Persons) Act placing a statutory responsibility on local authorities to house the homeless, it is estimated that in London alone 60,000 people are homeless at any one time and at the end of 1984 a further 10,696 households were in temporary or 'bed and breakfast' accommodation (Greve Report, 1986). Nationally only 45 per cent of the 185,000 who applied for re-housing under the Act in 1984 were accepted and many others, knowing the stringent conditions operated by local authorities and the appalling quality of much of the accommodation allocated, did not apply (Archbishop of Canterbury's Commission, 1985, 231).

Once again a central theme which recurs throughout this chapter is that of *inequality*. Despite the large number of people whose housing falls short of normal expectations, there is no absolute shortage of decent housing (Duncan, 1978). Rather, the problem is one of maldistribution. As is the case with any form of resource consumption, the abundance enjoyed by the privileged has its counterpart in the deprivation of the under-privileged, the effective disqualification of certain sections of the population from an adequate share of housing. Moreover, housing inequality also assumes a spatial dimension.

Just as access to employment is partly conditioned by the individual's location in space (Chapter 1) so too are housing opportunities. Although Holtermann (1975) has demonstrated significant variations between conurbation and non-conurbation housing conditions, it is at the intra-urban scale that such disparities are most manifest. Since the 1960s, studies (Robson, 1969; Herbert, 1972; Davies, 1978) have emphasized that housing poverty and abundance are spatially localized; and that their patterns of localization tend to recur from city to city. Especially salient is the contrast between inner city and suburbs (see in particular DOE, 1977a), though this is by no means the only important line of variation. Rural housing may still lack basic amenities (Chapter 7) and in Britain public-sector intervention has decanted part of the housing problem to the urban fringe. Although estates built in the 1960s and 1970s may appear satisfactory on physical criteria, they are commonly high-rise and/or system built, badly maintained, prone to condensation and damp and remote from valued services.

2.1.2 Houses as Symbols

Besides its practical uses housing also acts as a principal means of 'socio-economic status display' (Johnston, 1973, 78) or, as Dickens *et al.* (1985, 193) phrase it:

> People's identities and status as individuals and members of social groups are, in part, conferred and maintained in the home and in the process of housing consumption.

In an anonymous urban society, where the individual's personal identity is unknown outside an immediate social circle, the dwelling place assumes immense significance as a public badge of worth, a means of status affirmation. At the top of the social scale, the rich family's mansion is not simply a place of habitation. Its purpose is to proclaim the worth of the inmates (Galbraith, 1977). It is a visible monument to opulence, success and worthiness. Yet such proclamation is not limited to the extremely affluent. On a more modest scale, the villas and semi-detached houses which characterize the modern suburb throughout the English-speaking world are also designed to transmit signals about the social rank of those who live there.

By the same token, to live in a slum is a public admission of poverty, helplessness and dependence. On top of the material privations which it inflicts, the slum also stigmatizes and alienates, defining its occupants as non-members of the social mainstream. In effect it extorts a 'psychic' penalty, undermining self-worth and heightening feelings of social rejection.

Differences in housing *tenure* – whether one owns or rents one's shelter and

Table 2.2 Permanent dwellings started by type of authority and sector, United Kingdom 1971–1985

Year	Local authorities	New Towns	Housing associations	Total public sector[1]	Private sector	Total dwellings started
1971	122.1	10.4	11.2	146.4	212.2	358.6
1976	133.7	14.9	29.2	180.1	158.4	338.5
1977	96.8	10.6	28.4	136.9	138.6	275.5
1978	79.5	9.6	20.9	110.9	161.6	272.5
1979	58.4	8.4	16.1	83.5	148.2	231.7
1980	37.4	6.8	15.0	59.4	101.5	160.9
1981	26.1	1.9	11.9	40.3	118.9	159.2
1982	35.9	2.3	17.8	56.1	127.0	183.0
1983	36.9	2.0	16.0	54.9	148.0	202.9
1984[2]	35.0	2.0	17.0	54.0	158.0	212.0
1985[2]	29.0	1.0	13.0	43.0	154.0	197.0

Source: *Social Trends* 1985 Table 8.16; 1986 Table 8.11; 1987 Table 8.12.
Notes: [1] includes a small number of houses built by Government Departments and not included elsewhere.
[2] figures for 1984 and 1985 are rounded to the nearest thousand.

from whom – constitute a particularly potent source of symbolism in many societies. In Britain, despite an aggressive programme of council-house sales since 1980, one quarter of the total housing stock is rented from the state. Public opinion tends to attach superior ethical qualities to home ownership and to place a stigma on council tenancy (Berry, 1974; Association of Metropolitan Authorities (AMA), 1984a, 2), a form of occupancy which carries connotations of dependence, irresponsibility and willingness to rely on perceived subsidies from the public purse.

Even where council property meets all the requisite material criteria for satisfying its residents, it still imposes this burden of 'unworthiness' upon them. Historically, of course, the strength of this fallacious belief has varied considerably from one period to another, with the current period notable as one in which official British government ideology and practice is intensely hostile to council housing.

During the period 1979/80 to 1984/85 net capital spending on council housing has been cut by no less than 52 per cent; yet in the past decade total government expenditure has increased by 8 per cent in real terms (Harloe and Paris, 1984). This is even more clearly reflected in the number of public-sector housing starts, which fell from 180,100 in 1976 to 43,000 in 1985 (Table 2.2).

2.1.3 The Externalities of Residential Location

Any assessment of residential satisfaction would be incomplete if it considered only the dwelling house in isolation. Although the home itself is a crucial source of material and non-material utility, it is also the key to a range of further utilities which derive from the surrounding environment. The point is well explained by Bourne (1981, 14), who stresses the view of housing as

> a package or bundle of services – a view which recognizes that the occupancy of housing involves the consumption of neighbourhood services (parks, schools), a location (accessibility to jobs and amenities) and the proximity of certain types of neighbours (a social environment).

Here the dwelling place is conceived as a location in geographical space which in itself exerts a profound influence on the well-being and life chances of its occupants. Each location offers a special – even unique – combination of opportunities and restraints, exposing its inhabitants to a mix of assets and hazards in its immediate vicinity; and offering varying degrees of access to other points in the urban system (Merrett, 1982). For instance, a home in a deteriorating deprived inner-city area is likely to inflict on its occupants a range of physical dangers, psychological stresses, health hazards and social disadvantages irrespective of its *own* intrinsic qualities. The impact of this neighbourhood effect is of course highly variable from one individual and group to another, depending on age, class, race and gender. A high degree of personal mobility clearly reduces dependence upon the immediate vicinity but the ability to overcome the tyranny of distance is one that is denied to a surprisingly high proportion of the urban population (see further discussion in Chapter 6).

Housing then is both a major (for many people *the* major) consumption resource in its own right and a key to access to other opportunities. Given this, together with its obvious physical and spatial properties, its prominence in modern social geography is not to be wondered at. Yet over the past 20 years

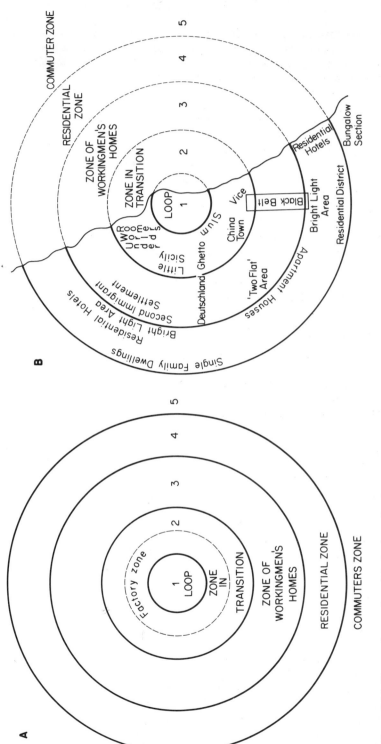

Figure 2.1 A Burgess's zonal model of urban structure.
B The application of Burgess's model to Chicago in the 1920s.
Source: After Park, Burgess and McKenzie 1925.

there has been not only an increasing concern with housing but also a profound change in the geographer's approach to the topic. Initially, with the rise of urban geography in the 1960s, geographers drew their chief inspiration from the pre-war Chicago School of Human Ecology, with its emphasis on the spatial segregation of residential areas as an indication of social differentiation – housing as a kind of social litmus paper, a visible clue to the identity of its inhabitants (Figure 2.1). This was grist to the mill of the new spatial analysis, affording limitless opportunities to apply ever more refined quantitative techniques to the measurement of population distribution at the urban scale.

Subsequently, however, increasing concern has come to be devoted to questions of why and how rather than what and where. This growing stress on the processes by which particular sections of the population come to acquire particular forms of housing in specific areas of the city has inevitably led to a sharper political awareness. New managerialist and neo-Marxist approaches have challenged many of the bland assumptions underlying the ecological school: the city as an orderly beehive where each member knows his functional and geographical place; segregation as a well-oiled mechanism for maintaining order between potentially conflicting interest groups. We must now at least admit the possibility that residential space patterns are the product of inequality, conflict and injustice.

In identifying this change of direction we do not imply that one approach has been entirely superseded by another. What has taken place is not a 'paradigm shift' but a change in the relative balance of the many theories which compete for the attention of the housing geographer (Bassett and Short, 1980, 4–5). The field of enquiry is now the scene of a healthy dissensus, which reflects the controversial nature of housing as 'political hot potato' (Bourne, 1981, 13) and the intense divergence of views which existed among the founding fathers of social science. It also stems from objective changes in the nature of capitalist society itself, which have required the social scientist to respond by modifying existing theory and developing new. Although in consequence the theoretical and methodological mix is a rich one (for a review of the relevant schools of thought and their offshoots see, for example, Bassett and Short, 1980; Bourne, 1981; Saunders, 1981), we shall, for manageability, reduce them to the following three broad perspectives: *Ecology; Managerialism; Marxism*.

2.2 The Ecological Approach: Burgess as Messiah?

The one Chicago School contribution which above all has captured the imagination of succeeding generations is the so-called *concentric zone model*, Burgess's (1925) urban prototype based on the city of Chicago. In the heady days of the 1960s, when human geographers first awoke to the possibility of reconstituting their discipline as a socially relevant science, the attractions of the Burgess model were irresistible. Here was a model of the spatial geometry of the city based on the premise that spatial differentiation is social differentiation made flesh: that spatial location denotes social location, that spatial relations express social relations and that spatial mobility signifies social mobility. Excitement was heightened by a further corollary of all this – urban spatial structures tend broadly to recur in all places where cultural circumstances are fundamentally similar and are, therefore, subject to some degree of predictability. In view of the interminable literature on the geographic and geometric properties of the model, there is no purpose in a lengthy reiteration (to refresh the memory, see any one of Mann, 1965; Robson, 1969; Johnston, 1971; Timms, 1971; Carter. 1980; Bassett and Short, 1980). Suffice it to remind the reader that, according to Burgess, the city is arranged as a series of concentric housing zones around a central node, itself dominated by commercial and industrial activities, with residential land use occupying the outer space. Residential space is internally differentiated, with social status rising with distance from the city centre.

What insights into the housing question are to be gleaned from the ecological perspective? Many, if we care to examine the assumptions underlying its elegant logic. It must be admitted, however, that much initial geographical research failed to probe at this level, being much more attracted to the visible and tangible (i.e. superficial) attributes of urban space. Much effort was expended in testing the universality of the postulated spatial structure. Did the concentric ring arrangement 'fit' cities in Britain, continental Europe and elsewhere or did some cities conform more to the 'sectoral' pattern asserted by Hoyt (1939)? One inevitable by-product of this orgy of empiricism was the application of increasingly advanced statistical techniques (factor, cluster, principal components and other forms of multi-variative analysis) to the measurement of spatial variance. By means of this *factorial ecology* it was possible to verify the essence of Shevky and Williams's (1949) hypothesis that social space is multi-dimensional – i.e. areal variation is not simply a matter of social class segregation but also of separation by family status and ethnic status (see Table 2.3 for fuller commentary and summary of selected works in this genre).

Table 2.3 Selected empirical findings from factorial ecologies

Author	City	Observed pattern of variance Social Status	Familism	Ethnicity
Rees (1970)	Chicago	Sectoral	Zonal	Clustered
Murdie (1969)	Toronto	Sectoral	Zonal	Sectoral
Timms (1971)	Brisbane Auckland	Sectoral	Zonal	Clustered
Robson (1969)	Sunderland	Zonal and sectoral	Weak	—
Herbert (1972)	Winnipeg	Sectoral	Zonal	Clustered
Davies and Lewis (1973)	Leicester	Sectoral	Zonal	Clustered
Johnston (1973)	Auckland Wellington Dunedin Christchurch	Sectoral	Zoral	Clustered
Weclawowicz (1979)	Warsaw (1970)	Weak	Weak	—

The majority of factorial ecologies in the Western world, and particularly North America, show distinct similarities, with social status, familism and ethnicity tending to display sectoral, zonal and clustered patterns respectively. In the United Kingdom the pattern tends to be disturbed by state intervention in the housing market and relationships tend to be weaker (see Robson (1969) and Herbert (1972) on Swansea and Cardiff). In the socialist city this lack of spatial differentiation leaves few discernable patterns (Weclawowicz 1979), while in non-Western cities data difficulties have made comparisons doubtful (Abu Lughod 1969, Berry and Rees 1969, Herbert and de Silva 1974).

Despite the welter of words devoted to splitting hairs with Burgess – zonality is distorted by topography (Mann, 1965), by state interference in the housing market (Robson 1969), by non-monetary motives (Firey, 1945), and so on – there is little fundamental dispute with his model *at the descriptive level*. Indeed many of his observations and classifications amount to pioneering insights into the urban condition. Throughout the English-speaking realm and in many other parts of the developed world, the inner city (zone in transition) most certainly *does* contain concentrations of inferior and decayed housing, inhabited by disadvantaged people and there most certainly *is* a link between these conditions and a whole pathology of crime, vice, deviancy, homelessness and every other imaginable form of social disorder. Almost every large Western city has residential areas *de facto* 'reserved' for those excluded from the social mainstream: whether these are central or peripheral, circular or triangular is immaterial to the spirit of the model. Elsewhere in the Western city, the remaining Burgess zones are invariably present – the stable working-class housing area, with its connotations of modest and deserving satisfaction, the middle-class suburb which sets standards of desirability for all to aim at, the extra-urban zone which screens the rich and powerful from the scrutiny of their 'subjects'.

As the geographer's technical virtuosity has come progressively to be tempered with political awareness, so attention has shifted towards the fundamental premises upon which the Burgess model is built. Although superficially the model is a neutral statement about spatial arrangements, it derives from a set of highly controversial postulates about the nature of human society, originally formulated by Burgess's colleague, Robert Park. Residential segregation (and hence housing inequality) are explained by reference to *biotic* processes, to the indisputable fact that humanity is part of the natural world and, as such, subject to instinctive drives, including the drive to acquire living space. Much as in the natural world plants compete with one another for soil and light, animals for territory, so in the city human species (in the guise of classes, races, ethnic groups and other interest groups) compete for space (Park, 1952). Consistent with Darwinian principles (see Saunders, 1981 on Darwin as Park's inspiration), the best locations are commandeered by the 'fittest' species – the most accessible central locations by big business, the spacious new residential land on the perimeter by the owners and top functionaries of big business. 'Lesser' species – the poor, the unskilled, blacks, immigrants – must adapt to the less favourable environments.

Park's vision of the city was one of a human community governed by the same principles as a natural ecosystem. Here we should underline that, despite all the emphasis on competition, this is overwhelmingly a vision of stability and equilibrium. Jungles, concrete or otherwise, are orderly places where the territorial struggle between individual plants and animals is tempered in the last instance by the mutual interdependence between species. This stress on mutual interdependence (derived in part from Durkheim's thinking on the division of labour – See Chapter I and Saunders 1981) is an important pillar of Park's theory, since it purports to explain why cities (and jungles) generally continue to function and reproduce themselves with minimal long-term violence. As in the ecosystem, so in the city each section of the population is indispensible to the functioning of the whole and therefore must be allowed sufficient space and sustenance for continued existence.

Most present-day social scientists would reject this reasoning as obscurantist, considering it self-evident that urban society is held together by *socialization* not by some inbred urge to cooperate. From infancy members of society are taught a knowledge of the duties and obligations expected of them and of the need to conform. Those who, like the zone-in-transition dwellers, fail to display a sense of loyalty to the whole are morally condemned and subject to judicial curbs and penalties. As a social theorist, Park explicitly recognized that among humans biological imperatives are expressed through legally and ethically defined codes of behaviour. Nevertheless, the essence of ecological theory is that ultimate primacy is accorded to the biotic rather than the cultural level.

Taken to its logical limits, Human Ecology is the ultimate and unanswerable justification of the status quo. The implication of its bio-determinism is that residential segregation, housing inequality, slums and homelessness are inevitable facets of the human condition, programmed into our very genes. All attempts to impose change whether by reform or revolution, are mere futile tilting at the unalterable laws of the universe. 'What exists exists because it must exist' (Westergaard and Resler, 1975, 12). The only kind of change envisaged is evolutionary change generated by the urban ecosystem itself. For example, just as natural species evolve into higher forms, so the urban community encourages the more enterprising poor to escape through social mobility from the slums of the inner city (see Chapter 5 on the theory of assimilation).

Stripped of its biological underpinnings Burgess's model loses almost all explanatory pretensions. Several attempts have been made by later authors to justify its spatial properties by recourse to non-biological analysis. Thus, for example, Wirth (1938), himself a Chicago disciple, used overtly sociological reasoning to explain segregation as a means of maintaining harmony through distance between potentially conflicting class and ethnic groups (for later writing on segregation as a functional necessity see for example Gans, 1962; 1979). In the 1950s a school of Land Economists produced a justification of concentric rings based on neo-classical assumptions and couched in the language of land values (Alonso, 1960; Muth, 1969; Evans, 1973). Still more recent attempts have been made to resuscitate ecological theory from within (see for example Berry and Kasarda, 1977) while even today its vocabulary retains a tenacious hold, with the term 'ecological' (meaning no more than 'socio-spatial') continuing to evoke a kneejerk response from otherwise non-believing urbanologists. Yet as long ago as the late 1960s it was evident to many scholars that some new more realistic perspective on the urban housing question was urgently required.

2.3 Social Realism and Political Awareness

2.3.1 Intoxicated by Growth: A Climate of Euphoria

It is easy to appreciate how ecological analysis could flourish in the ideological climate of the 1950s and early 1960s. A school of thought founded on assumptions about in-built social stability, order and consensus was ideally suited to the post-war 'boom' conditions which, on both sides of the Atlantic, had produced a widespread belief in the 'affluent society'. According to contemporary wisdom capitalism was changing its spots, becoming more benevolent as it achieved greater technological proficiency; creating unprecedented increases in the total stock of wealth and at the same time sharing out this wealth in a much more just fashion than formerly (Westergaard and Resler, 1975).

Housing change was seen as a classic symptom of all this. Especially in Britain, there seemed good grounds for believing that the chronic housing shortage which had bedevilled Western society from the earliest phase of industrialism was well on its way to a final solution. Advances in building techniques promised to deliver a steady expansion in the supply of new dwellings sufficient to meet the demands of growing numbers of socially mobile and newly affluent families. The whole process was accelerated by institutional changes, such as the growth of house-purchase credit through building societies and increasingly favourable tax allowances granted to mortgagors. In addition to all this, housing was part and parcel of the 'mixed economy'. The burgeoning free-enterprise housing market was complemented by an equally expansive public sector, through which the State (via Local Authorities) supplied subsidized housing designed to meet acceptable standards at a rent affordable by those unable to meet market rents. As a consequence of these twin trends acting in concert, the social geography of the British city appeared to be undergoing revolutionary change, with mass population decentralization from the slum captivity of the inner city to the promised land – both private and public – of the suburbs.

2.3.1.1 Hangover

Thus the ideological climate of the time was suffused with a belief in the efficiency of technology, growth, mobility (social and spatial) and welfarism as instruments of redistribution. Yet as the 1960s progressed, the mood of optimism began to turn sour: 'The late 1960s witnessed the end of the post-war economic boom – an event which unmasked the latent social conflict between different social groups and between capital and labour' (Bassett and Short, 1980, 3). With the benefit of hindsight, we can now appreciate that reports about the demise of inequality were greatly exaggerated, if not entirely wrong (Westergaard and Resler, 1975; see also Table 2.4). By the early 1970s, several studies of social welfare had demonstrated that poverty – existence below accepted thresholds of income and consumption and not merely underprivilege in relation to average levels of living – was a continuing experience for very large numbers of people (Abel-Smith, 1967; Titmuss, 1968; Townsend, 1979). Not unexpectedly, housing deprivation was a crucial symptom of this condition. At the popular level, the nation's conscience was repeatedly jolted by startling new reminders that its housing stock was not only monstrously inadequate but also a focus for acute and sometimes violent overt conflict – between slum landlord and tenant (Milner Holland, 1965); between Local Authority and tenant, as in the bitter strife between the residents of Ronan Point (a block of damaged and spectacularly unsafe high-rise flats) and the bureaucracy of Newham Council (Dunleavy, 1981); and between immigrants and established residents (e.g. Burney, 1967). These dramatic but sporadic events revealed merely the tip of a vast iceberg of housing scarcity.

2.3.2 The Inner City

Of special interest to geographers and planners was the emergence of the inner city as a *cause célèbre*. The discovery that poverty and housing deprivation were disproportionately localized in the ring of pre-1914 development which sur-

Table 2.4 The distribution of income in the United Kingdom, 1985

| Average per household (£s per year) | Quintile groups of households ranked by original income | | | | | All Households |
	Bottom fifth	Next fifth	Middle fifth	Next fifth	Top fifth	
Original income	120	2,720	7,780	12,390	22,330	9,070
+ Benefits in cash	3,260	2,580	1,190	790	680	1,700
Gross Income	3,380	5,300	8,980	13,170	23,000	10,770
– Income tax and National Insurance	(–10)	360	1,460	2,560	5,300	1,930
Disposable income	3,390	4,940	7,530	10,610	17,700	8,830
– Indirect taxes	790	1,420	2,050	2,640	3,840	2,150
+ Benefits in kind	1,370	1,390	1,430	1,490	1,590	1,460
Final Income	3,960	4,920	6,920	9,460	15,450	8,140

Source: adapted from *Social Trends* 1988, London: HMSO, Table 5.15.

Note: Based on a sample of 7,012 households for the 1985 Family Expenditure Survey. All figures rounded to the nearest £10. Original income is that of a household before taxation and benefits.

rounded the CBD in most large cities and conurbations fitted well with the belief that poverty was largely a *residual* feature of modern society. For the most part the poor consisted of that minority of the population stranded on the platform by the express train of economic, technological and social progress. Typically they could be placed in the following categories: the low-paid and unemployed (excluded by their inability to acquire the skills and qualifications demanded by the modernizing economy); large and single-parent families (victims of unorthodox personal circumstances). The sub-standard conditions endured by these minorities were shared by a fourth disadvantaged group, black immigrants, whose problem was one of adjustment to a new society (see Chapter 5).

The dominant image here is of the inner city as archaic relic, apart not simply in space but in time also: a throwback area with a population locked into the dying industries, occupations and buildings of a bygone age: a twilight life apart from the 'sunrise' world of hypermarkets, credit cards and high mass consumption. Even the black newcomers, with their origins in a 'pre-industrial' Third World, slot conveniently into this notion of a space–time lag in the diffusion of modernization. Although gloomy at first sight, such a notion does imply the possibility of benevolent change. If inner-city slum dwellers are simply those left on the platform, then what is to stop them catching the next train? Several contemporary writers, drawing their inspiration from the zone-in-transition concept, accordingly expressed confidence in socio-spatial mobility as a means of escape into the modern mainstream of the outer city (Abu-Lughod and Foley, 1960; Bell, 1958; 1968). Others, however, more in tune with the new spirit of realism, were beginning to question this, laying stress on the institutional mechanisms by which certain sections of the population were (seemingly permanently) denied this escape route. Especially influential was Rex and Moore's (1967) study of immigrants in Sparkbrook (Birmingham), which demonstrated the way in which racial discrimination (in this case notably the impersonal procedures of the Local Authority) acted as a block on black access to 'legitimate' forms of housing in 'conventional' neighbourhoods. Other writers, though not exclusively preoccupied with racial minorities, began to highlight the tendency for poverty and housing disadvantage to be persistently reproduced (Able-Smith and Townsend, 1965; Runciman, 1966).

2.3.3 Behaviourism – A Blind Alley?

For reasons difficult to fathom, geographers generally failed to respond immediately to the challenge of this changing climate. Although dissatisfaction with the ecological approach did lead to a search for new perspectives this did not take the form of questions about the underlying cause of the housing crisis. Among English-speaking geographers, the late 1960s and early 1970s produced a flurry of studies which attempted to explain residential pattern, structure and change by reference to the actors' own perceptions, especially their perceptions of urban space (Wolpert, 1965; Adams, 1969; Herbert, 1973; McCracken, 1975).

While a refreshing change from bio-determinism and 'number-crunching', this *behavioural* approach ran into the accusation that, in adopting a consumer-choice framework, it ignored the structural constraints to which most of the actors were subject. In cases such as black and immigrant groups, where residential choice was brutally limited by racism on all sorts of levels (see Chapter 5), any analytical emphasis on the resident as 'decision-maker' came dangerously (though unintentionally) close to mockery of the victim. It was scarcely credible that slums and ghettos could be explained by reference to the exercise of choice by their inhabitants. Gray's (1975) charge that social geographic analysis was in danger of becoming a vast exercise in 'non-explanation' struck a sympathetic chord among many disaffected practitioners of the discipline who were searching for new directions.

2.4 Managerialism

Far from a lone cry in the wilderness, Gray's critique echoed a growing dissatisfaction within the discipline itself. *Managerialism* represents one of the important alternative perspectives which were adopted as a result of this dissatisfaction. The outstanding attraction of the managerialist approach is its comparative realism in that it seeks to explain the unequal distribution of life chances by reference to those agencies whose responsibility and power it is to undertake that distribution – a simple equation which had previously escaped the geographer's notice. The origins of urban managerialism are to be found in Rex and Moore's (1967) Sparkbrook study which, besides representing a pioneering study of racism, also codified certain cardinal principles about the workings of the urban housing market. (See Saunders, 1981, 111–15 for a useful summary and critique.)

The first of these principles concerns *access*. Although the total stock of housing is not in short supply, *desirable* housing *is* and therefore access to it is rationed. It is vital to recognize here that housing is unlike any other consumer item in that it is ' . . . . expensive to build, difficult to modify and generally long-standing' (Bourne, 1981, 2). Indeed, it represents a huge long-term investment from the producer's point of view. At market prices – i.e. a level yielding an economic return to builder and landowner – it is beyond the reach of the vast majority of the population. As Castells (1977, 152) puts it, 'There is practically no private production of social housing, whereas one finds industries manufacturing consumer goods for the whole range of incomes.'

Hence in urban society, free market forces alone are unable to provide a key commodity essential to the very survival of the entire population. This very fundamental defect reached its apogee in the appalling squalor of the Victorian city where, according to outraged contemporary commentators, living conditions resembled those of a large-scale extermination camp (Engels, 1845; Booth, 1902). In a sense, the entire modern history of housing in Britain (and to varying degrees in most other Western nations) consists of a succession of Government responses to the inability of market forces to match housing supply to need.

2.4.1 Access to Home Ownership

One major thrust of Government policy has been directed at expanding home ownership. Quite correctly, Rex and Moore (1967) identify private home ownership as unquestionably the most desirable form of housing resource, conferring far more benefits upon its users than any other type (see also Table 2.5). Despite the common use of such terms as 'ownership' and 'private sector', house purchase is far from a crude matter of 'paying your money and taking your choice'. The bulk of home ownership is now financed by long-term mortgage loans and the purchaser/debtor is heavily subsidized by the State in the form of

Table 2.5 The Benefits of Home Ownership

a. Capital appreciation

Year	All house prices Average	Index	New house prices Average	Index	Retail price index
1954	1,990	100	2,110	100	100
1959	2,330	117	2,540	120	116
1964	3,420	172	3,570	169	135
1969	4,540	228	4,760	226	166
1974	10,200	513	10,570	501	274
1979	21,540	1082	23,420	1110	566
1984	32,810	1649	37,440	1774	852

Source: *House Prices Over the Past Thirty Years*, Nationwide Building Society, April 1985.
Note: the average mortgagor gained an average return of 17 per cent per annum on his/her investment in home ownership, plus the additional benefits conferred by the use and enjoyment of the property. This return rises to 19 per cent for new property. The average inflation rate is roughly half this (9 per cent).

b. Tax Relief on a New Average Mortgage in the First Year [a]

Year	Average house price	Average mortgage advance	Interest rate (end of year)	Tax relief
1973	8,830	6,680	11.00%	220
1975	10,130	7,970	11.00%	289
1977	11,390	9,350	9.50%	293
1979	17,680	13,150	15.00%	592
1981	18,830	15,530	15.00%	699
1983	23,230	18,830	11.25%	636
1985	30,550	24,800	12.75%	917
1986	33,280	27,990	12.25%	994

Source: calculated from data in House Prices in 1986, Nationwide Building Society, January 1987.
Note: [a] – figures refer to first time buyers; the average tax relief paid to a previous owner occupier in the first year after moving was slightly higher (£1037)

income tax relief. A lengthy history of Government subsidization has resulted in a national rise in home ownership from 9 per cent of households in 1914 to over 64 per cent by the end of 1985 (Boddy, 1980; Nationwide Building Society, 1986)[1].

It is clear that house ownership has become substantially 'democratized', no longer the privilege of a minority elite. Even so it retains a distinct degree of selectivity. Rex and Moore (1967) note that consumer access to this highly protected and highly valued sector is governed by *creditworthiness*, a factor which confers considerable power on those financial intermediaries – principally building societies but also local authorities and banks – whose task it is to

1 State subsidies to mortgagors rose from £135m. in 1965/6 to £1240m. in 1976/7, almost a four-fold increase in real terms (Boddy, 1980). In the mid-1970s the average subsidy to mortgaged owners (£140 per annum) roughly equalled that to council tenants. Over the past decade rapid increases in house prices and mortgage interest rates, a lack of new local authority building and the present government's attempts to fix public sector rents at 'full market value' have altered the picture dramatically. Subsidies to mortgaged owner-occupiers are now over four times those to tenants and for those with large and/or recent mortgages the discrepancy is even greater.

provide that credit. The nodal role of these agencies was first underlined by Pahl (1970), who coined the term 'gatekeeper' in recognition of their function as filters in the housing allocation system, with the power to decide the eligibility of applicants for housing credit. In the most literal sense, it is the gatekeeper who determines who gets what where. The gatekeeper concept inspired a healthy flow of geographical and other research, examining both the motives for and outcomes of building society decisions. For example, Boddy (1976), in a study of Newcastle, found that eligibility for credit was decided not only by objective criteria such as size of income but also by gender and by the manager's subjective judgement of the client's status – 'suitability', 'stability' and 'risk potential'. Manual workers were notably disadvantaged by this procedure. In addition to their income disadvantages they frequently failed to conform to the middle-class manager's stereotype of a potential property-owner (a finding replicated elsewhere, e.g. Foord, 1975; Williams, 1976 in the United Kingdom; Harvey, 1977; Dingemans, 1979 in the United States). Thus managerial decisions are seen as a major factor in the maintenance, perhaps even the promotion, of residential segregation. All this is very much at variance with the mobile and fluid socio-spatial system envisaged by most behaviourists.

A key theme in managerialism, then, is *constraint*, the manner in which scarce resources are rationed, access denied and social groups effectively directed to particular housing in specific areas. Though it is building society activity which has attracted greatest attention in this connection, a number of studies have also identified estate agents (Williams, 1976; Cater, 1981), property developers (Pahl, 1969; Ambrose and Colenutt, 1975) and surveyors (Lambert, 1976) as active agents in shaping the housing geography of the city. Almost invariably, the impression gained is one of bias, ideological as well as financial. Those who qualify for the benefits of owner-occupation are overwhelmingly those whose lifestyle conforms to the core values of a property-owning democracy (Saunders, 1984).

2.4.2 Publicly Owned Housing

In turning the spotlight on urban managers, Pahl and Rex and Moore were concerned less with suburban private property-owning and more with decision makers in the Local Authority housing sector. The second great twentieth-century change in the British housing market has been the growth of public-sector housing, which despite recent shrinkage (Table 2.6), accounts for over one in four households nationally and a larger proportion still in the conurbations and in Scotland. Through the financial backing and legal powers vested in them by the State, Local Authorities have until the late 1970s played an expanding role as builders and landlords of dwellings which are allocated on the principle of need at a rent often below market levels. In geographical terms, this large-scale public participation has produced settlement patterns very different from those predicted by the ecological model. In Britain, suburbanization has not been an exclusively private affair but has also spawned vast municipal estates as well, with a strategy of slum clearance and overspill rehousing breaking the middle-class monopoly of the urban periphery. Commenting on the rise of the council estate, Rex remarks drily, ' . . . .the working classes, having attained a measure of power in the town hall, have their own suburbs built for them. They are modelled on those of the middle-class people but are distinguished by the fact that once a week a man from the council calls for the rent' (Rex, 1968, 214). Implicit in this statement is the recognition that council housing *in principle* is another desirable and desired resource – a working-class right established through political leverage rather than a charity granted by patronage. It also represents a leap forward in material terms, an escape from the overcrowded, antiquated conditions of the old city to the kind of space and comfort which previously had been reserved for the middle classes.

In practice, there are many ways in which council housing falls short of this idealized role. Firstly, it is not openly available to every casualty of the private market. It is restricted to those who meet the criteria of 'need' as defined by rules laid down by individual Local Authority housing departments, criteria which despite some recent improvements may be so inflexible as to exclude many of the groups seeking housing (for an instructive example, see Table 2.7). In the Sparkbrook study Rex and Moore (1967) identified recent immigrants as a disqualified group, barred from entry by insufficient length of residence in the L.A. area. Elsewhere, further studies have highlighted the question of 'problem families', households excluded by housing managers as 'socially disruptive', 'troublemakers', 'social misfits' and potential rent defaulters (Gray, 1976; Skellington, 1980; Taylor and Hadfield, 1982).

Secondly, not all council housing is a desirable good: 'Most cities in Britain own housing on estates ranging from relatively desirable through to disgraceful and degrading' (Taylor and Hadfield, 1982, 249). For Rex and Moore the key distinction was between *legitimate* and *submerged* housing, the latter consisting mainly of old short-life property ('patched' houses) acquired by the council and used as temporary accommodation prior to demolition. We might add numerous other categories of submerged council estate, where standards fall far short of the minimum acceptable and where tenant dissatisfaction has become

Table 2.6 Housing tenure in Britain, 1914 to 1986

Year	Owner occupied	Public rented	Private rented
1914	10.0	1.0	89.0
1938	25.0	10.0	65.0
1945	26.0	12.0	62.0
1951	29.0	18.0	53.0
1956	34.0	23.0	43.0
1961	43.0	27.0	31.0
1966	47.1	28.7	24.2
1971	50.5	30.6	18.8
1976	53.3	31.7	15.0
1978	53.9	32.3	13.8
1980	56.2	31.2	12.7
1982	58.6	29.5	11.8
1984	60.9	27.9	11.1
1986	62.9	26.7	10.4

Sources: Boddy (1980), figure 2.5., p. 154; *Annual Abstract of Statistics* (1988a), table 3.7., Central Statistical Office.

Note: a – the private rented sector includes an increasing proportion of households renting from housing associations and co-operatives (2.5% of all households in 1986). Since 1986 the recent trends of an increase in owner-occupancy, a decline in council tenancy and a decline in renting from private landlords have continued.

Table 2.7 Housing waiting list regulations in West Midland Authorities (after Niner 1975)

Regulations	West Bromwich	Warley	Wolverhampton	Halesowen
Residential qualification	None	Must live in County Borough, or have lived in town for 2 years and not lived outside for more than 5 years, or have worked in Warley for 10 years	Must live in Wolverhampton	Must have lived or worked in Borough for preceding 12 months, or lived there most of life
Other qualification	None	Single person without dependants must be aged 25 or over	None	Over 18, and have established a *prima facie* need for housing
Eligibility for consideration	Must have lived in Borough for past 12 months, or 10 of the last 15 years	Applicant (or wife) must complete 2 years residence in Warley	As above	As above
Other qualifications	Must be in housing need: – lodgers – overcrowded families – living in flat after 5 years – living in house without bath after 5 years – ill health	Must not be, or have been in past 3 years an owner-occupier of self-contained accommodation. Must not be a tenant of adequate-sized self-contained accommodation with all amenities and in a good state of repair	If a private tenant of self-contained premises must establish significant evidence of housing need. If an owner occupier must have evidence of severe ill-health or social need	None, though an owner occupier may have to accept a nominated tenant for his property as a condition of rehousing

Source: taken from Short (1982), table 7.2., pp. 166–7.

increasingly audible – old, unmodernized, badly maintained estates, system-built properties, isolated, under-serviced overspill developments, estates destroyed by vandalism, petty crime, chronic rent arrears and other forms of community disorder. In many instances the categories overlap, producing night-marish living conditions for the inhabitants and mocking all the ideals of public housing as a social service (see Chapter 6).

Undoubtedly, the most highly publicized public-sector failure is the high-rise flat (Association of Metropolitan Authorities (AMA), 1984b). A classic symptom of the 1960s boom mentality, the high-rise flat represents unquestionably the purest expression of faith in technology as public benefactor. The system-built tower blocks were conceived as an outstanding cost-effective and technically brilliant solution to the problem of housing shortage. In practice they were expensive, badly designed and built and contributed significantly to the demise of the inner-city community (Dunleavy, 1981). After two decades of tenant protest and 'shock horror' media publicity it is now officially accepted that high-rise blocks are completely unsuited to family and community life. Worse still, it is clear that defects in construction have produced not only chronic problems of damp, noise and inadequate heating, but also a state of decay so advanced that a large proportion of dwelling units are a direct danger to life and limb (AMA, 1984b). Since 1975 almost all construction of multi-storey flats has ceased, but high-rise living remains a reality for many public sector tenants.

2.4.3 Conflict over Resources

Intertwined with the question of differential access is the further question of *class conflict*. A second cardinal principle enunciated by Rex and Moore (1967) was that the urban housing market can only be understood as an arena of struggle between various interest groups for a share of scarce resources. Class conflict differs fundamentally from the concepts of competition used by ecologists or neo-classical economists. It implies an unequal struggle between strong and weak, in which the former use their superior power to enhance their own resources at the expense of the latter – there are winners and losers. Although the losers – in this case those relegated to submerged housing – give every appearance of passive acquiescence, this does not denote satisfaction or agreement with the rules of the game (Dunleavy, 1977). Intermittently their awareness of themselves as victims bursts forth as organized political activity, sometimes legitimate (as with the emergence of residents' associations and community action groups), sometimes in a manner which transgresses the legal bounds of politics (as with the 'inner city' riots of the 1980s in Britain). More usually, however, dissatisfaction translates into *alienation*, a deep-seated sense of isolation, withdrawal and non-identity with society. Depending on the indivi-dual, this leads either to passive adaptation or anti-social behaviour (the expansive literature on alienation is selectively reviewed in Chapter 3).

Who are the new winners and losers? Rex and Moore's (1967) concept of class derives directly from Weber's (1947) proposition that stratification – the division of society into differentially rewarded groups – stems from the unequal possession of three resources: *wealth* (the economic dimension); *status* (social esteem, prestige); and *power* (the degree to which one dominates or is dominated). Until the Sparkbrook study, almost all class analysis presumed that the unequal distribution of all three resources was determined solely by production relations. For example, access to housing depends upon income/wealth, which in turn is determined by position in the labour market. Yet as we have now established, the housing market is not responsive only to income differentials. In the case of working-class entrants to decent council housing, their position in the market provided an effective addition to their real wage. In the past decade, however, the subsidy to public-sector tenants has been drastically eroded. Increasingly the housing market acts to redistribute real income in an extremely regressive fashion *from* low *to* high-income groups, with suburban owner-occupiers tending to gain extra rewards while the submerged housing groups incur extra penalties. This semi-autonomous operation of the housing market, the fact that one's relationship to the means of shelter may not be identical to one's relationship to the means of production, led Rex and Moore to propose a model of *housing classes*. Their model is designed to rank the urban population in terms of rewards, status and power in the housing market rather than in the labour market. Table 2.8 reproduces the original Rex and Moore classification.

2.4.4 Housing Class and Tenure

One important feature of the model is that housing class emphatically does not equate with tenure status. Even though suburban owner-occupation is naturally ranked top in terms of material and non-material benefits, tenancy of the better-built and more spacious outer council estates is seen as *superior to ownership of substandard inner-city property*. By and large public-sector housing is considered part of the legitimate housing mainstream, both materially decent and 'morally respectable'. Though obviously important, the private ownership of property is not the sole determinant of a household's shelter benefits nor of its

Table 2.8 Rex and Moore's (1967) Typology of Housing Classes

Housing class	Housing situation	Spatial outcome
I	Outright owner of a whole house	In the third and fourth zones from the centre (medium and high status residential) and outside the city in satellites
II	Owner of a mortgaged whole house	Predominantly in the fourth zone
IIIa	Council tenant in a house with long life	In the fourth suburban zone or, after slum clearance, in higher density council houses in the first and second zones
IIIb	Council tenant in a house awaiting demolition	After slum clearance often in higher density council property in the first and second zones
IV	Tenant of a whole house owned by a private landlord	Normally in the first and second zones
V	Owner of a house bought with short-term loans who is compelled to let rooms in order to meet repayment obligations	In all zones, but especially zones, one, two and three
VI	Tenant of rooms in a lodging house	Lodging houses, occupied by their owners and tenants, will be in zones one and two, but predominantly in zone two

Source: Rex and Moore (1967), pp. 274–5.

status and power in the housing market.

Twenty years on, however, there are those who would dispute the continuing relevance of this aspect of the model. For writers like Saunders (1984), recent government housing strategies – intensified financial and ideological support for ownership coupled with the impoverishment and progressive privatization of the public sector – are converting council tenancy into a residual and ever more undesirable form of tenure. The tone of this official hostility to the public sector can be gauged by a recent public pronouncement by a government minister: 'Council housing breeds slums, delinquency, vandalism, waste, arrears and social polarization' (Geoffrey Pattie, MP, reported in the *Guardian* 17 June, 1986). Throughout the present decade the entire concept and practice of adequately funded social housing allocated as of right according to need has been under threat and in consequence a yawning gulf has opened up between what Saunders calls the 'privatized majority' and the 'marginalized minority' – those who can achieve satisfaction through ownership and those who are left high and dry in a dwindling, decaying, under-resourced, stigmatized public sector of the last resort. It is notable that by 1984, the former category had extended well beyond the 'middle classes'. It now embraces no less than 66 per cent of skilled manual worker-headed households, 46 per cent of semi-skilled and 33 per cent of unskilled households (Central Statistical Office, 1987a).

Additional empirical weight is lent to Saunders's proposition by recent work (notably Hamnett, 1984), which finds that nationally the population in public-sector housing is: (a) declining numerically; (b) ageing; (c) declining in occupational status; (d) becoming increasingly dominated by the unemployed and other welfare dependants. This 'socio-tenurial polarization' (to use the egregious jargon favoured by these writers) is the inevitable result of the sell-off of the better dwellings on the more desired estates to the higher-income (or less badly-off) tenants. As time passes, so the council sector becomes systematically devalued, ever more equated with second-class housing status.

Or does it? While no one can dispute that council renting is under grave threat, what *is* in question is whether owner-occupation offers a genuinely superior option (or even a genuine option *per se*) in all cases. As we have argued elsewhere:

> In practice, the two supposedly distinct categories overlap . . . the 'privatized majority' actually contains a section of the 'marginalized minority', some of whom are living in near slum conditions in substandard environments. For us the division within home owners seems no less politically significant than that between owners and renters. (Cater and Jones, 1987, 200. (See also Dickens *et al.*, 1985 on the inadmissibility of equating tenure status with housing status).)

Or to paraphrase briefly, not all privately owned housing is good housing, notwithstanding the tendency for both official and popular wisdom to regard home ownership as synonymous with housing satisfaction. Indeed there are grounds for the belief that a not inconsiderable portion of owner-occupied dwellings should be included within the submerged housing category. Even today, a significant minority of owner-occupied dwellings are materially substandard in one or more respects (Table 2.1). More often than not these are aged inner-city dwellings, lacking the benefits of modernization and suffering from all the 'negative externalities' characteristic of their neighbourhood, though a substantial number of rural properties also still lack basic amenities.

Many home-owners in these and other dwellings fail fully to obtain the

Table 2.9 Household Tenure by Socio-Economic Group, 1983

Socio-Economic Group	Home owners:		Rented tenures:		Private rented
	Outright Owners	Mortgagors	Local authority[a]	Housing association[b]	
Professional	10	78	2	*	9
Employers & managers	17	70	5	1	8
Intermediate non-manual	12	67	10	1	11
Junior non-manual	16	51	20	2	11
Skilled manual	14	52	28	1	5
Semi-skilled manual	12	34	41	2	11
Unskilled manual	11	22	57	4	7
Economically inactive	42	5	41	3	9
Total	24	38	28	2	9

Source: Central Statistical Office (1987a), *General Household Survey 1985*, table 5.13.b.
Notes: [a] – includes those renting from New Town Development Corporations
  [b] – includes housing co-operatives
  all figures are percentages rounded to the nearest whole number. Based on a sample of 9607 households
  * – less than 0.5 per cent.

economic benefits normally associated with ownership (Short, 1982; Forrest and Murie, 1986; Cater and Jones, 1987). For example, cheap old inner-city property does not appreciate in market value as rapidly as suburban property and some-times fails even to keep pace with general inflation. Moreover, in the lower reaches of the private sector many owner-occupiers cannot obtain one of the prime assets of home-ownership, mortgage tax relief. This is because they are either non-mortgaged (cash purchase of very cheap property), mortgaged for only minimal sums and/or through informal arrangements or earning less than the minimum income-tax thresholds. These peculiar conditions apply strongly (though not exclusively) to black and Asian home owners (of which more in Chapter 5; see also Cater and Jones, 1987).

All in all then, although the detailed lines of the class-tenure system have become rearranged, sometimes forcibly so, the basic anatomy as charted by the Sparkbrook project remains intact. For example, racial-minority households may have exchanged the tenancy of multi-occupied slums for the ownership of inner-city terraces but they continue to occupy dwellings and localities which fall below the normally accepted standards of comfort and amenity (Rex and Tomlinson, 1979; Ward, 1982). As a further instance, many former council tenants now own houses which continue to be in need of improvement, which are situated in 'undesirable' neighbourhoods and which offer very poor prospects for resale. In response to these and the related considerations raised in this section, we have drawn up in Table 2.10 a modernized version of the housing class schema (with apologies to Rex and Moore) which attempts to summarize the rewards and penalties accruing to households in each of three major strata. The table under-lines the cardinal divide between conventional and sub-standard owner-occupation, and insists that the autonomy and self-determination (what Saunders, 1984 calls 'exclusivity') conferred by the former are relatively absent from the latter, except as a figment of the New Right imagination.

2.4.5 The Managerial Thesis Re-examined

If we accept that the housing market operates relatively independently as an allo-cator of rewards and penalties, then we place an even greater onus on the role of gatekeeping institutions. One is tempted to attach absolute importance to managerial decision-making as holding the final key to the mysteries of the housing crisis – the persistent scarcity of acceptable dwellings, the acute injustices which stem from the need for rationing and the yawning gap which exists between political promise and practical fulfilment (for historical com-mentary on official housing targets since 1945, see McKay and Cox, 1979; Dun-leavy, 1981; 1982). In addition to the specific forms of bias (perhaps unconscious on the part of the individual decision-makers) previously documented in this chapter, there is a more general sense in which the 'managerial culture' may be seen as hostile to many of the more vulnerable social groups. Particularly in the public sector, a great gulf exists between the goals, aspirations and perceptions of the managers and their expert professional advisers on the one hand; and those of ordinary people, their 'clients' on the other. Accordingly, in his initial develop-ment of the managerialist theme, Pahl (1970) took as central the premise that managers are the 'independent variable' in the urban allocation system:

Table 2.10 A modified version of Rex and Moore's housing classes scheme

Class	Tenure, occupants and typical location	Entry requirements	Utility derived by occupants: Material	Status	Power	Neighbourhood
IA	Owner-occupiers (outright or mortgaged) of legitimate housing; suburban (or sometimes conserved or gentrified)	Wealth or credit-worthiness; meet building society criteria	Acceptable shelter at a price subsidised by tax relief; appreciation of assets	Property ownership held in high esteem	Exclusivity; autonomy; mobility and choice	Local access to superior environment, schools and services
IB	Renters of legitimate council housing, usually suburban	Meet local authority 'points' and other criteria – prove need and worthiness. "Respectable" working class	Acceptable shelter; some rent subsidy	Stigma attached to council renting	Dependent on local authority provision and rules; limited choice and movement	Local services may be inferior; mundane and uniform design; poor location
2A	Owner-occupiers of low grade property; usually old and inner city	Failure to qualify for IA and IB – low income, unemployment, perceived deviancy (e.g. non-white, immigrant, single parent, "problem" family, mental illness, criminal record – all may act as disqualifiers)	Frequently unmodernised, overcrowded and lacking amenities; capital appreciation minimal and little tax relief	Pale imitation of IA	Independent, but often lack capital to maintain or improve	Old 'worn out' neighbourhood; high population density, mixed land uses; hazards and dangers include noise, dereliction, traffic, pollution, petty crime, vandalism filth; poor service provision; negative image etc. etc..
2B	Tenants of undesirable council property; inner city or overspill		Overcrowding, lack of amenities, disrepair	Second class version of IB; highly stigmatized	Dependency on often sub-standard local authority provision	
3	Private renters; mostly inner city	As 2A and 2B	Beset with similar problems to IIB, but higher rents, multi-occupation and shared facilities in many cases	Slum dweller	Dependent and vulnerable, though some legal protection	

Mainstream, 'Normal' or Conventional Housing (IA, IB)

Submerged, 'Subnormal' or Substandard Housing (2A, 2B, 3)

> . . . the task for research was to discover the extent to which these different gate-keepers shared common ideologies and therefore acted consistently with each other in generating and perpetuating definite patterns of bias and disadvantage . . . (Saunders, 1981, 119).

For all its attractions, this line of investigation poses serious problems. As Pahl himself was soon ready to acknowledge, he had attributed far too much power to a group of individuals and institutions who in reality are not so much *independent* as *intervening* variables, acting within constraints posed by larger more genuinely potent organizations. The point can be well illustrated by reference to the problem of overspill council estates, a category which contains many of the more spectacular public-sector housing failures. Here, 'pure' managerialist interpretation would tend to place maximum emphasis on the role of the local housing authorities concerned – on their managerial ideology, with its cult of 'the expert knows best', resulting in the choice of remote and unpopular locations and unsuitable design; on the lack of foresight and planning resulting in a shortage of shops and community facilities; on the defective selection procedures which have clustered many of the poorest and most economically helpless households in close mutual proximity.

While largely accurate on its own terms, all this takes insufficient account of forces entirely outside the Local Authority's control. At root, many of the inadequacies of the municipal housing system are attributable to financial constraints. As a broad generalization it is fair to say that rarely has any Local Authority enjoyed access to sufficient finance to ensure that every new estate is built according to high standards of dwelling and social provision in accessible areas and populated according to a sensitive and consultative rehousing procedure paying due attention to individual needs and preferences and to family and community ties. Once this point is conceded, it is clear that council housing shortfalls are at least partially traceable to *Central Government*, a primary source both of funding and of housing policy guidelines. Particularly pertinent in this context is the recent privatization of public sector housing (Table 2.11). This realization has led to a rich new field of research on relations between the national and the local state (Cockburn, 1977; G. Jones, 1980; Dickens *et al.*, 1985). How autonomous/dependent is the latter *vis-à-vis* the former? While not attempting a definite answer here, it is at least clear that the national state determines the political–economic environment in which the local state must operate.

But the buck cannot stop even here. In Britain, the Local Authorities with the greatest problems of council housing – in general the large Metropolitan District authorities – are not simply those which have suffered most heavily from the past decade of Central Government cutbacks (Harloe and Paris, 1984; see Table 2.2 on public sector completions). They are also those with the highest proportions of needy households, the unemployed and other marginalized groups with no means of housing themselves. Here again, Central Government may be seen as the independent variable, in its role as national and regional economic manager. Ultimately, however, the analytical chickens may come home to roost at a different level; as seen in Chapter 1, localized concentrations of marginalized workers may readily be explained as the product of the changing needs of capital, both national and international.

Table 2.11 The sale of council houses in England and Wales 1971–86

Year	Sales by local authorities	Sales by New Town DC's	Total sale
1971	17,214	3,157	20,371
1976	5,793	102	5,895
1977	13,020	365	13,385
1978	30,045	575	30,620
1979	41,665	795	42,460
1980	81,485	4,215	85,700
1981	102,825	3,660	106,485
1982	201,880	5,170	207,050
1983	141,615	4,835	146,450
1984	102,635	4,290	106,930
1985	92,295	3,110	95,405
1986	88,410	2,415	90,820

Source: *Social Trends* 1986 Table 8.15: 1988 Table 8.11.

2.5 Marxist Approaches to Housing

Not unexpectedly, the sharpest critique of managerialism has come from Marxists, who hold that managers are precisely what their name suggests – the executives of a higher will rather than a ruling elite in their own right. For geographers, the possibilities opened up by Marxist method were first seriously ventilated by Harvey (1973), who expressed some surprise that one of the great classical Marxist contributions to the urban question had apparently made no impact whatsoever in urban geography.

> The line of approach adopted by Engels in 1844 was and still is far more consistent with hard economic and social realities than was the essentially cultural approach of Park and Burgess . . . It seems a pity that contemporary geographers have looked to Park and Burgess rather than to Engels for their inspiration (Harvey, 1973, 133).

2.5.1 The Engels Model

As a direct statement about residential geography by a colleague of Marx, Engels's portrayal of Manchester in 1844 offers an appropriate introduction to this section. It is first of all apt because it forces us to concede that spatial patterns can convey only rather limited information about social relations. Space is suggestive, perhaps indicative, but never definitive. A reading of Engels suggests that the spatial geometry of early industrial Manchester was a somewhat cruder version of that discovered by Burgess 80 years later in Chicago – though so far as we are aware Burgess did not acknowledge this parallel. Engels describes what amounts to a central business district ringed by workers' housing, which in turn was flanked by the vastly superior housing of the bourgeoisie. The resemblance to the Chicago pattern is striking.

Yet the socio-political inferences drawn by each writer could not have been more dissimilar. For Engels the land-use geography of the city was the unambiguous product of class conflict. It was shaped entirely to the requirements of capital, the dominant class, which had commandeered the accessible heart of the city for its own production purposes and the perimeter for its own consumption. As befits a subordinate class, labour was forced to occupy the inadequate residual space at appalling densities in wretched conditions. There is no talk here of bio-determinism or ecological imperatives. The urban system was shaped by political and economic power. It embodied the expropriation of one class by another.

The model is also significant for our purposes in that it introduces a number of theoretical principles (only partially germinated here but later to come to maturity in Marx's *Capital*) which have become indispensible to a Marxian perspective on the city. First and absolutely pivotal is the concept of *class* and class conflict. As noted in Chapter 1 the Marxian tradition recognizes only two primary classes: *capital* (the bourgeoisie) which uses its power over the means of production to expropriate surplus value from *labour* (the proletariat). For Engels, Manchester's urban land-use pattern epitomized this polarity between the two mutually antagonistic classes, an arrangement designed to maximize capital accumulation at the expense of labour. Equally central to the housing question is the concept of the *reproduction of labour power*. Housing is one of the necessary consumption items by which the proletariat is enabled to live and to reproduce the next generation of workers. This is the principal material purpose of housing in the capitalist system. Yet, within the system, there is a constant tension (contradiction) between capital's drive to accumulate and the need to reproduce the requisite labour power. If too much surplus is extracted (through low wages and/or high rents) then workers' living standards fall below subsistence level and the supply of labour power is threatened by high mortality and disease. Conversely, if living standards rise too high, then profit levels and accumulation are threatened. The brutish housing conditions of the nineteenth-century industrial city seem to represent the minimum consistent with social reproduction. Any tendency for wages to rise or rents to fall was constantly countered by the recruitment of workers from the industrial reserve army in rural Britain and of course Ireland. As Preteceille (1981, 6) puts it, capital has ' . . . always tried to obtain the necessary labour without paying the price for it by means of structural unemployment, the mobilization of workers for other sectors (artisan production, peasantry) or other countries.' (See also Harvey, 1982 and Chapter 1 of this book.)

2.5.2 The Capitalist City Transformed

Non-Marxists would undoubtedly dismiss the above analysis as being no longer relevant. Since the age of Engels the transition from free market to welfare capitalism has been marked by a progressive improvement in mass housing standards, widening access to property ownership, ever-growing government involvement in housing and the growth of an intricate system of residential stratification which would seem to signify the breakdown of the traditional class polarity. Yet there is a sense in which this mutation is one of shadow rather than substance. Despite their very considerable variation in emphasis and direction, Marxist urban analysts (see in particular the influential work by Castells, 1977; 1978; 1983 and Harvey, 1973; 1982) are virtually unanimous in depicting modern urban change as being shaped *by* capital *for* capital. Capitalist expropriation has not waned nor has capitalist domination been supplanted by the power of new ruling elites. Instead, the conspicuous and extreme methods of expropriation favoured by the early industrialists and rentiers have been abandoned in favour of subtler, less visible processes, usually operating through a hazy web of intermediaries, frontmen and impersonal agencies. The outward style of this neo-capitalism (Castells, 1978) is to all appearances benign but its basic imperatives remain unchanged.

In the field of housing, Marxists find themselves in initial agreement with managerialists that the key development of the past century has been the expansion of state intervention. Bearing in mind the range of state housing-related activities – planning and land use management, as well as policies to stimulate housing supply – all would agree with Castells, who calls its contribution 'massive and decisive' (1978, 17) and 'indispensible for the survival of the system' (1978, 19). In varying degrees, all neo-capitalist societies depend on state aided (socialized) housing (Department of the Environment, 1977b). Together with education, health care and other public amenities which contribute to the social wage, housing should now be regarded as an item of *collective consumption* (Castells, 1977; see Bassett and Short, 1980 and Saunders, 1981 for a critique of this concept).

Where Marxism parts company with almost every other school is in its insistence that state policies are directed by the needs of capital rather than by electoral accountability, managerial power or some vague altruistic welfarism. Nor does public housing reform represent any kind of victory for labour (Preteceille, 1981, 8). On the contrary, modern Marxists remain broadly faithful to the dictum of the founding fathers that 'the executive of the modern state is but a committees for managing the affairs of the whole bourgeoisie' (Marx and Engels, 1848; 1969 reprint). Ownership of the means of production confers political power, not least through control of apparatus of state government.

2.5.3 State Housing Intervention and the General Interest of Capital

Although we stand by the above summary as a fruitful basis for an analysis of class–state–housing linkages, we are well aware of its inadequacies. Modern Marxist state theorists remain locked in debate as to how far the state may be said to serve the 'general interests' of capital and how far it enjoys *relative autonomy*; and on the manner in which state action to protect the interests of the dominant class as a whole may conflict with the specific interests of parts of that class. For example, British state policies to control rents and promote council housing after World War I were in the general interests of capitalist domination in that they were designed to appease housing dissatisfaction and working-class unrest but at the same time they brought about the long-term decline of private landlords (petty rentier capital), a specific fraction of the capitalist class. While detailed exploration of these subtleties is beyond the scope of the present book, the reader is referred to such texts as Crouch (1979) and Harloe (1981) for more sophisticated guidance.

At the material level, modern state housing intervention is primarily geared to ensuring the reproduction of labour power. Here the function of the state is to address itself to the 'great contradictory double bind' (Castells, 1978, 47) between accumulation and reproduction: to allow popular access to decent housing is to sacrifice the rate of profit but to deny such access is to threaten the reproduction process and, worse still, to court social upheaval and organized resistance on the part of labour. A central tenet of Marx's teachings was that primitive expropriation contains the seeds of its own destruction, generating a growing proletarian awareness of itself as an exploited class and leading to political action to overthrow capitalist rule ('class-in-itself' becomes 'class-for-itself'). That this prophecy has remained until now unfulfilled is due in no small measure to *appeasement*. As Castells (1978, 17) reminds us, the 'growing bargaining power of workers and popular movement' has continually obliged capital to 'cede to popular demands' for increased consumption (see also Preteceille, 1981, 6–7). Hence the very survival of capital as an unchallenged ruling class depends on the institutionalization of appeasement: ' . . . rising housing standards become incorporated into what labour comes to regard as the minimum subsistence wages' (Bassett and Short, 1980, 207).

We see here the intimate interconnection between the material, political and ideological levels of class conflict. State intervention is invoked because, as we have seen, private capital cannot profitably supply decent mass housing at an affordable price. But the primary object of this exercise is not to satisfy labour's housing aspirations but to secure the unchallenged political power of capital. Seen in this light, housing reform is first and foremost a counter-revolutionary strategy of social control. At the same time it is also a form of ideological control. Here we should emphasize that the essence of neo-capitalist power is *legitimacy* or rule by consent – the willingness of the majority of its subjects to tolerate the capitalist order as normal, unquestionable, routine, apolitical, even positively benevolent (Westergaard and Resler, 1975). Indeed, most subjects of capitalism are entirely unconscious of their subjection or even of the existence of a common class enemy. That such 'false consciousness' should be so widespread is hardly surprising in view of capital's control of all the organs of information (including the educational system and the media) which shape the entire culture (see Boggs, 1976 and Gray and Dickens, 1977 on Gramsci's concept of ideological hegemony). As always throughout history, the ruling class is literally in a position to write its own version of events but in doing this it does not stop at mere depiction: it continually offers concrete 'proof' of its own beneficence. Improved levels of collective consumption are among the more tangible means of winning hearts and minds.

2.5.4 The Residential Division of Labour

A further aspect of false consciousness is *class fragmentation* – the tendency for each member of the proletariat to identify with an immediate interest group defined by status or some other non-class factor. For example, it would be difficult in the extreme to persuade a home-owning Surrey bank clerk that, as a seller of labour power, he shares common cause with a Belgian miner or a black labourer from the Bronx. Consciousness of nation, region, race and, above all, status intervenes to stifle class consciousness. To a vast extent, fragmentation is a product of the manner in which capitalist social relations are organized. The housing market is a particularly potent source of divisiveness, with its residents differentiated into a finely graded hierarchy of status layers, each of which is perceived by its members as an interest group in its own right. Capital's material concessions are not awarded *en bloc* to the whole of labour. They are selectively directed and the resultant inter-group rivalries serve to deflect attention from capital itself.

At root, the key to class fragmentation is the *occupational* division of labour and numerous writers (usefully reviewed by Salaman, 1981) have examined the way in which the large corporation uses labour differentiation to stifle class consciousness and promote inter-worker rivalry. Citing Foster (1974), Salaman refers to ' . . . . a deliberate employers' strategy of dividing the working class and, at the same time, bribing its upper layers into political acquiescence, into identifying with the interests of the employers' (1981, 177). Whether deliberate

Table 2.12 The Residential Division of Labour and its Purposes for Capital

Mode of housing	Purposes for capital: Material	Political	Ideological
IA Mainstream Owner-Occupation (Privatized)	State-subsidized high level reproduction of privileged labour. Expanded accumulation for housing market capitalists. General accumulation aided by consumerism.	Co-opt of class allies. Stake in preserving status quo	Subsidised individualism and anti-collective support for capitalist proper relations. Social distance from rest of labour
IB Mainstream Council Renting (Socialized)	State-subsidized reproduction of manual labour; lowering of wage costs to employers	Class appeasement; direct bureaucratic control of labour in their living space. Appeasement by conceding shelter demands	Moral and spatial isolation from owners; collectivist values
	CRITICAL CONSUMPTION CLEAVAGE		
2A Submerged Owner-Occupation (Privatized but Marginalized)	Degree of self-reproduction for sections of labour and surplus labour	As IA, but individualism only minimally subsidised	
2B Submerged Council Renting (Socialized and Marginalized)	Increasing degree of self-reproduction of surplus labour	Moral and spatial isolation from rest of labour	
3 Private Renting (Marginalized in a Free Market)	Self-reproduction of surplus labour	Moral and spatial isolation from rest of labour again commonplace	

or unconscious on the part of capital and the state, such practices are clearly reproduced in the housing market, with its creaming off of non-manual workers and 'labour aristocrats' (Salaman, 1981, 177) into relatively privileged forms of housing and its physical segregation of each layer of the working class into discrete spatial compartments. To a very great extent, the residential division of labour is the counterpart of the occupational division of labour extending its principles from the workplace into the home.

Turning now from high abstraction to rather more concrete issues, Table 2.12 is an attempt to summarize the major social divisions of the British housing market in terms of their role in the residential division of labour. The table employs the same five modes of housing as identified earlier (Table 2.10) but this time we assess their utility for *capital* rather than for the *consumers* of housing. Each of the five modes of housing is evaluated for the part it plays in helping to secure (a) *capital accumulation* (its *material* purpose); (b) *capitalist domination* (its *political* purpose); (c) *capitalist legitimacy* (its *ideological* purpose, rule by consent.) No doubt this approach will be criticized for being unduly functionalist, but there are precedents for it (for example, Dunleavy, 1982) and in any case some attempt at qualification will be made in the ensuing text.

2.6 Towards a New Model of Housing Classes

2.6.1 Good Quality Owner Occupation (1A)

As will be recalled (Table 2.6), owner-occupation has expanded decisively since the 1920s, to a point where it now accommodates a clear majority of British households. Throughout this period it has been promoted by the central state through such instruments as financial support for building societies together with income tax relief for home-buyers (since 1951) and capital gains tax exemptions for home sellers. While not surprisingly the initiative in this respect has historically been taken by Conservative governments, it is also true that since the 1960s the Labour Party, 'a reluctant champion of owner occupation' (Saunders, 1984, 218), has shown a similar practical (if not ideological) commitment to this form of tenure (Short, 1982).

Because of its many paradoxes, this remarkable rise of petty property owners has posed acute problems for class analysis. In the first place there is a sharp discordance between the political rhetoric and popular perceptions of home ownership on the one hand and its true nature on the other. Like any other private individualistic market-place institution, home ownership is automatically equated with ' . . . saving, thrift, hard work, independence – the very best qualities for a hard-working stable population . . . In contemporary Britain owner-occupation is more than just one of three tenure forms, it is an emblem, an image rich with meaning and full of wider social and political significance' (Short, 1982, 119). Yet this 'sturdy independence' imagery overlooks the obvious fact that owner-occupation could never have taken off as a major tenure without very substantial state aid. Short (1982) estimates that in 1978/9 owner-occupation cost the state no less than £2½ billions in relief on income tax and capital gains alone. The holder of an average new mortgage of £29,622 receives income tax relief of £945 in the first year of repayment (June 1986 values; see also Cater and Jones, 1987). So massive is the level of subsidy that even members of royalty and the Church have been driven to publicly advocate the withdrawal of tax relief on mortgage repayments (Archbishop of Canterbury's Commission on Urban Priority Areas, 1985). In contemporary Britain home-ownership is thus a curious hybrid tenure, a form of 'subsidized individualism' (Forrest and Murie, 1986, 6a) which plays a very real though heavily disguised part in the process of collective consumption. Yet even when this point is established, owner-occupation poses a further analytical dilemma centring on the complete lack of fit between labour market and housing market positions which it represents. The typical Briton now appears simultaneously as a seller of labour power (manual or mental) and as an owner of petty home capital. Needless to say this was not a problem for Engels writing at a time when labour was propertyless in both spheres, subject to expropriation in the workplace and secondary expropriation by rentier capital in the home. Subsequently of course the widespread growth and 'democratization' of owner-occupation has completely destroyed this close correspondence. Taken at face value this development offers powerful support for the suggestion that capitalist class conflict has been superseded by some form of pluralistic society in which access to valued consumption resources is no longer governed by one's relationship to the means of production and in which property-ownership is no longer critical in defining a dominant class.

An uncompromising retort would be that this is precisely the kind of myth which home-ownership is *intended* to promote: and that the expansion of this tenure is one of a number of developments in late capitalism which have served to mystify the true nature of class relations (see Harloe, 1981 and Short, 1982 on the assertion that house-ownership fosters 'false consciousness'). Recent Marxist analyses of home ownership tend to stress its utility as a bastion of social stability, a 'bulwark against Bolshevism' in the words of a pre-war Tory electioneering slogan. Here the role of owner-occupation is more than simply to paint a vague halo of benevolence around the capitalist system. More than merely a form of appeasement or crude bribery, owner occupation functions as a process of *incorporation*. By this is meant the co-option of sections of labour as loyal class allies who positively identify their own interests with those of capital. Part of this has to do with vested interests. As Castells (1977) points out, any set of debt-encumbered petty property owners stands to make great losses from the over-turning of property relations, which helps to explain the great enthusiasm shown by twentieth-century conservative policy-makers in Britain and elsewhere in 'giving them a stake in the system' through widening access to home ownership or, as Agnew (1981, 457) puts it, 'promoting the expansion of commitment to the prevalent social order by the development of personal stakes in its survival'. (For further comments see Boddy, 1976; 1980; Merrett, 1982; Short, 1982; Dickens *et al.*, 1985.)

While this may not be entirely convincing – after all many tenants may also be locked into the status quo – Kemeny (1980) argues that to fully appreciate the 'conservatizing effects of owner-occupation' we must go beyond narrow vested interests into the realm of ideology, to assess the effect which proprietorship exerts on the political consciousness of these owners.

> . . . a central characteristic of owner occupation is that it is founded upon the privatization of housing consumption. As such it is anti-collectivistic and is structured in such a way that it discourages cooperation and the communal organization of social security (Kemeny, 1980, 373).

While any satirical reference to privet hedges and lace curtains is by now a worn cliché, it is still the case that the privatized life-style of the home-owner fosters a sub-culture which is the very antithesis of the solidary working-class 'cultures of resistance' discussed in Chapter 1.

A corollary of all this is that widespread home ownership also generates and fosters class fragmentation. It is at the very heart of the residential division of labour, whereby conflict becomes channelled along *intra*-class rather than *inter*-class lines (Harloe, 1981). At the ideological level it is not simply a question of home-owners identifying with capital and lending legitimacy to its core values such as the sanctity of private property – important though this may be. It is also a question of non-identity with the propertyless. Possession of property, however small, arguably tends to nourish self-worth, even social superiority and to lead to negative steretyping of the less privileged. The simultaneous but separate development of home-owning and council renting has starkly highlighted the popular moral distinction between the 'sturdy independence' of the former and the client status of the latter. As the prolific literature on urban residential geography attests, this moral divide is further deepened by residential segregation, minimizing inter-group contact and socializing group members into separate 'class cultures'. It is no coincidence that the rise of owner-occupation is a kind of residential confirmation of work-place status and a means by which these groups are enabled to put distance between themselves and those of 'lesser' status who are reproduced in council housing. Moreover, increasing diffusion of tenure to the more affluent manual workers may also have weakened the sense of identity between owning and renting sections of the working class. According to Darke and Darke (1979, 21–2) this widening cleavage has been developing since the 1930s with the first major migrations of manual workers into private suburbs: ' . . . the interests of the property-owning worker began to diverge from those of the Local Authority tenant or the worker still living in an inner-city slum'. Stedman Jones (1983, 218) traces the suburban exodus of skilled London workers to an even earlier period, the 1870s when it became a 'mass phenomenon'.

Although we have concentrated heavily upon the political and ideological 'functions' of this tenure for the reproduction of capitalist hegemony, we should note briefly that owner occupation also has direct implications for the capital accumulation process itself. Aside from those branches of capital directly concerned with the provision of private housing (building and property capital, finance capital, exchange professionals), it might be argued that employers across the whole spectrum stand to gain from the heavily subsidized reproduction of key labour power in a relatively privileged form of shelter. It has also been shown that the general process of private suburbanization acts as a major stimulus to demand for various kinds of consumer goods (e.g. cars, household durables – see Harvey, 1973, 270–5).

2.6.2 Submerged Owner-Occupation (2A)

It goes almost without saying that many of the above comments do not apply with full force to owner occupiers in the lower reaches of the private sector, in what we have called submerged housing. We have established that, for these households (among whom ethnic minorities are heavily 'over-represented' – Chapter 5), ownership is no escape from housing disadvantage (Karn, 1977/8; Ward, 1982). Even more significantly, they are deprived of many of the 'normal' benefits of this tenure, being in practice the victims of regressive fiscal and financial transfers, and at first sight it is hard to reconcile their position with the conventional logic of owner-occupation.

> Typically their housing situation represents little more than a token nod in the direction of material appeasement, not sufficient one would imagine to engender a sense of satisfaction or of loyalty to a system which in effect punishes them for 'good' behaviour. . . . On the surface this differential allocation of rewards would seem to pose a serious threat to social stability by creating . . . a mood of dissatisfaction among the losers (Cater and Jones, 1987, 199).

In practice, however, this form of ownership provides a highly effective answer to many of the present-day requirements of the capitalist state, in particular the need to cut public expenditure by shifting the burden of their own reproduction on to the consumers themselves. A mode of housing which provides its occupants with relatively low, indeed minimal, state benefits but at the same time conceals this by conferring 'psychic' benefits (i.e. the illusion of independence, security, respectability) is evidently a very potent instrument for ensuring social peace without paying the full price for it. Predictably, then, owner-occupation in the inner city and other localities, which yield very poor returns for landlords and property capital, has been relentlessly encouraged by recent governments.

All we have tried to do here is sketch a theoretical guideline for future research. As yet the question of marginalized owner-occupation has received very little attention, and what data is available tends to be aggregated at a very coarse scale. For example it has been shown that, while capital appreciation from house-price inflation in the South East of England (18 per cent per annum between 1975 and 1985) has been almost double the increase in the Retail Price Index, in Northern Ireland the average rate of return has failed to match the annual rate of inflation (9 per cent compared with 10 per cent (Nationwide Building Society, 1987)). Similarly it has been shown elsewhere that, while the rate of appreciation of older terraced property nationally has hovered around the inflation rate, newer, suburban and exurban, semi-detached and detached property has shown a much higher rate of return (Cater and Jones, 1987). More detailed research is clearly needed to discover the precise extent to which certain consumers 'lose out' on benefits like tax relief and price appreciation, and also the extent to which this is compensated in other ways, e.g. by improvement grants.

2.6.3 Mainstream Local Authority Sector (1B)

We should note at the outset that any attempt to itemize the 'functions' of municipal housing for capital and the state or to shrug such housing aside as mere reformist appeasement conceded primarily on the terms of the dominant class should be strictly qualified. Indeed there are good reasons for presenting the rise of the public sector as a response to organized political pressure, a genuine concession won by labour on its own terms, even a shift in the balance of class forces – at least in particular places at specific times. Some knowledge of the historical geography of council provision in Britain can shed light here. Ever since the last century there have been immense local and regional variations in the scale and growth of local government housing intervention, variations which to some degree have persisted up to the present and which can be linked to geographical differences in the power of organized labour (as discussed in Chapter

1). Dickens *et al.* (1985) argue that these areal variations are more than simply local impacts of national housing policies: the very fact that local authorities (the 'Local State' (Cockburn, 1977)) have always acted as the agent for central state housing provision has given to the local state a certain *relative autonomy* – i.e. a capacity to influence (though not in the last instance to determine) housing outcomes (Dunleavy, 1981). From the 1890s onwards an increasing number of (mainly urban) local councils entered an era of long-standing dominance by the Labour Party, for whom the provision of social housing was a major issue. Dickens *et al.* (1985) use the example of one such locality, Sheffield, to argue that the local state cannot be dismissed as a passive recipient of central state policies. Among the many imprints of a half-century of Labour control of Sheffield council ('coupled with a very strong trade union movement and a strong labourist political culture in the city generally' (Dickens *et al.*, 1985, 161)) has been a very high level of municipal housing provision, far in excess of the national average. In Britain, then, the local state has traditionally enjoyed significant leeway in its interpretation of central state housing policy together with some discretion in the allocation of central state funding. Where this has been combined with decisive and long-term socialist control of councils and localized pressures from powerfully organized labour using legitimate democratic channels it has produced a strong commitment to council housing provision (see Melling, 1980 and, for the contrary view, Preteceille, 1981, 7–8).

Yet while this view of social housing as a positive reward for the labour movements's own initiatives has much to commend it in cases like pre-war Sheffield, it completely fails to fit many other cases. For example, Dickens *et al.*, show that in rural North Norfolk council house building was extremely well advanced in the 1930s despite an overwhelmingly Conservative local council and a concomitant absence of direct leverage for the labour movement. Here the rise of social housing provision can only be explained as an initiative by landowners and large agricultural capital anxious at the threat to their supply of labour power posed by a critical shortage of housing. (Dickens *et al.* 1985; see also Saunders, 1984, 209–10 on the convergence of multiple interests to create social housing.)

On a broader level too, much of the literature concerning the post-1919 emergence of a formal state commitment to council housing conveys a firm impression of a capitalist state using the provision of social housing very much on its own terms as an instrument of social control. According to Harloe (1981, 28) '. . . . it was concern with having a "guarantee against revolution" which above all inspired the first large-scale council building programme in 1919.' Certainly the housing crisis of the period appeared to pose a serious, even a revolutionary, threat to public order which needed to be urgently defused (see the literature on the Glasgow rent strikes – e.g. Dickens, 1977; Melling, 1980; Castells, 1983) but it is debatable how far the state's response can be construed as a satisfying long-term gain for labour.

Social housing in Britain has turned out to be an extremely variable consumer good, with periods of high-quality building to strict standards being interspersed with phases of shoddy cost-trimming such as that which produced the hated high-rise flats. Bureaucratic allocation procedures have ensured that the poorest families end up in the worst of these dwellings, in effect creating a cruel parody of the private housing stratification system (Taylor and Hadfield, 1982, 258) and mocking the original utopian ideal of decent housing according to need (Williams *et al.*, 1986). The full irony of the situation is highlighted when we note that some of the greatest housing disasters of the 1960s and 70s have been perpetrated not by 'uncaring' Tory local authorities but by the very guardians of labour themselves in cities such as Liverpool (Manion, 1982) and Glasgow (Gibb, 1983; Wannop, 1986).

Yet it is not solely the tenants of the 'dump' estates who are entitled to feel betrayed. There is a sense in which council tenancy itself, irrespective of its shelter satisfaction, has always constituted an infringement – invisible for most of the time, no more than a minor irritant for some tenants, but still palpable – of human liberty. That infringement has been encapsulated by Ward (1974) in the phrase ' . . . that awful tradition of paternalism which has been inherited by one third of the households in Britain'. He continues: 'And if you live on the reservation [sic] . . . you are the victim of a thousand small humiliations unknown to the owner-occupier and foreign even to the private tenant, who is never expected to feel that the landlord is doing him a favour' (Ward 1974, 12).

From its very inception municipal housing administration has always reduced its residents to client status, passive recipients of public welfare, dependent upon outside agencies for repair and maintenance and subject to petty rules and regulations concerning access, mobility, house upkeep and even trivial matters of day-to-day living such as the keeping of pets. Castells's observation that daily life in the community increasingly resembles the 'cadences of assembly-line work' (1978, 33) must ring true for many council residents for whom the regimented residential experience is a direct extension of that in the workplace. Although the 1980 Housing Act removed a few of the petty restrictions on public-sector residents, there is still a great deal of truth in Salaman's (1981) assertion that the tenant is 'controlled in overt and onerous ways which clearly articulate the "untrustworthiness" of the manual workers'.

It is difficult to fathom how this relationship represents a victory for labour or a shift in the balance of class forces. On the contrary, the council housing system appears as a highly beneficial arrangement for capital and the British state. More than simply a means of reproducing manual labour power, it also provides for the direct control of workers through bureaucratic regulation (Dunleavy, 1981; 1982). Moreover, it harmonizes perfectly with the ideological logic of the residential division of labour. The irresponsibility, dependence and passivity which client status imposes is virtually guaranteed to reinforce the negative stereotypes which private-sector dwellers project on to those in the public sector (Berry, 1974, 105–7; Ward, 1974, 12–13). There can be little common cause between the two nations of the housing system.

2.6.4 Council House Sales and Privatization

If social housing so effectively accords with capital's need to divide and rule, why then is a British Tory government (explicitly devoted to capitalism as an ideal) currently selling off the public housing stock at a discount to sitting tenants? The sale of council houses has been proceeding for a lengthy period but it is only in the present decade that the rate of privatization has significantly exceeded that of new additions to stock (Table 2.11). Naturally this vexed question has recently received close attention in the housing literature, notably Harloe (1981), Harloe and Paris (1984), Saunders (1984) and Forrest and Murie (1986).

A central point stressed by all the above writers is that while council housing may well have played a valuable role for the capitalist order in the recent past, this role has been contingent upon specific historical circumstances and has become less and less essential in the past two decades or so. Both Harloe (1981) and Saunders (1984) present the council estate era (roughly 1920–70) as representing an interval (almost an aberration) in the long-term domination by private housing. Broadly, there have been three overlapping historical phases of housing provision, which Saunders dubs 'market', 'socialized' and 'privatized'. We have now entered the third phase, a swing back to private housing provision, though now in the form of owner-occupation and not renting from landlords as in the first phase. The relatively brief period of social housing arose in answer to a peculiar set of conditions at a particular time: namely the political imperative for the state to deliver a quantity and quality of shelter beyond the capacity of the private landlord to supply or of the average worker to afford. But now those conditions no longer wholly apply. Increased and more stable earnings for the *majority* of workers in recent years has made mortgage credit a feasible proposition for them while the fiscal crisis of the state has made it necessary to cut public expenditure and shift the burdens of reproduction on to labour itself. In Harloe and Paris's (1984, 89) words, housing policies have become 'hostages to the central imperatives of monetarism'.

As Forrest and Murie (1986) point out, however, this last point is self-contradictory, since in most cases council homes are sold at a discount, implying varying degrees of financial loss to the state. In addition, where former council tenants become owner occupiers with significant mortgages they usually represent an increased charge to the state. Perhaps once again it is the ideological arguments which carry more weight than the materialistic considerations. At the risk of repeating previous points, we would observe that the sell-off of public housing helps to incorporate still more workers into the capitalist property ethic, to create new divides between tenants and former tenants and to further erode a housing mode which, because of its non-commodity form and the collectivist ethic which it symbolizes (Kemeny, 1980), exists as a permanent thorn of potential subversion in the capitalist body.

2.6.5 Submerged Local Authority Housing and Private Renting (2B + 3)

These two strata are considered together because jointly they contain most of the nation's marginalized households and the extremes of residential deprivation. The council 'failures', hard-to-lets or 'dump' estates have by now received sufficient attention to need no further introduction (Berry, 1974; Merrett, 1979). Many of their definitive attributes – the unfitness of their dwellings and the poverty and social disadvantages and powerlessness of their dwellers – are also the hallmarks of the private renting sector. Quite correctly Taylor and Hadfield (1982, 242) describe this latter small and dwindling tenure as ' . . . a residual set of housing resources, a last resort for those failing to qualify for other housing modes (Harloe, 1981), a type which includes most of the remaining multi-occupied dwellings in the inner-city "twilight zones".'

Since they meet little more than the barest needs of their occupants, these two housing strata might seem to pose a real threat to capitalist order, but in practice the residentially deprived are rarely able to mount effective resistance. Frequently this is because they themselves are internally fragmented, for example by racial divisions, since blacks and immigrants tend to form a good proportion of this residential sub-proletariat (see Castells's (1978) case study on the effects of fragmentation on residential resistance).

More to the point, however, they are isolated from the mainstream of the working class, alienated from them in the literal sense of being estranged. For Harloe (1981) the primary rift within the working class is reproduced in the housing market. This rift is between the traditionally unionized majority who enjoy a strong labour market position and political representation via the Labour Party; and marginal labour, 'those who have limited or no utility for the purposes of production', the unskilled, the elderly, the disabled and those excluded by racism. The urban housing system is a classic instance of Harloe's assertion (1981, 21):

> In political terms this can lead to outcomes which reflect the interests of the more organized sector of the working class while excluding or even disadvantaging those of more marginal groups. 'Normal', 'respectable', 'decent' housing is designed for the virtually exclusive use of the organized sector of the proletariat.

We might conclude this argument by observing that, ideologically, the function of the residentially subnormal is to legitimate the position of others in society. As Galbraith (1977) has it, part of the pleasure in being rich lies in having what others have not, and so the poverty of the slum dweller serves as a constant reminder of the comparative plenty enjoyed by the mass of labour. At the level of popular myth the victims of the housing system are an articulation of the ethical distinction between the 'deserving' and the 'undeserving' poor, the latter being a category of humanity who are, according to conventional wisdom, innately self-destructive. In this way any highly publicized outbreak of riot, vandalism, mugging or other slum disorder fuels reactionary opinion, vindicating the allegation that poverty is self-inflicted. This alienation of slum dwellers from the rest of labour is one of the most critical ruptures in class solidarity and as such has always made a major contribution to the stability of the capitalist order – providing of course that it is correctly 'managed' (see Chapter 6).

2.6.5 Alternative Tenure Modes

Despite their comparatively small contribution to the British national housing stock, the minority tenures, housing associations and housing cooperatives, are well worth consideration. In principle they offer an opportunity to burst out of the straitjacket imposed by the two dominant housing modes. In Britain approximately one million people live in homes provided by housing associations, non-profit making organizations with their origins in the nineteenth-century voluntary housing movement. In the last three decades, and especially since the 1974 Housing Act, they have played an increasing role in the provision of housing, especially for the elderly and in inner-city locations, obtaining funds centrally through the state-financed Housing Corporation. They thus have more in common with the public sector than the privately rented category of which they form part, acting as small-scale and, currently, ideologically acceptable, providers of 'primarily' good-quality welfare housing. Their size and diverse character nonetheless limits their impact, and they have been criticized as being expensive, inefficient, uncoordinated, lacking in accountability and potentially undemocratic (Short, 1982, 188–93).

Housing co operatives make up only a miniscule proportion of the total stock, but are of interest in that they seem to represent a radical alternative to the heavily subsidized individualism of more dominant modes and the regimented paternalism of the other. By definition they imply an entirely collectivist form of ownership, with each cooperator owning a nominal share and the total assets being owned by the whole group – the exchange of client status for active responsible control. In practice, they have often been frustrated by institutional and legal barriers (Ward, 1974), and the most innovative co operative experiments frequently fall foul of partisan politics at the local and national level. The Housing Corporation has been reluctant to fund council tenants who wish to develop socially-owned, self-managed housing on the grounds that council tenancy by definition constitutes adequate housing. Secondly, little funding or support has been forthcoming from many Labour-controlled local councils dedicated to the building of municipal housing on traditional lines. Whatever the underlying reasoning here, we can conclude that the long tradition of centralized municipal housing control is one which dies hard and constitutes one of the two immovable barriers to the growth of a third housing arm. The other is the equally doctrinaire obsession with owner-occupation at the national level. Currently alternative tenures are caught in an unholy alliance between the New Right and the Old Left.

2.7 Crisis, Change and Resistance

Up to this point we have deliberately emphasized the counter-revolutionary processes of the housing market. It is above all necessary to explain how it is that, despite glaring injustices and the continuing existence of shelter poverty, the capitalist order can continue to count on the effective consent of its members; as Dunleavy (1977) has highlighted, acquiescence is the dominant theme. It is, however, appropriate to remind ourselves that this is not the *sole* theme, that despite the comprehensiveness of bourgeois hegemony the situation is by no means a static equilibrium, with labour politically neutralized and posing no threat to the established order. In reality, the historical development of capitalism constantly generates fresh crises which must be resolved (or at least displaced) if the system is to survive.

It is in the urban milieu that the contradictions of capitalist economic growth are most visible in the literal sense of the word. At the heart of the city, the perpetual growth of economic activity has exerted inexorable pressures for outward expansion, rising land values and displacement of residential areas. Already a problem in the last century (see Stedman Jones, 1971 on the erosion of workers' living space in Victorian London), the process has continued, rising to a modern crescendo in the post-war orgy of city-centre redevelopment, office building and speculative property dealing which has occurred in many advanced capitalist countries (Boddy, 1981). Numerous studies have documented this unequal conflict between the needs of local residents and the desire on the part of commercial interests to realize the full exchange value of land, either through retail or office development (Ambrose and Colenutt, 1975) or through *gentrification*, the substitution of up-market housing for low-income workers' accommodation (Hamnett, 1973; Williams, 1976; Castells, 1978).

While this 'urban reconquest' (Castells, 1978) may accord with the short-term economic interests of those capitalists directly involved – builders, landowners, developers and financiers – its political impact is potentially one of destabilization. Partly this is because of its high visibility and its socially destructive consequences. Even where redevelopment is presented as a planning exercise justified by such catchwords as 'progress', 'modernization' and 'efficiency', it may be difficult to persuade informed public opinion that speculative private profits are less than extortionate; or to disguise the manner in which these profits are won by destroying settled communities, scarce housing and environmental amenities to create space for office blocks, urban motorways and car parks.

In Europe since the late 1960s overt resistance to the injustices and oppressions of the capitalist housing market has become increasingly manifest, stimulating a body of research into all aspects of urban protest across a spectrum ranging from the fairly low-key but widespread rumblings of residents' action groups to sporadic instances of premeditated damage and violence (Castells, 1978).

Such collective and organized responses by groups of consumers are what Castells (1977; 1978; 1983), the foremost exponent of research into consumption struggles, calls 'urban social movements'. Castells's argument is that territorially based issues over housing/environmental redevelopment can unite all the residents of the threatened neighbourhood, whatever their perceived class or status. Urban social movements, then, are 'essentially locally based protest groups linking together disparate social groups' (Harloe and Paris, 1984, 91). And even in exclusively working-class areas Castells (1978) claims that alliances with non-resident middle-class interests cannot be ruled out. This he illustrates with a case study of a poor area of inner Paris scheduled, like so much of that city, for clearance, commercialization and gentrification. Here the presence of university students played some part in the organization of a resistance movement among the tenants.

The significance of urban social movements in Britain and North America has been the subject of intense debate (see, for example, the articles and responses by Hooper, Duncan and Goodwin, Saunders, and Dunleavy in *Political Geography Quarterly*, 1982). The most widely accepted view (Kirby, 1982; Pinch, 1985) is that protest movements in Europe are 'more radical in their demands, more disruptive in character and working-class in orientation' (Pinch, 1985, 148). In the United Kingdom and the United States neighbourhood organizations tend to be predominantly middle-class, defensive, confined to small spatial units and disinclined to link their protests to broader consumption issues. Although it is possible to cite a wide range of examples of community action (see Wallman, 1982 on residents' organizations and Dickens et al., 1985 on squatting), their impact is typically limited and transitory, with class-based self-interest re-emerging to transcend such 'artificial' spatial communities (Cockburn, 1977; Saunders, 1979).

Many consumption struggles are primarily concerned with issues other than housing, and we return to this issue in Chapters 5 (Racial and Ethnic Minorities) and 6 (The Urban Neighbourhood). Nonetheless there are considerable overlaps between consumption issues and housing issues and both are almost invariably present in some form. In view of all that has been written in this chapter about housing's prodigious economic, political, social and psychological importance, such centrality is hardly to be wondered at.

Recommended Reading

There is an extensive literature on housing problems and housing policies, penned by both geographers and non-geographers. Two recent examples in the former category are Larry Bourne's *Geography of Housing* (1981), a wide-ranging review strong on spatial patterns and processes and angled towards the North American market, and John Short's accessible *Housing in Britain* (1982). A more detailed exposition of theoretical approaches to the housing market can be found in Bassett and Short's *Housing and Residential Structure* (1980), which contains a thorough analysis of structuralist perspectives, while an article by Peter Saunders (1984) in the *International Journal of Urban and Regional Research* informs some of the debate on consumption cleavages in this chapter.

Two substantial studies by Steven Merrett, of *State Housing in Britain* (1979) and, with Fred Gray, of *Owner-Occupation in Britain* (1982), provide the most comprehensive surveys of the principal sectors of the British housing market, while Balchin's recent book, *Housing Policy* (1985) provides an up-to-date introductory review of the role of the state. A brief and coherent summary of government intervention in the housing market until the late 1970s can be found in Chapter 2 of McKay and Cox's *Politics of Urban Change*. Most recently, a collection edited by Peter Malpass, *The Housing Crisis* (1986), contains several up-to-date and pertinent papers on contemporary Britain, emphasizing the neglect of the past decade and predicting an increasingly severe housing crisis as we move into the 1990s.

3
Crime and Disorder

In the academic study of crime the question 'Where?' has always commanded attention, so much so that it seems at times to have obscured the much more profound question 'Why?' Although it was only in the 1970s that geographers themselves began to develop a serious interest in the field (Scott, 1972; Herbert, 1976b; 1979), spatial analysis had been pursued by scholars from other backgrounds. Moreover it had been pursued for a very long time. The study of the geographical incidence of crime and delinquency is generally thought to have begun a century and a half ago with the work of the 'cartographic school', pioneered in the 1830s by the French writer Guerry and imitated by others in Britain and elsewhere (Herbert, 1982). The sharp regional variations in offence rates disclosed by these early map-makers supplied the first systematic evidence that criminality is space-specific. In mid nineteenth-century Britain this was evident both at the inter-regional scale, where law-breaking was strongly concentrated in the most urbanized and industrialized areas, and at the intra-urban scale, where offenders tended to be heavily concentrated in small pockets within the poorest

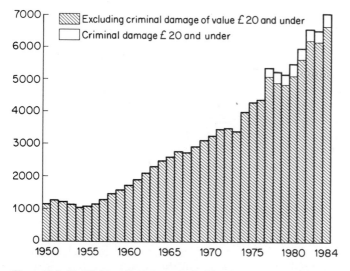

Figure 3.1 Notifiable offences recorded by the police per 100,000 population 1984, England and Wales.
Source: **Home Office 1984.**

Table 3.1 Notifiable offences recorded by the police, England and Wales, 1950–1986.

Year	No. of offences ('000s)	No. of offences (per '000 population)
1950	479.4	1094
1955	462.3	1040
1960	800.3	1742
1965	1243.5	2598
1970	1568.4	3221
1975	2105.6	4283
1980	2520.2	5119
1981	2794.2	5630
1982	3088.3	6226
1983	3071.0	6191
1984	3313.8	6674
1985	3611.8	7261
1986	3847.4	7708

Sources: Home Office 1984: Criminal Statistics, England and Wales. London: HMSO.. Central Statistical Office 1988: Social Trends. London: HMSO, Table 12.2.

working-class neighbourhoods (Mayhew, 1862; see Herbert, 1982 for a useful summary).

Subsequent developments in crime patterns have to a remarkable degree retained the basic elements outlined above. It goes almost without saying that there have been colossal changes in the law, in policing and in sentencing policy. There has also been a most visible and well-publicized rise in the total of (officially defined and recorded) indictable offences, culminating in the marked escalation of recent years (Fig. 3.1 and Table 3.1). And yet the time-honoured areal variations persist, at least in basic outline.

The apparent upsurge of criminal behaviour charted in the above figure is widely viewed as a critical social problem: a 'law and order crisis' for the political right, part of the 'crisis of late capitalism' for the radical left. Whatever kind of crisis it may be, it is evidently one whose evolution is geographically uneven. Just as in Mayhew's time, there are clearly identifiable localities – specific urban neighbourhoods within the largest cities – which exist at the 'sharp end' of the crime crisis. Both offenders and their victims are highly localized, often, though not always, in the same areas. Although not for a moment could we claim that every other location is crime-free, there is nevertheless a relative freedom which, in an age of allegedly high mobility, borders on the remarkable.

This last point is worth underlining. Contrary to popular perceptions that it has become a ubiquitous problem that threatens to get out of hand, crime is less than commonplace. One particularly vivid example of the gap between image and reality is provided by Clarke's (1984) study of the London Underground system. Although 'common knowledge' holds that the Tube is an exceedingly dangerous place, recent statistics show that only one in 173,000 passengers per year is actually robbed. While this is one too many, it demonstrates the need for a sense of proportion. Although it has long been recognized that crime is seriously under-reported (Hood and Sparks, 1970; Nettler, 1984), criminal behaviour (as legally defined) involves only a small fraction of the population. Moreover, we reiterate, it is only a small proportion of the total inhabited space which houses criminals in significant numbers and suffers severe and frequent criminal acts.

This vulnerable portion of inhabited space is overwhelmingly urban. Not only do urban areas as a whole record higher crime rates than rural areas; most forms of law·violation are also positively correlated with city size and hierarchical rank (Harries, 1974; 1976). At the micro-level it is the inner city (zone in transition, twilight area) which, in a succession of American research studies dating from the pre-war Chicago School to the present, has been shown to bear the brunt of urban crime (Shaw and McKay, 1942; Schmid, 1960; see Brantingham and Brantingham, 1981a for a summary of the literature). In Britain (and other states with strong public-sector policies of slum clearance and rehousing) this simple geometry has been succeeded by a bi-polar pattern, with peripheral 'problem estates' of public-sector housing displaying relatively high levels of criminal activity.

3.1 Human Ecology and its Derivatives

As in so many other spheres of geography, so in the geography of crime the Chicago tradition offers an obvious launching pad. Of the various competing perspectives within criminology it is the ecological school which has addressed itself most directly to the spatial dimension of the problem, being concerned with the areal co-variance between criminality and various aspects of the built and social environments. Judging by the prominence accorded to it by subsequent authors (for example Baldwin and Bottoms, 1976; Baldwin, 1979; Brantingham and Brantingham, 1981a; Davidson, 1981; Herbert, 1982), the definitive product of the Chicago School was that of Shaw and McKay (1942), who mapped the incidence of juvenile offences in Chicago and five other United States cities, concentrating on Burgess's concentric-zone model as a correlate of delinquency.

How did the geography of delinquency fit into the geography of the urban system? Very neatly, it seems, with a steady downward gradient from central city to outer perimeter. Shaw and McKay found the highest rates of both crime commission and criminal residence to be in the area adjacent to the Central Business District (CBD), where criminality appeared to be strongly correlated with areal concentrations of poverty, disease and broken homes. In Chicago the offender rate in the innermost zone was found to be three times that of the outermost zone, a pattern broadly replicated in Denver, Philadelphia, Cleveland and Seattle. Not only is this spatial pattern recurrent, with the inner city featuring as the principal crime zone in a host of studies on both sides of the Atlantic: it has also proved highly resilient to the passage of time. For example, the Chicago pattern persisted right through to the 1960s and later. As Brantingham and Brantingham (1981a, 14) note, 'all subsequent American criminological theorizing appears to accept Shaw and Mckay's findings as the facts requiring explanation'. Subject to the modifications noted earlier, this would also seem to apply to British crime geographers, whose task has been to explain both the inner-city crime concentrations and the outer council-estate concentrations which are associated with slum clearance and the residential relocation of inner-city populations (see Fig. 3.2).

What kind of relationship have the ecologists uncovered? Have they established a genuine link between criminality and space *per se*? Or have they simply conflated space with some other variable? The 'other variable' could well be class. Areal concentrations of offenders tend also to be areal concentrations of

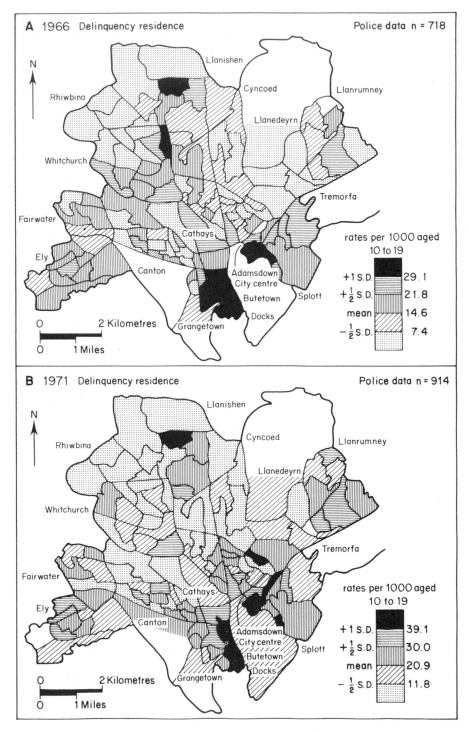

Figure 3.2 Residential distribution of known delinquents in Cardiff 1966 and 1971.
Source: **Herbert 1976b, 478–9.**

low-income households, unskilled workers and the unemployed. Indeed for some writers this link is utterly self-evident: 'Today it is those families with annual incomes below the poverty line which fill the police stations' (Platt, 1981, 21). While this inverse correlation between crime and social class is certainly spatially manifested, this does not mean that spatial arrangements themselves hold the power to influence criminality. Spatial variations in offences and offenders may be no more than the passive result of urban processes which segregate the poor – and those predisposed to crime – into discrete portions of the urban space.

We cannot, however, dismiss the role of space quite so easily. For example, there is considerable dispute among criminologists as to the association between criminality and poverty (Tittle *et al.*, 1978). While material deprivation may be a motive for some crimes, especially burglary and theft, acceptance of the link is far from universal. At one extreme, writers like Erlich (1973) have attempted to analyse criminal behaviour in terms of neo-classical utility theory. This finding has been supported by Braithwaite (1981) who, in a survey of 53 studies in 12 countries of juvenile crime from 1941 to 1979, discovered that no less than 44 of them had found lower-class juveniles to be more crime-prone.

At the same time a number of area-based studies, some dating back to the pioneering crime surveys of the last century, assert that the poverty-crime link is by no means deterministic and is mediated by several intervening variables. In early Victorian England Mayhew (1862) was arguing that the main explanation was the socialization of children into crime as a profession (Herbert, 1982). Poverty *per se* was not the basic cause, even though it provided a general context and was certainly present in the criminal 'rookeries'. A more recent and frequently cited source is Lander (1954), whose multivariate analysis of cities in the USA seemed to establish that localized peaks of juvenile delinquency were only weakly related to the socio-economic status of these areas. Arguing on a more abstract plane, Clinard (1963) adds weight to this theme by reminding us that the law-abiding poor far outnumber the law-breaking poor; and further, that there are all manner of additional behavioural sources of deviance – family, marital, sexual, alcohol-related – which are not necessarily linked to poverty. We will return to this question later in the chapter, but in the meantime we might note that class/wealth is one of a range of socio-economic attributes which, together with racial/cultural and demographic variables (sometimes interrelating, sometimes not) tend to predispose individuals towards law-violation. Herbert (1982) lists many of the attributes which are typically, though by no means inevitably, associated with offenders, and these are summarized in Table 3.2.

For the geographer, the fascination with Shaw and McKay's work stems from its evident discovery that space itself is an active variable, one capable of asserting considerable influence on criminal and deviant behaviour. The essence of their discovery was that the inner city stands out as a high crime area and an area of criminal residence even when socio-economic and demographic characteristics are held constant. This implies that crime is to a significant degree the product of locality and not solely the character of residents. Quite simply, the same people tend to behave differently in different locations and environments. For Shaw and McKay, clinching evidence is provided by their further finding that the zone in transition has maintained its high crime rate throughout a lengthy period in which its population composition has been altered by successive waves of new immigrants. While residing in the inner ring, each immigrant group appears to

Table 3.2 Common Attributes of Known Offenders

Category	Indicator		Sub-group at risk
Demographic	age	→	young
	sex	→	male
	marital status	→	single
	ethnic status	→	minority group
	family status	→	broken home
	family size	→	large
Socio-economic	income	→	low
	occupation	→	unskilled
	employment	→	unemployed
	education	→	low attainment
Living conditions	housing	→	substandard
	density	→	overcrowded
	tenure	→	rented
	permanence	→	low

Source: Herbert 1982, 40.

have generated high crime rates, but also to have drastically reduced these rates upon migration to a less central part of the city.

To disentangle the cross-cutting links between location, class and crime is no simple task. One possible inference to be drawn from Shaw and McKay is that *neighbourhood social composition* is the crucial component. Support for this reasoning comes from several more recent studies, one of the most widely quoted being the Sheffield crime survey which found that lower-class individuals were more likely to commit crimes if they lived in lower-class neighbourhoods than if they lived in mixed or higher-class neighbourhoods (Baldwin and Bottoms, 1976). Seemingly the collective class character of the locality may either amplify or reduce the criminogenic propensities of individual class membership.

Yet this is by no means the whole story. Another consistent theme to emerge from several decades of areal crime surveys is that given types of social area do not produce given crime rates or offender rates: on the contrary, rates may fluctuate wildly, even when social class composition is held constant. In Britain both Herbert (1977a; 1977b) and Baldwin and Bottoms (1976) find that spatial variations in offender rates are associated with variations in household tenure. It is neighbourhoods dominated by owner-occupied dwellings which exhibit the lowest rates. Evidently this tenure dimension is one which, in Britain, cuts across the social-class dimension. We shall return to this important finding later in the chapter.

In the United States, where tenure is not a powerful source of social differentiation, it is nevertheless also the case that the geography of crime is partially independent of the geography of poverty. The marked bias towards the inner city, an eternal feature of North American crime geography, is still present even when social-class variations are controlled for. This non-class-related locational effect is highlighted in recent work by Sampson and Castellano (1982; see Table 3.3), who show that offender rates in high-status central city areas are roughly on a par with those in low-status suburban areas and actually exceed those in low-status exurban areas. Whatever may be the effect of poverty and low status on cri-

Table 3.3 Annual offence rates in North American cities by location and area social composition, 1973–1978.

| | Area social composition (% of total families below $5000 annual income) | | |
	Low (27–29%)	Medium (11–26%)	High (0–10%)
SMSA central cities			
Theft	4187	2368	1397
Violence	5426	4671	3909
Balance of SMSA			
Theft	1422	1159	912
Violence	4154	3959	3340
Areas outside SMSAs			
Theft	443	464	462
Violence	2117	2682	2599

Source: after Sampson and Castellano 1982.

minal propensity, it is clearly one that is critically modified by relative geographical location within the urban system. Inner city and zone-in-transition crime levels are not simply a product of the relative poverty there.

What then can explain them? What is the missing element? In addressing this question ecologists have relied heavily on two basic theories – *social disorganization* and *neighbourhood subculture* – or derivatives and variations of these. Each theory attempts to identify the conditions specific to certain portions of the urban space which produce a criminogenic (i.e. crime-creating) environment. Each of these theories warrants a full discussion in its own right.

3.1.1 Social Disorganization

For the ecologists, one of whose intellectual precursors was Durkheim, the essence of the zone-in-transition was social disorganization, a concept closely related to *anomie*. Anomie signifies normlessness or 'moral deregulation' (Saunders, 1981, 45), and was presented by Durkheim as a product of urbanization and the collapse of the old rural order. In the intimacy of the traditional village where, in Frankenburg's (1966) words, virtually the whole population works, plays and prays together, there is likely to be (or to have been?) a unitary value system – a moral right and wrong – together with strong social pressures against deviant behaviour (Frankenburg, 1966; Bell and Newby, 1971; see also Chapter 7). With mass urbanization, however, all this is broken down, with the majority of people now living in large urban settlements where anonymity permits greater freedom of action and where rule-breaking is far less conspicuous. In Wirth's (1938) view, the collapse of traditional morality is further accentuated by rapid population turnover, by the transience and superficiality of many relationships, and by the heterogeneity factor, the recruitment of migrants from a wide range of divergent national and regional cultures and the division of labour into a multitude of differentiated occupational strata. Deviant urban behaviour, then, is not simply a case of rule-breaking, it may reflect an absence of generally agreed rules (normlessness) or even a multitude of conflicting sets of rules (competing norms).

Hence, as Saunders (1981) notes, the modern city has acquired an extremely negative image in popular folklore and even in certain academic circles. Urbanism is 'associated in the most vivid way with the pathological aspects of modern society . . . with the erosion of collective morality. . . . Given the role of the city as the primary force for change, it is mostly in the cities that the anomic character of modern societies becomes most evident' (Saunders, 1981, 44–5; see also Clinard, 1963).

If this reasoning is accepted it goes some way towards illuminating rural–urban crime differentials. Even so, it tells us nothing about *intra*-urban patterns since, if anomie is a by-product of urbanism, then it should logically be universal throughout urban space. Yet in practice this is far from the case. As we have seen, every available shred of empirical evidence shows crime to be acutely biased towards certain definable portions of the urban system. Whether defined by offences or offenders the lawless realm is a minority phenomenon, encompassing only a relatively small fraction of urban population and territory.

The practical truth is of course that the connection between urban life *per se* and cultural disintegration is by no means clear-cut. There is no linear relationship between increasing urbanization and anomie. Indeed a central feature of the theories of both Durkheim and Wirth is a recognition that the destructiveness of modern urbanization is counterbalanced by its capacity to generate new institutions of social order to replace those of traditional society. In Durkheim's terminology the *mechanical* solidarity of the old order is substituted by *organic* solidarity, a sense of belonging to a large-scale non-local society whose members are held together by mutual self-interest. This reasoning was also taken up by Wirth (1938), who saw the intimate controls of the old rural order as being ousted by new forms of social integration more appropriate to a changing modern society. Impersonal forms of control (e.g. the police force and the law) replace personal moral sanctions; mass media disseminate values, replacing word of mouth; wider loyalties to city, nation and state replace parochial loyalties (Wirth, 1938; see also Morris, 1968; Smith, 1980; Saunders, 1981).

While these generalizations are far too sweeping to be applicable to every aspect of the urban system, organic solidarity must nevertheless be seen as one very important contributor towards urban social stability. At the risk of stereotyping, we might suggest that it finds its apogee in middle-class suburban neighbourhoods and commuter satellites, where privacy, anonymity and individualism are no bar to law and order. Here, in truly Wirthian fashion, order is maintained by voluntary conformity to the formal rules of the larger society. Since there is a near universal consensus about the values embodied in these rules, the traditional community-based control mechanisms – from prying neighbours to the church – are redundant.

For other classes in other districts, however, this logic tends to break down. In many stable working-class areas of the city, community is far from redundant, with law and order still seemingly based on the personal ties of kinship and neighbourhood. While urbanization has certainly removed people from traditional rural communities, it has also facilitated the creation of new forms of community. This is evidenced by the extensive body of literature on the working-class community (see, for example, Hoggart, 1957; Stedman Jones, 1983) and the immigrant/ethnic community (see, for example, Boal, 1978; and Chapter 5), which establishes that, far from existing in a state of anomie, many urban groups have reconstructed their own intimate social networks within the city itself. While obviously much looser in character, the urban neighbourhood community retains many of the essential properties of its rural forerunner and provides a source of personal identity and mutual support for its members. Whereas the suburban middle class may derive material and moral satisfaction from a career and from a sense of participation in the affairs of the larger society, many sections of the working class turn to the community to supply these needs. In the context of criminology the end product of these two contrasting social arrangements is not dissimilar: both are conducive to order, stability and harmony. They keep people satisfied.

If this reasoning is accepted then the ecological connection between delinquency and the inner city becomes clear. 'Criminal neighbourhoods' must be regarded as abnormal areas in which (for whatever reason) the essentially urban institutions of social control – organic and communal solidarity – are absent or imperfectly evolved, areas whose inhabitants are not fully integrated into the dominant moral order of the city. This is precisely the point made by ecological theory with its stress on adaptation and on social disorder as maladaptation. Moreover it is transitional groups who are most likely to be maladapted and here the role of migration assumes critical importance. This process, by transferring people from one culture to another, places them outside the moral order. For example, migrants from remote rural regions bring with them norms and values which may be at odds with those of urban society, a particularly potent factor in the case of the foreign migrants who made up a substantial part of the migration to the great American cities.

Plausibly enough, then, delinquency in the inner city – the principal zone of first generation migrant settlement – was seen by Shaw and McKay (1942) as an inevitable consequence of immigrant alienation from the conventions and laws of their adopted society. Hence the very high recorded correlation between ethnic areas and high crime rates in the zone in transition. But ethnic social disorganization is no more than a transitional condition. In accordance with ecological theory, over time each incoming group becomes adapted to its adopted society, acquiring the job skills to advance occupationally and the social skills to assimilate culturally. These adaptations are accompanied by residential movement out of the slums and into the 'respectable' parts of the city, where the behaviour of group members takes on a law-abiding character indistinguishable from that of the population at large. Meanwhile their place in the inner city is taken by the newest wave of incomers, who will undergo the same sequence of changes (see Chapter 5). In this way the consistent crime rates recorded over time by Shaw and McKay are the function of location not of ethnicity nor of social class *per se*.

Empirical testing has provided some degree of support for the disorganization hypothesis, although this has been far from unqualified and more recent findings have tended to be even less conclusive. In one of the earlier works, Lander (1954) applied multiple regression to crime rates in Baltimore, finding some association between population instability (transience, high turnover) and juvenile delinquency. His methodology has, however, been heavily criticized (Baldwin and Bottoms, 1976; Baldwin, 1979). In the United Kingdom, a study of Leicester (Jones, 1958) similarly equated delinquency with instability, this time in the shape of newly built public-sector housing estates. Although the dislocating effect of slum clearance and rehousing is now a well-worn theme, the Leicester study was one of the first to draw attention to the resulting breakdown of community controls. It is probable that these new and/or relocated council tenants were undergoing a similar process of displacement to that experienced by new arrivals in the inner city – the loss of one set of standards and references and the absence, in the immediate term at least, of satisfactory substitutes. This reasoning led Bagley (1965) to argue for the creation of artificial substitutes in the form of enhanced social provision. His study of council estates in Exeter found a significant negative correlation between youth-service provision and juvenile delinquency. Thus a key variable would appear to be the relative poverty of life in the new estates compared to the richness and intimacy of life in the slum.

While the above writers appear to have made out a clear case for disorganization as a predisposing factor in crime, their conclusions are disputed by other empirical work. Baldwin and Bottoms's (1976) Sheffield study, for example, finds no correlation between either population turnover or social provision and crime rates; and further that some of the highest crime areas are contained within pre-war rather than recent municipal estates, a finding echoed by Herbert (1976b; 1981). These pre-war estates are not only settled communities but also, like most estates of their type, tend to exclude potentially anomic lifestyles – for example, the unmarried, the transient, the recent immigrant – and would therefore qualify as stable rather than disorganized areas. Given this and the contradictory evidence thrown up by other studies (see for example Schmid, 1960; Davidson, 1981), it is clear that we need to look beyond disorganization as a simple cause and beyond factorial ecology as an investigative method. As Davidson (1981, 74) warns, 'ecological analyses . . . have continued to develop statistical sophistication without providing much theoretical clarification'. (See also Brantingham and Brantingham, 1981a.) Indeed we might at this point remind the reader of the general critique of multivariate analysis and particularly the fact that the end product is entirely dependent upon the selection, exclusion and weighting of variables. It is likely that those factors identified as predominant in their relationship to crime and deviancy – race for some authors, economic status for others, anomie for yet others – are less a social reality than a statistical construct, an arbitrary mathematical caprice.

3.2.2 Sub-Cultural Theory

Aside from certain well-rehearsed pitfalls of ecological theory – the ecological fallacy (Brantingham and Brantingham, 1981a) and the temptation to attribute causation to correlation (Baldwin and Bottoms, 1976; Baldwin, 1979; Davidson, 1981; Herbert, 1983) – Shaw and McKay's reasoning also appears self-contradictory in at least one important respect. Alongside the theme of disorganization they develop a further theory of *cultural transmission*, based in effect on the premise that deviant behaviour is anything but disorganized or anomic. Far from resulting from inadequate socialization, deviants are individuals who have been socialized all too well into an alternative sub-culture many of whose values are diametrically opposed to mainstream morality. Herbert (1981, 23) refers to a 'delinquent tradition "nurtured" among some sections of society', and relates this to Mayhew's (1862) description of the Victorian delinquents of London's slums, 'children born and bred to the business of crime'. In order to accept this, we must also accept that delinquency (and indeed other forms of deviancy) are not born of *dis*organization but of *counter*-organization, a highly cohesive community structure (family, neighbours, peers) in which children are allegedly conditioned into a sub-culture antithetical to the dominant morality.

For geographers, sub-cultural theory holds obvious appeal because of the role that locality may be held to play in shaping such deviant organization: 'sub-cultural theory . . . can be theorized in spatial terms, groups can be identified with places' (Herbert, 1981, 24). There are of course dangers in overstating the case for a neighbourhood effect. As argued in Chapter 6, the links between neighbourhood and community attain real significance only in rather special circumstances and are often tenuous or even non-existent. Nonetheless, there is a certain promise in the argument that the local urban neighbourhood may frequently act as a territorial base for community life built on mutual proximity and shared space. This will apply most strongly when community cohesiveness is heightened by some common bond such as shared ethnicity, religion or work experience and where contact with other groups and places is limited by segregation and low mobility. By and large, well organized community life is responsible for generating and transmitting conventional morality; it is, as we have seen, one of the classic bulwarks of social order and 'normality'. At the same time it is equally capable of socializing its members into deviancy, as the problem neighbourhoods of inner city and council estate will testify.

At this point we are left with a number of unresolved questions. What are the abnormal conditions which cause a minority of neighbourhoods to produce lawbreakers while the rest produce law-abiders? And why in even the most lawless communities are lawbreakers invariably in a minority?

On the first of these questions classical ecology lays substantial weight on immigration and on immigrants as displaced from one culture and not yet absorbed into another (See 3.1.1 above). Here the high crime rates of American blacks and, less conclusively, of Irish and West Indians in Britain, appear to confirm the proposition. We might also cite the case of certain Italian Americans who brought with them as part of their 'cultural baggage' an ancient Sicilian self-help organization which was then successfully converted to the business of organizing urban crime. Even here it goes almost without saying that only a small minority of this population (or indeed any other) are associated with the lawless activities which popular myth wrongly attributes to entire ethnic groups.

In sharp contrast to this, however, the experience of many other migrant communities past and present on either side of the Atlantic conspicuously denies any connection between immigrant status and criminal deviance. Rather the reverse, since much of the literature on, for example, the Chinese (Light, 1972), Japanese (Bonacich and Modell, 1980) and Jews (Rosentraub and Taebel, 1980) in America and on South Asians in Britain (Robinson, 1986) argues for community organization as a force for channelling energies into 'positive' and 'constructive' outlets such as entrepreneurship, property ownership and self-advancement through education and the professions. Elsewhere, ethnic cohesiveness has been presented as a force for legitimate political advancement (see the classic study of New York by Glazer and Moynihan, 1963). Hence, while we might accept that some neighbourhoods may transmit and reinforce deviant values, these are not necessarily immigrant or ethnic neighbourhoods. On the contrary it is not unlikely that rural–urban migrants may bring cultural traits which are even more strongly disposed towards order than those of the receiving society. In Britain, Indians and Pakistanis are frequently cited as law-abiding communities whose behaviour is ordered by deference to the authority of family, religion and culture (Rex, 1982a).

None of this means that sub-cultural theory must be rejected out of hand. Several recent writers (among them Herbert, 1977a and Bottoms and Xanthos, 1981) have found it to be a useful tool for explaining criminality on particular council estates. But they have also demonstrated that localized deviant sub-cultures need to be assessed in their entire context, not simply treated as if they were internally self-generated and virtually independent of the dominant social structure.

3.2 Opportunities for Crime

The research reviewed so far in this chapter has been concerned primarily with offender motivation: or, more properly, with the socio-spatial environments which might predispose individuals towards lawbreaking. However, most criminologists would now accept that while motivation is obviously a *necessary* condition, it is not a *sufficient* requirement. The other essential condition is opportunity. Most forms of crime can hardly occur in the absence of available targets – property or human victims – and therefore the criminologist or crime geographer must accept what Brantingham and Brantingham (1981a) call 'an opportunity/motivation rubric' for explaining the pattern and incidence of criminal acts. It is, of course, a matter for the individual investigator where he/she puts the emphasis. In recent years an increasing number of writers appear to have been placing the emphasis on opportunity.

In a useful review of opportunity-oriented work, Davidson (1981) demonstrates that space is once again an influential agent, not just a passive receiver of human actions. Two aspects of space in particular – distance and architectural form – have received close attention. On the effects of distance, Brantingham and Brantingham (1981b) note a pronounced distance–decay effect, with the bulk of crimes against property being perpetrated close to the offender's residence. This is due not only to the time/money costs involved in any 'journey to crime' but also to spatial awareness biases. Information about potential targets is richly available in the immediate vicinity of the home but tapers off with distance. A behavioural approach, using classic concepts such as mental maps, activity space and awareness space, thus has considerable relevance. The Brantinghams argue that, no matter how inexperienced or disorganized he or she may be, the would-be criminal actively 'engages in a search behaviour' (1981b, 29), which is constrained by considerations of distance and information. It is also constrained by the need for personal security, which tends to confine offending activity to familiar territory: the offender is 'unlikely to penetrate into totally foreign areas where he will feel uncomfortable or stand out as different' (Brantingham and Brantingham, 1981b, 29).

Implied in the above reasoning is the notion that the geography of offences is closely tied to the geography of offender residence. Yet the Brantinghams are careful to deny that the two patterns are identical. On the contrary, they criticize earlier writers in the ecological tradition for their failure to distinguish between the geography of crime and that of criminals. By no means all criminals will be confined to their 'home patch'. Among the factors that influence their willingness and need to travel to 'work' are their motivation, type of crime, experience and level of organization. Also, because of their sheer attractiveness as targets,

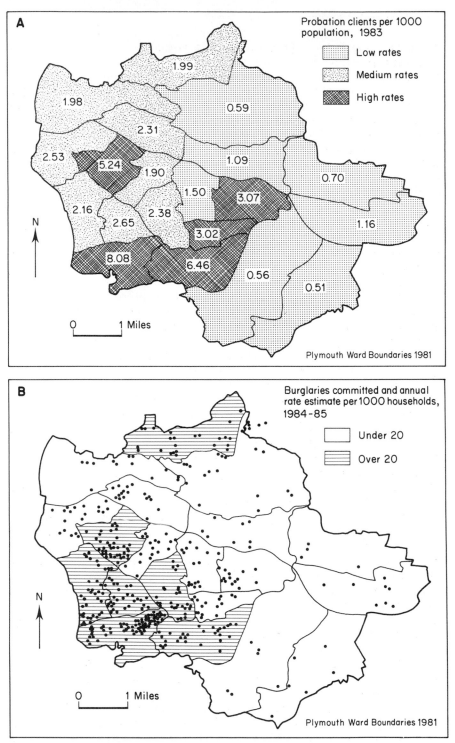

Figure 3.3A Residential distribution of probation clients and B the geography of burglary in Plymouth, England.
Source: **Mawby, 1986, 61, 63.**

certain areas such as the CBD will draw offenders from a wide radius. Even so, the tendency for crime to be committed close to home is most pronounced, reflecting the limited organization and experience and the lack of mobility of the young males disproportionately represented among offenders. Evans (1986) finds that in a town in North Staffordshire, England, almost half the burglaries are committed within 0.8 km of the convicted burglar's home (see also Evans and Oulds, 1984), while the Sheffield Crime Survey found three-quarters of the city's burglaries to have occurred within two miles of the offender's residence (Baldwin and Bottoms, 1976). Mawby's (1986) work on Plymouth similarly indicates a significant coincidence between the geography of burglary and the residential distribution of probation clients (Fig. 3.3). In comparison with the rest of the city, south-west Plymouth is both more crime-prone and more likely to house lawbreakers (see also Mawby, 1981 on Sheffield). This substantial overlap between crime location and criminal residence is broadly confirmed by more wide-ranging reviews of the 'journey to crime' literature (Phillips, 1980; Davidson, 1981).

According to Brantingham and Brantingham (1981b), selection of targets by the potential offender will be governed not only by proximity but also by an assessment, however crude, of the risk of being identified and/or apprehended. Since Newman's (1972) influential work, architectural design and layout has been identified as one of the main factors determining whether or not a potential target is readily accessible at low risk to the criminal. Newman's theory of *defensible space* thus represented a bid to link crime directly to the form of the built environment. As Davidson (1981) observes, Newman's ideas harmonize with popular feelings that modern cities are dangerous, threatening places and that it is often their physical design which heightens this insecurity by making it easier for burglars, muggers and rapists to operate, secure both from the public gaze and from the long arm of the law. If this is indeed the case, then the problem can be easily (though not cheaply) rectified by progressively redesigning the built environment. To paraphrase Newman, the architect armed with some knowledge of the structure of criminal encounter can simply avoid providing the space which supports it. For the average citizen the diagnosis is even more simple. You do not need a knowledge of 'criminal encounter' to know that gigantic depersonalized built structures full of dark empty spaces belonging to no one and screened from public view are threatening and likely to be dangerous.

Newman's work, then, was in some senses the scientific articulation of a widespread perception. His research in American cities identified large-scale high-rise apartment projects as disproportionately prone to vandalism, assault and burglary, with the suggestion that the higher the building the higher the crime rate within it. This is not only because architecture on this scale is inherently threatening and alienating; it is also because buildings of this type are the very negation of defensible space. The label 'defensible space' essentially denotes areas over which local residents can exercise some sort of proprietorship or guardianship, either individual or collective. In physical terms it is space with which residents can identify as virtually an extension of their own homes and which is designed in such a way as to facilitate surveillance (see Mayhew, 1981 on the surveillance effect). This is precisely the kind of space that is lacking in American public housing projects or in British high-rise council flats. Here, defensible space is virtually confined to the dwellings themselves. Corridors, stairs, lifts, entrance halls and fire-escape exits are a kind of territorial limbo over which residents exercise little responsibility. In principle these areas 'belong' to all residents but in practice they belong to no one and to anyone at all. Furthermore, physical design ensures that these are unobserved spaces. For Newman, observation is a key requirement in defining space and protecting it against those who would misuse it. Equally important are the symbolic properties of space. Structures should be designed to transmit unambiguous signals about who the rightful occupants are and to eliminate ill-defined 'no man's land' spaces.

Since Newman's original essay, numerous writers on both sides of the Atlantic have attempted to develop or apply his ideas; much of this research is synthesized by Davidson (1981). At the time of writing, the most widely cited British example is Coleman's (1985) painstaking study of the relationship between physical form and 'social malaise' in London. This study emphasizes architectural determinism in an extreme fashion. A vast exercise, involving over 4000 blocks of flats and a further 4000 houses, the study attempts to correlate such symptoms as vandal damage, graffiti, litter and excrement with 15 physical design variables. On one level, this is an invaluable condemnation of architectural failings and (another) final nail in the coffin of the high-rise flat. In other respects, however, its approach is disturbing. In a highly critical review, Dickens (1986) castigates Coleman for confusing correlation with casuality. While vandals, for example, may be aided by a lack of security, visibility and communal responsibility for 'public furniture', these design features do not necessarily explain the propensity for vandalism.

Dickens also criticizes Coleman's failure to compare high-rise in the public and private sectors, and indeed it does seem that form of tenure, with all that this implies for ownership, control, individual sense of responsibility and autonomy, is obviously a very relevant variable here (see Mawby, 1981; Table 3.4). Moreover, the variations in crime incidence both within and between building types are so great as to suggest that design factors exert far less impact than Newman or

Table 3.4 Offender and offence rates in nine residential areas of Sheffield

Area	1	2	3	4	5	6	7	8	9
Predominant Housing Type	OO	PR	PR	PR	CH	CH	CF	CF	CF
Offender Rate	5.2	141.3	26.5	23.1	96.7	46.6	32.4	76.7	22.2
Rank	9	1	6	7	2	4	5	3	8
Offence Rate	9.5	113.8	27.4	18.2	85.1	37.0	23.7	31.2	20.2
Rank	9	1	5	8	2	3	6	4	7

Source: adapted from Mawby, 1981.
Notes: Offender and offence rates are expressed as the number per 1000 households.
OO denotes owner occupied
PR denotes privately rented
CH denotes council houses
CF denotes council flats
While the above table does not prove causality, it certainly suggests two marked tendencies: a. a high degree of overlap between offender residence and offence commission; b. a tendency for offenders to both live and operate within the same type of residential environment.

Coleman would have us believe. Reporting on the Sheffield Crime Survey, Bottoms and Xanthos (1981) reveal that, within the council sector, the worst areas for crime incidence are all pre-war estates built decades before the advent of the tower block. They therefore conclude that high-rise construction has little connection with spatial patterns of crime incidence – in Sheffield at least.

These narrowly defined points should be seen as part of a more far-reaching critique of architectural determinism, one that has been levelled at Newman in the past (for example Mawby, 1977) and is equally applicable to Coleman today. For all its virtues, architectural determinism is an easy way out, a pragmatic theory which prescribes purely technical solutions for problems which are at root social or even political. Thus Newman exhorts that living environments and public spaces be constructed so as to make them more secure for residents and other legitimate users. The present authors would certainly not dispute the literal need for such improvements – on the contrary we would support any drive for substantial design modifications which would make public housing more liveable as well as more secure. But we are under no illusions that such measures would solve the crime problem. As Allatt's (1984) studies of experimental council-estate security schemes suggest, security measures and redesign may suppress burglary, theft and vandalism in the treated areas themselves, but such measures do nothing for the criminogenic conditions which predispose people towards lawbreaking. All they do is displace such activities to other places and other forms of expression.

3.3 The Radical Critique

From the foregoing it is evident that the spatial analysis of crime has been extremely strong on empirical data and statistical method. It thus embodies the dominant virtues of the methodology which for much of the 1960s and 1970s held sway throughout geography: an uncompromisingly positivist stance and a technical sophistication reflecting the scientific pretensions of its practitioners. Yet to some critics it was extremely weak on explanation – like much human geography (Gray, 1975). We have already seen that in empirical terms the spatial approach tends to turn up rather inconclusive findings; correlations and associations certainly, but often with no clear proof of causality (Baldwin, 1979). In other cases, the work seems to have been little more than a time-consuming exercise in proving the geographers were actually unwilling to face up to the 'real' causes of crime, preferring instead to dabble in superficial phenomena.

This point was particularly forcefully argued by Peet (1975) in a paper which was both a specific critique of Harries's highly empiricist approach to crime geography (Harries, 1974) and a general challenge to the entire field. Peet's basic point was that the kind of criminality which is the geographers' staple diet, 'the geography of lower-class crime; "crime in the streets" or black crime' (Peet, 1975, 277), is little more than a skin-deep manifestation of far more deep-rooted social ills.

> Crime is a surface expression of discontents which lie deeply embedded in the social system. Like any surface manifestation, crime can provide clues to the particular forces which cause it; these, in turn, may be traced to deeper contradictions which churn in the guts of the social and economic system. . . . A study which starts, continues and ends at the surface cannot possibly deal with cause (Peet, 1975, 277).

Deviancy and many other forms of 'social malaise' are symptoms rather than causes themselves. As Peet argues, to concentrate on managing symptoms rather than attacking the real problem at its roots is to lend uncritical support to the judicial machinery of the state and to accept at face value the premise that offenders must be suppressed, coerced and punished. Indeed Harries is quite explicit about the policy implications of criminology and crime geography: 'a much larger range of expertise must be brought to bear on issues of crime control and prevention' (1976, 384). In this way academic, work can act as ideological justification for the status quo: 'geography comes to be of use only in preserving the existing order of things by diverting attention away from the deepset causes of social problems and towards the details of effect' (Peet 1975, 277).

Though addressed primarily to a geographical audience, Peet's admonitions were an echo of a wider radical upsurge within mainstream criminology, both in the USA and in Britain. Up to the 1960s it is fair to say that criminological theory, for all its many variations, had been based squarely on the premise that criminality is an aberrant form of behaviour. Whether produced by their own deficient personalities or by defects in their surrounding environment, lawbreakers are antisocial misfits, people who in Peet's words 'are not well meshed with the system' (1975, 279). The implication here is that the system itself is above question and all would be well on the law and order front if the minority of troublemakers could somehow be 'meshed' into that system. The onus is on the individual, not on society, and the keynote is conformism.

It is this very view which radical criminology emerged to dispute: the assumption of a just society whose benefits, rights, duties and obligations are fairly apportioned to all; a society in which order is maintained by a consensus in which the vast majority of members accept the authority of the state as legitimate (Young, 1979). For Friedrichs (1980) the first purpose of radical criminology is to challenge traditional assumptions about the legitimacy of the state and its laws:

> In the traditional view, law secures and enhances civilization; law is based upon the universal good and democratic consensus. Law settles disputes, defines relationships, educates, promotes desirable behaviour and protects the idea of justice . . . a 'lawless society' is a contradiction in terms (Friedrichs, 1980, 42).

But the radical view sees law as 'supportive of elite interests' (Friedrichs, 1980, 42) and based on coercive power, the reverse of an expression of popular will. Hence it is inherently illegitimate (see the discussion in Young, 1979).

Before proceeding to some of the applications opened up by these ideas, we should note that both radical geography and radical criminology have met with serious objections from those who would prefer to remain within a more conventional framework of research. For example, Herbert (1982) finds Peet's critique to be based on a rather superficial account of poverty and deprivation as criminogenic conditions and to offer very little in the way of positive new directions for research. At the same time, Herbert is far from dismissive of many of Peet's propositions, taking very seriously assertions that crime is symptom not cause, that the geography of crime is too often that of lower-class crime, and that positivist approaches may act as a 'think-tank' for the interests of the ruling class. For a truly dismissive account of radical criminology, the reader should consult Toby (1980). In a paper entitled 'The New Criminology is the Old Baloney', Toby asserts

> The New Criminology is not new. It draws upon an old tradition of sympathy for those who break social rules. . . . The notion that the poor are compelled to steal by the unbearable misery of their lives reflects over-identification with the underdog (1980, 24).

It is hoped that the following sections may allow us to decide whether these are the crusty sentiments of a dyed-in-the-wool reactionary or a salutary caution against over-indulgence in ultra-left utopianism.

3.3.1 Labelling and Labellers

Few geographers of crime have heeded Peet's cry for a radical re-direction paralleling that of the new criminology, an omission provoking Lowman's remark that 'in the geography of crime the positivist demon still awaits an epistemological exorcism, (1982, 325–6). In an attempt to perform the necessary ritual, Lowman has advocated the use of labelling theory and has attempted to demonstrate its possible applications. The attraction of the labelling approach is that it attempts to place deviant behaviour not simply in its immediate neighbourhood/community context but also in its larger 'socio-legal context' (Lowman, 1982, 307). By this Lowman means that rule-breaking is to be understood as a response to the social control mechanisms – ranging from informal social disapproval to the formal agencies of the state – which exist to define, deter, detect, punish and correct crime and deviancy.

Prior to this approach (developed by writers such as Becker, 1963 and Lemart, 1967; see Taylor, Walton and Young, 1981, 139–71), criminologists almost invariably presented the 'commonsense' version of social control agencies as necessary *responses* to crime, as society protecting itself against anti-social behaviour by the use of the police, the judiciary and the penal system. With a fine sense of irony, labelling theory neatly inverts this causation by concentrating instead on the way in which criminal behaviour may actually be a reaction to attempts to control it. According to Lowman, there are two ways in which the 'crime control industry' distorts the rate of crime: indirectly, by the production of statistics which misrepresent the true nature of criminal behaviour as well as its magnitude (Lowman, 1982, 308); and directly, through stimulating and reinforcing criminality (Lowman, 1982, 309). Each of these points warrants a full discussion in its own right.

3.3.1.1 *Official Statistics and the Manufacture of Crime*

From the researcher's viewpoint, 'crime and delinquency . . . have some advantage in that an elaborate set of laws enforced by a highly professional judicial system exists to define what constitutes a criminal act . . . certain acts . . . are clearly labelled as offences against society' (Herbert, 1983, 75–6). Thus, for practical purposes, the researcher benefits from a massive, meticulously assembled data bank (police and court records), enabling him/her to count, classify and locate offences and offenders. Unlike Herbert, however, radical criminologists see this as arbitrary and misleading. Even on its own definitions, the official recording procedure fails to give an accurate picture of the real number and type of offences committed, a problem now well known and exhaustively discussed (see Hood and Sparks, 1970; Taylor *et al.*, 1981; Nettler, 1984).

Among the pitfalls of official procedures are:

1 the extensive under-reporting of crime by victims and witnesses (Nettler, 1984). Actual crime comfortably exceeds recorded crime and, despite the useful attempts to quantify the experience of crime (see 3.4), 'the dark figure of crime' (Lowman, 1982, 308) remains a mystery.
2 the extensive social bias in the reporting, recording and prosecution of offences. The police exercise a great deal of discretion as to whether a particular act should be defined and processed as a crime. Nettler (1984, 46) refers to the operation of a 'judicial sieve' and emphasizes that whether or not a detected offender is arrested and prosecuted depends upon the nature and circumstances of the crime, the characteristics of the offender, the offender's attitude to the officer and whether or not there is access to a lawyer. 'The police force, the police officer, the victim and the offender do make a difference – up to some limit – in determining what is recorded as a crime' (Nettler, 1984, 46). In particular the well-documented race and class bias in the criminal labelling process produces a false over-representation of certain categories which is sufficient to question the credibility of much of the positivist conventional wisdom. Any correlation between crime and poverty/minority status/slum areas is actually based on a rather arbitrary definition of who is an offender.

Orthodox crime geography has done little to challenge these definitions. Officially defined crime has been accepted at face value and, as Peet (1975, 278) reminds us, 'no mention is made of upper-class, middle-class or "white collar" crime'. Quite patently, this near-exclusive concentration on working-class offenders paints a partial and biased picture, a point which can be readily grasped by reference to Young's (1979) distinction between what he calls *voluntaristic* criminals and *pathological* (or 'determined' – i.e. shaped by forces outside their control) offenders. Voluntaristic offenders are those who actively embrace crime as a professional enterprise and Young cites such cases as the property speculator, the professional burglar and the fraudster, highly organized operators 'voluntaristically embracing the cupidity of capitalism' (Young, 1979, 18). By contrast, he sees pathological offenders as in a sense driven into criminality by the unequal social order which penalizes, impoverishes and excludes them. For example, some types of petty thief might be seen as 'mere artifacts of their wretched environment' (Young, 1979, 18), victims driven by circumstances into victimizing others. (On this theme see also Platt, 1981; Box, 1983.)

There is, of course, no clear-cut distinction between voluntaristic and determined criminals. Nor is all working-class crime necessarily determined: some of it, such as workplace pilfering, may well be calculated, organized, and even fairly large-scale. This nevertheless does not overturn the central argument that working-class street crime is hugely over-represented in official statistics. It is over-represented firstly because the disorganized lower-class offender is highly vulnerable to detection, secondly because street crime is typically more visible ('transparent' as Young, 1979 calls it) than fraud or commercial swindling, and finally because this type of crime is 'a common enemy; all sections of the population (except perhaps the left-idealist criminologists) are united in their opposition to street crime' (Young, 1979, 20). Whatever the cause, the researcher should allow for what Greenberg (1981, 58) calls 'class differences in immunity from law enforcement', rather than simply reproducing them without qualification.

3.3.1.2 The Impact of Labelling on Criminal Behaviour

The central assertion of labelling theory is that the individual's experience of law enforcement agencies (and indeed of any other agency with the power to attach value-laden definitions) acts to reinforce and amplify deviant behaviour. The actor may accept the label and adopt deviancy as a career. As a result of the perceptions of others, there occurs a change in self-perception, an aggressive assertion of deviancy. Lemert (1967) suggests that having been given the stigma, the individual may use it as a positive sense of identity, embarking on a deviant or criminal career to 'fend off the painful involvements' (Taylor *et al.*, 1981, 151) with 'straight' society. At the same time of course, labelling removes one of the basic disincentives for deviancy: the individual no longer fears falling because he/she has already fallen. Paradoxically, then, social control produces an effect directly opposite to that intended. 'The amplification of deviance is the unintended consequence of the effort to contain it' (Lowman, 1982, 309).

Evidently there are numerous attractions in a labelling approach. Unlike many of the orthodoxies which have engaged geographers in the past it is highly dynamic, concentrating as it does on the active interplay between law enforcement and criminal behaviour, thus allowing for the human element of free will. In this way it is also far less deterministic than those perspectives where local environment or individual psychology are held to dominate behaviour. Moreover, for the geographer an added attraction lies in the fact that labelling is applied to places as well as individuals. Whole neighbourhoods (in some cases entire cities) become stigmatized and their inhabitants wear a collective badge. In such circumstances, negative labelling acts as a self-fulfilling prophesy; whole neighbourhoods actually do deteriorate as a result of a process stimulated by the creation of an initial negative image. The definition produces the thing defined (Taylor *et al.*, 1981) as the residents of a stigmatized district adopt negative, defensive and/or hostile postures ('reactive antagonism' (Lemert, 1967, 43)).

At this point a word of caution is in order, since to push the labelling logic to its limits would be to enter into the realms of the absurd. For Taylor *et al.* (1981, 145) the arguments advanced by Lemert come close to claiming that there can be no crime without an audience to label it so: 'It is almost as if without labels there would be no deviance'. A related and more basic criticism is that labelling hangs too heavily upon certain questionable psychological assumptions and is not a social theory in the full sense. Deviance should not be accepted as an initial random act which is then repeated in reaction to societal disapproval.

Such an acceptance would neglect 'the contingencies and circumstances of everybody's lives . . . and . . . the larger social inequalities of power and authority' (Taylor *et al.*, 1981, 154). Thus criminal and deviant responses need to be located in their proper social context, since it is only in specific circumstances that negative definitions may produce a negative response. Most frequently these circumstances will be those of material deprivation and powerlessness.

3.3.1.3 Urban Managers as Labellers

Many of the better recent attempts to apply labelling to the spatial aspects of crime have successfully surmounted this last criticism. A case in point is Gill's (1977) readable study of the decline from respectability to criminality of a council-estate neighbourhood in a northern English city. The central argument of the study is that negative labelling is the critical determinant of the decline of 'Luke Street', but that this needs to be placed in its full historical, political, socio-economic and indeed physical context. Faithful to the tenets of the labelling approach, Gill is primarily concerned with the way in which external definitions 'create' criminal neighbourhoods: 'By "created" I mean . . . the way they are created in the minds of people whose residential location is far removed from such places' (Gill, 1977, 1). Here he is adopting the notion of a 'social audience' (Taylor *et al.*, 1981), outsiders (individuals, groups and institutions) who act as arbiters and standard-setters.

In the case of Luke Street this social audience is far from monolithic and by no means consists solely of law-enforcement agencies. It contains first of all the public, residents of other areas of town who react with fear and disapproval to the mention of Luke Street. Local residents themselves are sharply aware of this social rejection – after all it is frequently articulated in the local press (Gill, 1977) – and in this they are fairly typical of slum and ghetto dwellers elsewhere. For example a study of a 'problem estate' in Knowsley, Merseyside, reported residents' fears of 'the shame of having people visit' and the failure of estate dwellers to obtain jobs or credit because of the bad name of the area (Liverpool Polytechnic, 1982; for parallel experiences see Manion, 1982).

For Gill, however, the crucial component in the social audience is the local authority housing department, since it is their policies which at first unwittingly initiated and subsequently confirmed a downward status spiral. Built in the 1930s, this remained a highly desirable council estate until the 1950s, with pride taken by inhabitants in maintaining their homes and with a long waiting list of prospective tenants. The rot set in with the construction of modern peripheral estates and the consequent fall in the relative status of Luke Street. This fall in demand led to the filling of vacancies by families stigmatized as 'problems'. The result was cumulative – an exodus of 'good' residents and a concentrated lumping together of a mass of every imaginable vulnerable social category – the unemployed, the welfare dependent, the large family, the single-parent family, all defined by housing managers as rent-arrears risks with low standards of household management.

In cases such as Luke Street a further labelling process occurs in that hard-to-let estates tend to be the least well maintained, with housing management taking the tacit view that scarce resources would be better employed in estates with more 'reliable' tenants. Certain estates are effectively labelled 'unworthy'. Here it is the council rather than the residents themselves which might be held responsible for the lowering of standards. Irrespective of the tenants' own predispositions, a policy of neglect, of poor and tardy house maintenance and failure to repair the public environment is at best a poor model for the residents and at worst an open invitation to juvenile vandalism and criminal damage. Once again there is a feedback in that, once the vandalism sets in, the council is even less inclined to allocate investment to such troublesome areas. Fig. 3.4 sets out some of the interlocking and cumulative interactions.

Gill is by no means alone in this emphasis on local authority decision-making and bureaucratic allocation procedures. This managerialist approach to the creation of delinquent council estates has been adopted by several other British criminologists and represents one attempt to break with American-dominated theory and replace it with an approach more appropriate to an urban setting which differs decisively from the American city. For example, state intervention

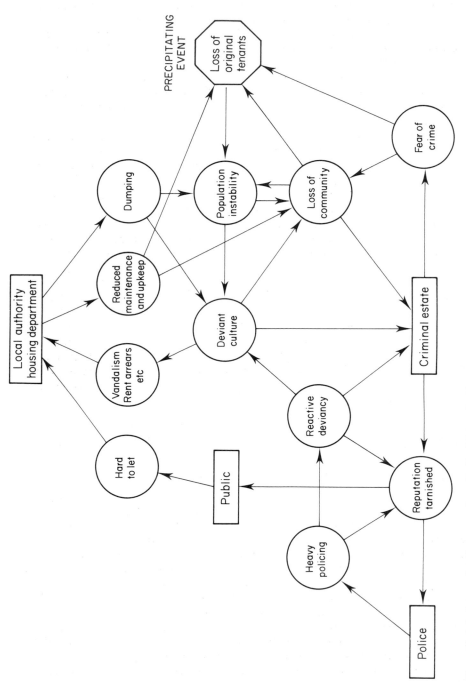

Figure 3.4 Actors and actions on a hypothetical Council estate.

has had a major impact on the British housing market, paradoxically acting to redistribute life chances in a regressive fashion (Pahl, 1975) and to create 'housing classes' which do not correspond to social divisions created by the division of labour (Rex and Moore, 1967). Hence, insofar as poverty does exert a crime-creating influence, then this will be greatest in areas where housing class (rather than simply occupational class) is lowest. Consistent with this reasoning Bottoms and Xanthos (1981) and Herbert (1976b; 1977b; 1982) have shown that:

1 areas dominated by owner-occupied housing exhibit lower offender rates than would be predicted by their socio-economic composition. Here, we may assume, the status of property owner exerts a stabilizing influence upon the social behaviour of all classes

2 Crime concentrations are localized within neighbourhoods of submerged rented housing, whose occupants tend to be not only economically marginalized but also penalized by a form of housing which gives little material, financial or psychological satisfaction.

As we have argued in Chapter 2, both housing and neighbourhood act to isolate them morally from the respectable mainstream of the population.

Whatever the explicit weight which they attach to this kind of social labelling, all the above authors are interested in pinpointing the managerial processes which create these neighbourhoods, most of which are in the public sector. By and large the writers concur with Gill as to the broad processes involved, though with some differences in emphasis. They all point to housing allocation procedures which act to place the most disadvantaged people in the most disadvantaged places: unlike Gill, however, the Sheffield study finds no evidence of direct labelling or grading of tenants by the local housing department (Bottoms and Xanthos, 1981, 207). There is also a general agreement on the cumulative self-sustaining nature of housing deterioration (see also Manion, 1982). In addition both Herbert in Cardiff and Bottoms and Xanthos in Sheffield attach very great weight to the operation of deviant subcultures in such estates. Thus, in one of the Sheffield case-study estates:

> Criminality as such is not taught to children or friends, but attitudes consistent with criminality are transmitted. Children may learn at a young age that 'dad fixes the electric meter' . . . or that 'mum bought a carpet cheap from someone in the pub' (that is, the carpet was obviously stolen). Such behaviour comes to be seen as not particularly wrong, but a necessary way of obtaining desirable goods, or of solving certain material problems (Bottoms and Xanthos, 1981, 213).

At first sight this may seem to be harking back to an increasingly dated and tarnished approach. Managerialism, however, lends subcultural theory a more progressive thrust by focusing on the external forces which create this collective local deviant mentality. It cannot be denied that deviant attitudes are present or that they are actively transmitted within a cohesive solidary community which, moreover, is physically enclosed and spatially isolated to an inordinate degree (Bottoms and Xanthos, 1981, 212). Nevertheless these attitudes are not simply internally generated by the pathological inadequacies of the local residents themselves: they are constructed by defective management procedures which ghettoize a collection of disadvantaged people in a bad environment, thereby reproducing and enlarging the social inequalities to which these residents are prey. Like any other ghetto, this is one imposed from without, not constructed from within.

3.3.1.4 *Policing*

The third element in Gill's social audience are the police. Many labelling studies see law enforcement as absolutely crucial, the prime cause of reactive deviancy. In his study of prostitution in Vancouver, Lowman (1982) places heavy emphasis on policing and upon judicial procedures as a means firstly of criminalizing a section of the prostitute population and secondly of displacing it spatially within the city. Geographical displacement rather than eradication is a further unintended consequence of enforcement, a treatment of symptoms rather than cause.

In other aspects of labelling too, policing is given great prominence. In particular many studies of black crime interpret criminality as at least in part a reaction to aggressive and intrusive policing in ghetto areas. There is a widespread tendency for black individuals and black residential space to be labelled and targeted for heavy police activity, both because they are easily identifiable and localized areas and because their inhabitants are highly visible. Many writers add to this the further factor of alleged police racism (Scraton, 1982; Gutzmore, 1983), which automatically equates a black skin with potential criminality. Whatever its origins, the intensity of police activity in many ghettos on both sides of the Atlantic can be held to inflate the level of black crime, both in the sense that it ensures that more blacks than whites are apprehended for any given crime, and that it provokes the kind of aggressive deviant responses noted by Lowman (1982). Add to this, allegations (many of them well substantiated: see Takagi, 1977; Platt and Takagi, 1981; Marable, 1984) of extorted confessions and police brutality and it is difficult to resist the conclusion that the ghetto is a classic example of the paradox of reactive deviancy. Certainly Scraton (1982) in his analysis of the Liverpool 8 riots of 1981 is unequivocal in his condemnation of policing in that area, highlighting the role of the Chief Constable himself in creating a mood of racism within his force. Alternative versions of such events are provided by Lea and Young (1982) and indeed Scarman (1981) himself on the Brixton riots.

Black criminality is only one aspect (though an important one) of the labelling question. In general terms, geographers should be aware of the central thesis about police labelling; that differential policing concentrated in social areas stigmatized as trouble spots produces differential responses and differential crime statistics. Because police manpower resources are limited they must be heavily deployed in areas of greatest criminal potential and there is thus a need for *a priori* labelling of certain areas.

Wiles (1975) argues that spatially differential patrolling is a cumulative process and he illustrates this by reference to the history of policing in British cities. From its earliest beginnings in Britain, policing tended to be concentrated in those areas thought to pose the greatest threat to social order. Thus concentration automatically produced extra knowledge of crime in these areas and artificially inflated their crime rates. In turn, these rates could be used as 'evidence' that such neighbourhoods needed still heavier policing. There is more than a suggestion here that recorded crime patterns are not the 'true' geography of crime. To a significant degree they are the geography of police patrolling, reflecting a tendency to discover and record more crime in the most intensively patrolled localities. Crime is where you look for it. Recorded patterns may also reflect the differential use of police discretion, as for example in a willingness to prosecute for relatively minor offences in some areas but to caution such offenders in other areas. Here spatial bias may well express class bias, with working-class offenders in working-class localities more vulnerable to prosecution. In addition, as Mawby (1981, 142) suggests, 'police definitions of suspected persons might vary by area, such that police would be more likely to question suspects from certain areas rather than others'.

While all this amounts to a cogent argument which should be taken seriously, it has not won universal acceptance and has even been categorically denied by some researchers (e.g. Mawby, 1979; 1981). Even so, geographers should be aware of possible social, racial and spatial biases operating at every stage of the managerial and judicial processes.

3.4 Victimology: The Injured Parties in Crime

3.4.1 Victim Research

Apart from the labelling and managerialist pathways, there have been other attempts over the past decade or more to break away from old-style positivism. One such approach which is now beginning to influence geographical work is victimology which, as its title declares, aims to shift the focus away from crime as officially recorded and towards crime as seen through the eyes of its injured parties. Among the many advantages of victim studies is that they present a more human picture of crime, one which is lost in the pages of official statistics.

From the researcher's viewpoint, the major gain from victimology is that in many respects it provides a more accurate reflection of the experience of crime than that furnished by police records. This is certainly the case in the USA, where surveys based on victims' own evidence suggest that the true figure of national crime (excluding murder, corporate and white-collar crime) may be as much as four times that indicated by the FBI (Platt, 1981). As we have seen, before any act can be recorded as an offence it must undergo the judicial sieve, which means that many acts deemed to be unlawful by their victims simply disappear. Even before that, any act must first be *reported* to the police and it is at this stage that much of the discrepancy arises between actual and recorded offences. According to Platt under-reporting stems largely from public distrust of the police:

> The primary reason for not reporting crimes is the belief that the police are either incapable of solving crimes or are likely to aggravate the situation by brutalizing or intimidating the victims. This distrust of the police is realistically based on the extensive experiences of working-class communities, especially racial and national minorities, with police brutality and ineffectiveness (1981, 15).

Whatever the truth of this assertion, there is no substitute for research based on first-hand reporting by victims themselves, and the victim survey has begun to make its mark as an important alternative source of crime data. In Britain, the first British Crime Survey was directed at a sample of 16,000 people who were asked about their experience of crime victimization (Hough and Mayhew, 1983; 1985). Since then this nationwide survey has been repeated and complemented by a number of local studies in Merseyside (Kinsey, 1984), North Central Birmingham (Smith, 1982; 1986) and Scotland (Chambers and Thombs, 1984). Though none of its practitioners would present the victim survey as a flawless substitute for official data (Mayhew and Smith, 1985), the general feeling is that its pitfalls are outweighed by its advantages.

As yet the technique has not been extensively applied to the spatial patterning of victimization. Broadly it is fair to say that, like official statistics, victim surveys once again highlight the inner city and the local authority ghetto as peak locations. In this respect they underline an issue of great criminological significance – namely that it is poor people in poor areas who run the greatest risks. Whether or not we accept that criminal behaviour is itself a direct reaction to poverty and oppression, we must recognize that it is certainly members of the most oppressed classes who are the most victimized by crime. This 'poor preying on the poor' syndrome is underlined by several American writers who have drawn attention to the phenomenon of 'black on black' crime, a now widespread scourge of the ghettos, which heaps fear upon the pile of burdens which the black poor must in any case shoulder, (US Law Enforcement Assistance Agency, 1975; Shin, Jedlicka and Lee, 1977; Marable, 1984). Such 'parasitical criminality' (Platt, 1981, 16) poses painful intellectual problems for radical-left analysis, which rejects crude and unqualified condemnation of the criminal. At the same time such analysis must recognize that 'crimes do have victims, whose suffering is real . . . a radical criminology which appears to deny this will be seen as callous and rightly rejected' (Box, 1983, 3: see also Greenberg, 1981; Platt, 1981). This dilemma will be addressed again in the final section of this chapter.

Despite the undoubted risks to which many deprived groups are exposed, most studies reveal that victimization patterns are not a simple function of class or income. Particularly illuminating on this point is the North Central Birmingham Crime Survey which examines the cross-cutting impact of class, demography, lifestyle and geography on victim risk (Smith, 1982; 1986). Some of its findings – for example that young males are more prone than either females or the elderly – are confirmation of now widely established relationships. The victim-proneness of young males is related to lifestyle in that it is the most mobile and active sections of the population who are most exposed to risk (Smith, 1982). Other findings begin to test new ground. As one of the few victimologists to try to isolate spatial effects, Smith (1982, 395–9) finds significant variations by area type within the inner city, with residents in Housing Action Areas and General Improvement Areas being less vulnerable than the admittedly small sample living in non-designated areas.

3.4.2 Fear of Crime

Closely associated with victimology is the study of the fear of crime, an issue which has begun to stimulate research in the USA and latterly in Britain. In its social and behavioural impact fear of crime may be as potent as victimization itself:

> In recent years it has become recognized that people's anxieties about crime amount to an independent problem in their own right: people may change their lifestyle as much or more because of their fear of crime as of their actual experience of it (Merseyside County Council, 1986, 1).

In other words, many citizens tailor their movement patterns to avoid times and places seen as dangerous. As Smith (1986) emphasizes, this can be extremely debilitating, subjecting people not only to emotional stress but constricting their movements and imposing all manner of personal and even financial costs. This is of major impact on the elderly and other groups most prone to this feeling of vulnerability. In effect their activity space is reduced, distorted and confined to dwindling portions of secure time and space. Fear is socially isolating and a force undermining local community life.

At the same time, fear appears in certain respects to be not entirely rational. Much recent work – including Hough and Mayhew (1983; 1985), Kinsey (1984), Maxfield (1984) and Smith (1986; 1987) – suggests that the fear of crime is not necessarily related to the experience of victimization itself or to the statistical probability of victimization. While it is naturally the case that crime victims express more anxiety than non-victims, there are many categories of non-victim for whom the possibility of victimization is a real worry. Indeed, as Smith observes, 'many of those who seem most anxious are actually least at risk' (1986, 37; see also Warr, 1984). Among Smith's 'vicarious' victims, women and the elderly are invariably over-represented, despite the statistical truth that it is young males who run the greater risk.

It would of course be misleading to analyse any emotion, particularly fear, in terms of calculative rationalism. Most of the recent victim surveys make this point forcefully, as for instance Kinsey:

> Merseysiders in general worry a lot about crime but those under greatest social and economic pressure also suffer most from crime: they worry more, perhaps too much, but they do have the most to worry about (1984, 24).

As on Merseyside, so elsewhere in the United Kingdom, the fear of crime is greatest in the poorest and most crime-ridden areas, notably the inner city and the most deprived council estates. Once again there appears to be an aggregate neighbourhood effect, which helps shape the individual residents' perception almost independently of personal experience. Among the various 'non-crime origins of fear' (as Smith, 1986, 41, calls them) are processes which originate outside the local area and are then projected inwards upon it. For example residents of racially mixed areas tend to be disproportionately fearful in response to the criminal image of black youth – an image which many would argue has been falsely created by police and media (see Gutzmore, 1983). Other more localized origins are traceable to a decaying, derelict physical environment, with its implied threat as a kind of visible symptom of disorder; and to the breakdown in local community life, which has left individuals without support and hence a prey to all manner of anxieties. Indeed fear of crime might be best seen as a kind of 'displaced anxiety' in many instances (Smith, 1986).

Clearly these investigative pathways are as yet largely untrodden. We believe, however, that there is much potential here for valuable research to which social geographers could make an important contribution, building on the openings created by writers like Smith.

3.5 Political Economy and the Geography of Crime

For most radical geographers, the approaches reviewed above would not go far enough towards answering Peet's appeal for an excavation beneath the superficialities of crime symptoms. To be sure, there has been definite progress in this direction. Victimology represents a long-overdue shift in focus, while managerialism and labelling are altogether more realistic than ecological positivism. Nevertheless at the time of writing there appears to be a certain lag between mainstream radical criminology and the geography of crime. Whereas the former has already given rise to a strong body of work in the Marxian tradition of political economy (for example Quinney, 1977; Fine *et al.*, 1979; Hall, 1980: Greenberg, 1981; Platt and Takagi, 1981; Box, 1983; Lea and Young, 1984 and some of the papers in Inciardi, 1980), the latter has conspicuously failed to apply any of this to the question of criminal space and environment. Indeed Herbert's latest summing up of the work of crime geographers is obliged to dismiss 'political economy' in one short paragraph and to conclude that, although lip service is occasionally paid to Peet's critique, it has never been centrally applied as a means of gaining real insights (Herbert, 1986, 7).

It is clearly beyond the scope of the present authors to attempt to write a 'Marxist geography of crime' starting from scratch and in the space available here. We do, however, propose to set out some of the basic elements of the modern Marxist perspective on crime as a context in which geographers might work. Following on from this, we also offer a pointer as to how this perspective might guide research on the specifically spatial dimension of criminality.

3.5.1 Marxism as Context

Peet's principal quarrel with orthodox crime geography is that, with its heavy stress on working-class street crime and strategies to cope with it, it swallows capital's own definition of the problem without a murmur. It unquestioningly accepts, even justifies, bourgeois ideology. This ideology strips the crime issue of its class character and presents the capitalist state as presiding over the common good by disciplining and correcting that minority of the working class who jeopardize the common good by breaking social rules, rules which are specifically designed to ensure peaceful and harmonious co-existence. Capital as class is invisible in this scenario, an inert and passive entity which presumably plays no part, either in creating crime or in reacting to it (Box, 1983).

Yet, as Young (1979) points out, this cosy view of a just society in which order is maintained by consensus, with the vast majority of people accepting the state as legitimate, has been under challenge for some time. Marxist criminology had its origins in the late 1960s, when it became the voice of marginalized groups such as blacks, feminists and gays, groups who had increasingly come to see themselves as legally repressed merely because they were black, female or gay. Their protests were articulated by radical academics, 'the marginal middle class' according to Young. Essentially the task these new criminologists set themselves was to invert the established reasoning and replace it with a picture of criminals and deviants as members of a victimized class reacting against oppression. As Friedrichs (1980, 38) has it, 'It is not human nature itself but capitalism that produces egocentric, greedy and predatory behaviour'. In all essentials this assertion is entirely true to the spirit of classical Marxism, as the following extract from Engels illustrates:

> When one individual inflicts bodily injury on another, such injury that death results, we call the death manslaughter; when the assailant knew in advance that injury would be fatal we call it murder. But when society places hundreds of individuals in such a position . . . its deed is . . . [also] . . . murder'

Hence, in defiance of conventional thinking, deaths and injuries caused by exploitation in the workplace, by poor living conditions and by all the other human sacrifices on the altar of profit should be ranked with the most serious of crimes. The magnitude of capitalist crime in our own time is emotively documented by Quinney (1977, 142) who writes of:

> Crimes of the government, the capitalists and the corporations against the rest of the population. These offences take a daily economic toll that is beyond estimation. For example there are the losses caused by sickness and death resulting from the pollution of our food and environment; injuries from police brutality; and the loss of jobs and income through unemployment, racism and sexism (See also Platt, 1981; Box, 1983).

Quinney's classification of crime in capitalist society is reproduced in Table 3.5.

To illuminate this inversion of the 'normal' perspective on crime – capital perpetrating the *real* crimes thereby eliciting a response from labour which it then may define as criminal – a historical perspective is essential. In this case the past is a remarkably copious source of information about the present. This is especially so in view of the work of radical social historians on the way such concepts as law and order, riot and hooliganism, have been understood in earlier formative periods of capitalism (Thompson, 1963; Hobsbawm, 1969; 1971; Pearson, 1978). Initially we should grasp here that the birth of European capitalism was facilitated by 'larceny on the grandest scale imaginable' (Greenberg, 1981, 39). Not only did primitive accumulation involve the piratical plunder of what is now known as the Third World; at home it also involved the forcible driving of the peasantry from the land by enclosures, the destruction of traditional livelihoods and the passage of laws restricting common access to land.

Table 3.5 A classification of crime in capitalist society

Bourgeois crime (Crimes of domination and repression)	Working class crime (Crimes of accommodation and resistance)
Crimes of economic domination	Crimes of resistance
Crimes of government	Personal crime
Crimes of control	Predatory crime

Source: taken from Quinney (1977).

As is now documented by the above historians, this capitalist conquest met with popular resistance at all stages. Rural displacement was marked by widespread turmoil and the food riot was a persistent interruption in the history of eighteenth-century England (Pearson, 1978). In rural Wales a local variation on this theme were the 'Rebecca' riots in protest against road tolls which threatened to prevent small farmers bringing their produce to market (Williams, 1985). Early urban-industrialism was also marked by revolts, this time against the factory system, and taking the form of factory arson and machine-wrecking ('Luddism') by purposive well-organized bands of workers.

One of Pearson's (1978) core themes is that the conventional treatment of these events is 'history-from-above', a ruling-class version of events designed to justify its own crimes. In this way capital's theft from labour is called 'progress' and is legitimized and actively promoted by the state through repressive legislation against all forms of worker resistance. Then as now, people are treated as culpably deviant 'when they disrupt capitalist social relations' (Greenberg, 1981, 58).

Deriving from this are a number of propositions which may be used to illuminate the pursuit of crime geography:

1 Crime is historically specific. The kinds of behaviour which preoccupy the crime geographer are by and large creations of the capitalist epoch.
2 'Law and order' is no recent issue. 'Moral panics' have been a recurrent feature of social life since the birth of capitalism.
3 Crime and civil disorder frequently result from class struggle over space.

3.5.1.1 Crime as a Product of Capitalism

To Marxist commentators, much crime is thus 'endemic to capitalism' (Platt, 1981, 18). It is produced by capitalist social relations and represents a reaction against class oppression. One view propounded in some ultra-left circles (and sharply criticized by Young, 1979; Box, 1983) is that criminals are 'proto-revolutionaries', whose actions are a (misplaced) challenge to authority. While this may be justified when applied to the early Luddites or to present-day workers criminalized by repressive anti-strike legislation, it utterly fails to describe street crime. As we saw in the previous section, much street crime is predatory and individualistic and victimizes members of the same class. 'Criminality as an effective, though limited, method of waging class warfare began to decline with the development of industrial capitalism' (Platt, 1981, 25) and most modern working-class crime is anything but revolutionary in intent, appearing more as a kind of reactionary method of accommodating to existing class relations (Hirst, 1972). Even so, such criminality may still be seen as the actions of a sub-class who harbour very real grievances against a social order which dispossesses them but who have no legitimate outlet for such dissatisfaction.

3.5.1.2 Crime as an Enduring Feature of Capitalism

The 'law and order' issue is not one which is peculiar to our own time. Not only do we see parallels between the social disorders of yesterday and today, but also we see no particular mystery about the motivating processes operating at either period. As in early industrial capitalism so in the post-1970 period there are forces in motion which threaten labour's livelihood, liberty and dignity. No specialized knowledge of psychology is needed to predict that such a stimulus can evoke a reaction. The machine-wreckers of the early nineteenth century have been presented as blind, unthinking, mindless thugs in much conventional history, but a history written from the viewpoint of the rioters themselves would tell an entirely different story. As Pearson (1978, 123) enquires, 'What is blind about an action which strikes out to destroy something which threatens one's own destruction?'

What is different about the present period is that the threat takes a different form and that the target at which to strike out is less visible or accessible. Firstly, what is the threat? It is clear that people are no longer striking out against loss of land or of independent artisan livelihood. Rather they are reacting on a material level to a loss of jobs (capitalist restructuring) aggravated by a simultaneous reduction in state welfare support (the decollectivization of consumption: see Chapters 2 and 6). In this context the crisis of law and order can be seen as entirely consistent with the model of the capitalist state put forward throughout this book. Modern capitalism, notably during the early post-war period, was distinguished by extensive state intervention to underwrite a rising social wage for labour (collective consumption). Using the state in this way, capital has been able to rule largely by consent, by being accepted as legitimate by most sections of labour. A kind of 'social contract' obtained whereby, in return for law-abiding behaviour, the citizen receives guaranteed civil rights and a decent standard of living. Hence the mass of public opinion supports the law. Like every other aspect of capitalist structure, it is accepted as legitimate, so that coercion is only necessary to deal with those minorities deemed as a threat – apolitical in the case of criminals, political in the case of strikers and revolutionaries – to social order.

Since 1970 this stable edifice has progressively crumbled, almost as if it had been no more than a temporary, almost illusory, interval in the conflictual history of capitalist social relations. One way of looking at this is to say that the proportion of marginalized or surplus labour – i.e. those excluded from the social contract – has risen so sharply that it can no longer be peacefully contained (Quinney, 1977). As Platt (1981, 23) puts it, 'Monopoly capitalism emiserates increasingly larger portions of the working class'. This marginal population is always present under capitalism; it constitutes a major element in the population of slums, ghettos and other submerged areas, including, of course, crime areas. But in manageable numbers it poses no serious threat to social stability because of its political and moral isolation from the 'respectable' working class: such people can be dismissed as unemployable, as racially handicapped, or as wallowing in a culture of poverty. Here it is hard to resist charging traditional criminology with supporting this ideology by its portrayal of criminals and crime as pathological. Even Marx and Engels were decidedly contemptuous of this 'lumpen proletariat', those whom they called the 'social scum'. (See Greenberg, 1981 on the way this influenced early Marxist thought on crime.)

Our own view is that this sub-class is indeed crime-prone, largely because it has been impoverished and rendered politically impotent by forces not of its own making. The crux of the present problem (from the state's perspective) is that, with one-sixth of British households living below the poverty line, the marginalized have become too numerous to ignore or to isolate. It is hardly facile to equate rising marginality (especially youth marginality) with rising recorded crime rates. Under present conditions, petty crime assumes an enlarged role as a self-help survival strategy (see Marable, 1984 on ghetto crime as a function of increasing marginality) and as a means of self-assertion on the part of the utterly powerless (Dunning et al., 1988). Rising crime must also reflect the state's own politico-legal response (Hall et al. 1978). As the army of impoverished grows, so the legitimacy of capital and the state becomes open to challenge and the balance begins to slide from rule by consent towards authoritarian repression. Hence tougher laws, which criminalize the hitherto non-criminal, and tougher law enforcement, which brings forth its own counter-violence, most dramatically in the shape of the British inner-city riots of the 1980s (Cowell, Jones and Young, 1982; Hall, 1985; Sivanandan, 1985; see also Chapter 5).

None of this answers the question of why so much working-class crime victimizes members of the same class, or indeed of the same sub-class. That the surplus population should turn against its own poses extremely difficult problems for radical analysis and brings to light dangerous inconsistencies in Marxist thinking. For example, many radical criminologists worry about a tendency (common among labelling theorists and Marxist writers in the 1960s and not yet extinct) to romanticize the criminal, the victim of capitalist oppression valiantly striking out in reprisal (see the criticism by Lea and Young, 1984). Much street crime is anything but heroic and to present it as such is 'over-simplified and derives from a shallow and highly selective reading of Marx' (Greenberg, 1981, 12). Moreover common criminals are widely feared and detested by members of their own class (see 3.4 above), a feeling so obviously based on real experience that it can no longer be dismissed as 'false consciousness' (Young, 1979; Box, 1983).

In seeking a rational answer which does not stoop to the derogatory labelling of offenders, radical writers have invoked various forms of logic. Quinney's (1977) response to the victimization of the poor is to categorize such offences as acts pursued by those who are already brutalized by the conditions of capitalism. In this case the notion is of individuals desensitized by their own poverty and rejection, an echo of Engel's writings on the demoralization of the working class. While this is not to be dismissed, it should be seen in the wider context of a society in which there are no longer any 'real' targets for the disaffected to hit out at. Nowadays few factories remain in the inner city; even fewer have relocated to the peripheral council estates. Even if they were there, they would be unlikely targets for attack since, in contrast to their historic predecessors, modern working-class youths are not resisting proletarianization – what they object to is their forced *de*proletarianization and the lack of jobs, prospects and identity which this implies (Braverman, 1974). But how can this be resisted? What are they to strike out at? Capital? Present-day capital is infinitely elusive, geographically remote, physically intangible, masked by state apparatus and therefore virtually immune from any direct reprisal by the new dispossessed. Not surprisingly their reprisals appear disarticulated and directed at random targets. By and large the only substantial, accessible and visible targets are those front-line agents of the state, the police. Interaction between the police and the local community figure prominently in the ensuing sub-section on criminality as a product of territorial class struggle.

3.5.1.3 Crime and Class Struggle over Space

The early events charted by radical social historians can also be seen as the outcome of a class struggle over spatial resources, as capital asserting a right to expropriate rural land which had formerly supported the people. In the teeth of violent popular opposition, this expropriation was only accomplished by extensive legal changes explicitly designed to eliminate all but *capitalist* property relations (Pearson, 1978; Greenberg, 1981). The imperatives of capital accumulation demanded that rural land be commercialized (transformed from use value for labour to exchange value for capital) and that labour be displaced to the new urban centres of industrial capital.

This capitalist conquest of space was not a temporary or finite process necessary simply for the establishment of industrialism. On the contrary it continued into the new urban milieu and in some form or another has remained up to the present. Urban space itself had to be fully incorporated into capitalist social relations. Here Cohen's (1979) work on the policing of the working-class city is of prime relevance. In his studies of inner London in the late nineteenth and early twentieth centuries, he argues that the establishment and expansion of the metropolitan police force had a great deal to do with the process of spatial incorporation – in effect the agents of the state establishing territorial control on behalf of capital. To use his own phrase, the police were engaged in regulating working-class 'usage of time and space so that it did not obstruct the traffic of industry and commerce' (Cohen, 1979, 120). The management of local community living space was usurped, so as to bring it into line with the needs of capital. This can hardly be equated with the Highland clearances or the enclosure movement, but it nevertheless represented an undermining of established forms of life and livelihood. In Islington Cohen (1979, 120) represents police activity as 'consisting of general physical intimidation coupled with systematically arbitrary arrests', the object of which was to suppress 'street cultures and their irregular economies'.

To be sure, Islington was a notorious high-crime area during the period under review, but a significant proportion of this 'crime' consisted of redefinitions of behaviour which was sanctioned within the community itself – the criminalization of traditional street pastimes, as Cohen terms it. This is of obvious relevance to the analysis of 'criminal' areas in the 1980s. How far do such areas reflect not the pathological deviance of their occupants but the outlawing of aspects of local community lifestyle?

In a Marxist analysis we must ask further questions about why it is in the general interests of capital to impose (via the state apparatus) standards of bourgeois morality on every nook and cranny of urban space. After all, deviant communities may serve useful purposes. In certain respects they may even act to stabilize the social formation. As Marx himself noted, petty crime is a means of subsistence for certain members of the surplus population who otherwise might get up to worse mischief. The 'Arthur Daleys' of this world are profoundly conservative. Criminal and deviant sub-cultures also give a sense of identity and purpose to their adherents. Moreover, local criminal areas also act as convenient containers – rather like the 'sink estates' perpetuated by local housing managers – easily monitored by officialdom and ensuring that the rest of the city is relatively trouble-free. Why meddle?

There are all kinds of answer to this question. We might invoke the economic needs of capital accumulation and argue that during periods of expansion there is a need to incorporate the residents of every backwater of the space economy as regular producers and consumers. Like any other form of petty entrepreneurial activity, petty crime tends to lock up valuable resources. A more serious motive for state intervention is that criminal communities cannot be spatially contained. Those urban enclosures which we call 'criminal areas' do in practice contain a great range of individuals from the hardened career criminal to the completely law-abiding: the latter tend to comprise the great majority in many areas. And it is precisely these sections of the local population who suffer most from the fear of crime and as victims. Once again the issue is one of legitimacy. The state needs to ensure that the law-abiding working class continue to pin their faith in its laws and rules. Such faith is likely to be undermined unless the state is seen to be acting to quash those who refuse to acknowledge these rules. As several Chief Constables have recently observed, there can be no 'no-go areas' in the British city. While it may perform a stabilizing function in certain respects, in the last instance the criminal enclosure poses an unacceptable threat to state legitimacy.

3.6 Conclusion

In asking us to view crime as part of a social totality instead of as a specialized malfunction, Marxist method enables a re-evaluation of the 'dreadful enclosures' which are the focus of crime geography; especially it helps to demystify the people who live in them. These are the people who Platt and Takagi (1981, 35) claim 'can be best understood in the context of the capitalist labour market', victims of what Braverman (1974, 280) describes as the 'clearing of the market place of all but the "economically active" and "functioning" members of society'. In the present crisis, these are the front-line casualties of the mass unemployment and low-wage casualized activities which are now such a prominent feature.

While criminality may certainly not be equated with poverty in any simplistic way, it is nevertheless true that high-offender areas house a large part of that social stratum referred to as the marginalized or surplus population. This stratum is subject not only to economic dispossession but also to the erosion of its living space by urban renewal (Harvey, 1978; Cox, 1981), by intrusive land uses (Bunge and Bordessa, 1975) and, as we have seen, by repressive policing and criminalization within that living space. Moreover, it is politically unrepresented as well as economically outcast, having no legitimate voice within the democratic social contract (Harloe, 1981; Lea and Young, 1982). Petty criminality is just one response to this outcast condition. When right-wing penologists such as van den Haag (1975) advocate banishment for certain crimes, they might reflect whether the inner city is not already functioning as a latter-day Elba.

Recommended Reading

The work of Sue Smith has been well received by geographers in this field, and her *Crime, Space and Society* (1986) is a substantial, up-to-date and comprehensible review. This book succeeds in combining research from both spatial and non-spatial perspectives and usefully reviews recent work on emerging fields of study, such as the fear of crime (see also, for example, Maxfield's research for the Home Office, *Fear of Crime in England and Wales* (1984)). Two earlier texts, *The Geography of Urban Crime* (1982) by David Herbert and Norman Davidson's *Crime and Environment* (1981) are both sound and comprehensive reviews of research set firmly in a geographical framework. An occasional paper compiled by David Evans (1986), *The Geography of Crime*, provides a useful, if understandably brief, overview of research and supplements this with papers from four leading practitioners; a book edited by Evans and Herbert is scheduled for publication in 1989.

No geographer wishing to gain a comprehensive review of research in this field can ignore the work produced by official agencies and those from other disciplinary backgrounds. Much of the traditional data required to undertake empirical research is collected by official sources and is subject to *caveats* concerning its accuracy, nature and objectivity. For example, data provided by the Home Office and published in the usual reference texts (e.g. *The Annual Abstract of Statistics*) gives an indication of the level and nature of crimes recorded by the police, but the *British Crime Survey* (Hough and Mayhew, 1983; 1985) and a number of local studies (e.g. Kinsey's (1984) work on Merseyside) give a far more accurate and illuminating reflection of the experience of crime and its effects. The majority of the theoretical debate has also been located elsewhere in the social sciences, and the competing perspectives act as the main organizing focus for the preceding chapter. Advocates of each approach and a synthesis of at least some of their interpretations can thus be obtained by referring back to the text; for more detail it is then essential to consult the research referenced therein.

Finally, in addition to the sample of texts listed above there are a wide range of journals publishing research on crime, the most well-established in the United Kingdom being the *British Journal of Criminology*. Studies of the fear of crime are best developed in the United States, and, among others, the journal *Victimology* contains many significant contributions.

4
Gender

4.1 A Woman's Place? Gender and Disadvantage

> Everywhere women are worse off than men: women have less power, less autonomy, more work, less money, and more responsibility. Women everywhere have a smaller share of the pie; if the pie is very small (as in poor countries), women's share is smaller still. Women in rich countries have a higher standard of living than do women in poor countries, but nowhere are women equal to men (Seager and Olson, 1986, 7).

In this quotation, part of the introduction to a pioneering international atlas, Seager and Olson graphically demonstrate the pervasiveness and persistence of gender-based disadvantage and the commonality of women's experiences in both West and East, North and South. The atlas provides a valuable and visible testament to the extent of inequality and a worthwhile starting point for readers beginning this chapter. Our focus is somewhat narrower and more theoretical than empirical in its detail. However, before we attempt to explain the subjugation of women in contemporary Western society it is appropriate to provide some empirical evidence to highlight the extent of this disadvantage. We start by looking at income and employment.

4.1.1 Women and Paid Work

We all work, in the sense that we undertake tasks and duties that involve some form of effort, mental or physical, and which take time that we would prefer to use for other purposes. Indeed, preliminary results from the 1986 United Kingdom Labour Force Survey indicate that on average men spend almost 43 hours per week at work (Department of Employment, 1987). In contrast, non-married women in employment devote an average of nine hours per week fewer than their male counterparts; even more noticeable is the fact that the mean number of hours spent in the workplace by an employed married woman is less than 28 per week, more than 50 per cent below the equivalent figure for men. But this survey incorporates only a proportion of the time expended; it is an analysis of *paid employment*. This distinction may not be particularly important for the majority of employed males. Just 6 per cent of working men were not 'full-time' (i.e. were employed for fewer than 35 hours per week), and for almost all working males paid employment clearly constitutes the bulk of the time they devote to the tasks and duties which can be covered by our initial definition; indeed for many men there is little (if anything) to add to the effort expended in the office or on the factory floor.

For women the picture is different. Even among non-married working women 30 per cent were employed for less than 35 hours per week. Most married women in employment worked part-time (58 per cent), a figure which rose to 69 per cent for working women with dependent children (Department of Employment, 1987). Yet in contrast almost all time-budget surveys, in which the respondent keeps a diary of his/her daily activities, demonstrate that women, and particularly those with family commitments, have significantly less discretionary time than men. For example, a survey undertaken by the Henley Centre for Forecasting for the British government found that women in full-time paid employment had nine hours per week less free time than similarly employed men; for those working part-time the comparable difference was eleven hours. Indeed, even a full-time employed male had 4 per cent more leisure time than a supposedly 'unemployed' housewife (Table 4.1).

The key to explaining this is *domestic labour*. According to Oakley (1974, Table 5.4), the average time spent on housework by a sample of urban housewives in 1971 was 77 hours; and some of these housewives also had paid employment! Nor is the work rewarding. There is no direct renumeration and no pension rights. Most of it is undertaken in isolation, away from the camaraderie and solidarity of the workplace, and much of it is monotonous in the extreme. It is not regarded as a voluntary undertaking, nor one shared by both partners. Although a minority of men believe that household tasks should be equally shared, fewer still put this into practice, and even in the mid-1980s women are primarily responsible for washing and ironing in 88 per cent, for cooking in 77 per cent and cleaning in 72 per cent of married households (*Social Trends 1986*, 36 (Central Statistical Office, 1986c)). In married households men took primary reponsibility for the above tasks in just 1 per cent, 5 per cent and 3 per cent of cases respectively.

Given the relative absence of women from paid work it follows that female-headed households are likely to have below-average income levels. Precise data are not made readily available, but 89 per cent of single-parent households and 80 per cent of single adult households over retirement age are headed by women. The average weekly expenditure of a household in the United Kingdom in 1985 was £161.87, but comparable figures for low-income households in the above two categories (which would include the vast majority of those headed by a woman) were just £66.84 and £46.51 respectively (*Social Trends 1987*, 108 (Central Statistical Office, 1987c)). In part this is a consequence of the lower participation rates of women in paid employment, but even women in full-time work are significantly disadvantaged. Table 4.2 (a), shows the average gross weekly earnings for men and women, with the former being over 50 per cent higher in 1986, a ratio which has remained constant since the mid-1970s. Even after allowing for the fact that men work longer hours than women, their hourly rate of pay is more than one-third higher (Table 4.2 (b)). Despite the Equal Pay Act of 1970 and the 1984 equal value amendment (designed to assess pay relativities when jobs cannot be directly compared) in no occupation did wage rates for full-time female employees equal those of their male counterparts. The 1986 New Earnings Survey indicates that women's earnings averaged 74.1 per cent of those of their male counterparts, with a range from 91.8 per cent for police officers to 65.0 per cent for sales supervisors (Equal Opportunities Commission, 1987).

Despite (or perhaps because of) this, female participation rates in the labour

Table 4.1 Time use in a typical week by sex and economic status, Great Britain 1985

Hours

Weekly hours spent on:	Full-time employees[1]		Part-time employees[1]				Retired people
	Males	Females	Males	Females	Housewives		Retired people
Employment and travel[2]	45.0	40.8	24.3	22.2			49.8
Essential activities[3]	33.1	45.1	48.8	61.3	76.6		60.2
Sleep	56.4	57.5	56.6	57.0	59.2		58.0
Free time	33.5	24.6	38.5	27.5	32.2		7.9
Free time per weekday	2.6	2.1	4.5	3.1	4.2		9.1
Free time per weekend day	10.2	7.2	7.8	5.9	5.6		

Source: Equal Opportunities Commission 1987, 50.
[1] Excludes the self-employed.
[2] Travel to and from place of work.
[3] Essential domestic work and personal care. This includes cooking, essential shopping, child care, eating meals, washing and getting up and going to bed.

Table 4.2

(a) Average gross weekly earnings including the effects of overtime, full-time employees aged 18 and over, 1970–1986.

Pounds per week

	1970	1974	1975	1976	1977	1978	1979	1980	1981	1982	1983	1984	1985	1986
Men	29.7	47.7	60.8	71.8	78.6	87.1	99.0	121.5	137.0	150.5	163.3	177.0	190.4	205.5
Women	16.2	26.9	37.4	46.2	51.0	56.4	63.0	78.8	91.4	99.0	108.8	116.4	125.5	136.3
Differential	13.5	20.8	23.4	25.6	27.6	30.7	36.0	42.7	45.6	51.5	54.5	60.6	64.9	69.2
Women's earnings as a % of men's	54.5	56.4	61.5	64.3	64.9	64.8	63.6	64.8	66.7	65.8	66.6	65.8	65.9	66.3

(b) Average gross hourly earnings, excluding the effects of overtime, full-time employees aged 18 and over, 1970–1986.

Pence per hour

	1970	1974	1975	1976	1977	1978	1979	1980	1981	1982	1983	1984	1985	1986
Men	67.4	104.8	136.3	162.9	177.4	200.3	226.9	280.7	322.5	354.8	387.6	417.3	445.3	481.8
Women	42.5	70.6	98.3	122.4	133.9	148.0	165.7	206.4	241.2	262.1	287.5	306.8	329.9	358.2
Differential	24.9	34.2	38.0	39.5	43.5	52.3	61.2	74.3	81.3	92.7	100.1	110.5	115.4	123.6
Women's earnings as a % of men's	63.1	67.4	72.1	75.1	75.5	73.9	73.0	73.5	74.8	73.9	74.2	73.5	74.0	74.3

Source: Equal Opportunities Commission 1987, 39 and 38.

market are increasing; indeed this has been one of the most evident trends in Britain and much of the Western world as we emerge from the depths of the recession of the early 1980s. Government ministers extol the benefits of employers 'developing and realizing the full potential of their women workers and of getting more women into industry' (Mr Ian Lang, Parliamentary Under-Secretary of State for Employment, cited in the *Employment Gazette*, September 1986, 349). The same journal lauds the fact that 'more women find work', pointing out that 'all but 2000 of the 170,000 additional employees in the employed labour force in 1985 were women' (1986, 348).

However, before we conclude that women are being liberated from the domestic sphere, we should note that:

1 Female activity rates are still 25 percentage points below male activity rates. Although this compares with a 37-point gap in 1971, a significant part of this is due to a six-point fall in male rates (*Employment Gazette*, July 1985, 259).

2 Furthermore, 'nearly nine out of every ten women joining the employed labour force (in 1985) entered part-time employment.' (*Employment Gazette*, September 1986, 348).

3 Women are still subject to considerable industrial ('horizontal') and occupational ('vertical') segregation, being particularly concentrated in semi- and unskilled employment in retailing, the clothing industry, catering and cleaning. Women in white-collar employment are primarily in lower-grade clerical and secretarial posts, while the relatively small proportion of professional women employees are concentrated in the 'caring' professions, notably education, welfare and health (Murgatroyd and Urry, 1985).

4 As the evidence above indicates (Table 4.2), many of the sectors in which women are over-represented are notoriously under-paid, and even in those sectors in which some measure of apparent equity has been achieved women tend to be concentrated in lower-paid grades (Equal Opportunities Commission, 1987).

5 Female participation in the labour market is still subject to major discontinuities, notably pregnancy, childbirth and child care, and moving house due to marriage or husband's job change. As Dex (1984, 545–9) demonstrates, such discontinuities and ruptures in women's career histories often lead to downward occupational mobility (see also Joshi, 1984; Martin and Roberts, 1984). Nor in times of industrial recession are women any less likely to suffer redundancy (Martin, 1985); indeed, given their interrupted work record, many women are particularly vulnerable, especially when jobs are shed on a 'last-in, first-out' basis.

6 Women experience far greater spatial constraints on access to employment, leading to an increased dependence on homework (Allen and Wolkowitz, 1986; 'the homeworker is a casual labourer exploited on the basis of her ascribed role as a woman/wife' (Hope *et al.*, 1976, 103–4, cited in Bilton *et al.*, 1987, 162)). Those who are able to leave the home tend to operate in a limited spatial field, constrained by children's school times and holidays (Tivers, 1985). Women are also twice as likely to rely on public transport for their journey to work, and twice as likely to journey on foot (Equal Opportunities Commission, 1987, 50). For those in rural areas or on peripheral estates these constraints are particularly acute (Little, 1986).

It seems clear from the evidence above that women's employment, and in particular its recent expansion, can hardly be seen as solely, or even primarily, for the benefit of women themselves. Massey (1984a) identifies the way in which women have been used to restructure local labour markets in south Wales, providing a cheap, flexible and compliant (often non-unionized) source of labour for the new electronics and service industries which are replacing the old male-dominated heavy industries like steel and coal, where capital was forced to compromise with entrenched unionized resistance. On the same theme Williams (1984, 258) writes of the sexual recomposition of the Welsh labour force, where

> of the core working population, 45 per cent are women. Of these some 42 per cent are part-time, moving in and out of work according to the rhythms of child-birth and rearing. The pressure on them mounts. Women go on working beyond retire-ment age in larger numbers than men. . . . In the shadow land of the informal economy . . . no one knows exactly what is happening.

The same author goes on to show a similar trend to that identified for the whole of the United Kingdom, an increase of 10 percentage points in the female activity rate between 1968 and 1982 coinciding with a 7½ per cent fall in the total active population. Once again the superficial opening-up of the job market to women barely disguises their role as exploitable replacements for redundant males.

From these examples it is clear that female employment plays a critical role in capital accumulation, not least in the way in which capital uses space and locality in its struggle to maintain its level of accumulation and its control over labour. Another instance of this is Murgatroyd and Urry's (1985) case study of capitalist restructuring in the local Lancaster economy. This region has experienced considerable industrial disinvestment as industries such as chemicals have rationalized, concentrating more and more of their capacity in a declining number of large plants and pulling out of smaller peripheral locations. Since the 1950s this has resulted in the local sexual division of labour becoming even more skewed. Women by 1971 were heavily over-represented and concentrated into clerical and retail occupations and the clothing industry, even more so than at the national level. Once again an erstwhile manufacturing region was being reconstituted as a service economy, opening up usually low-quality jobs for women.

> Thus the expansion of women's employment did not imply a greater similarity between the positions of men and women in the local labour market, nor indeed in the division of labour more generally. A high degree of occupational segregation persisted in Lancaster. . . . Far from women entering the traditional domains of male activity, it was the feminized industries that expanded . . . there is no evidence that the types of work and the conditions of employment of women noticeably improved and there was an increasing level of female unemployment (1985, 49).

4.1.2 Women in the Housing Market

In the housing market, even more than the labour market, women have long been invisible, or, more accurately, have been made invisible by the policies and processes of those that allocate housing. Only when she is an independent wage earner does a woman receive serious consideration from the providers of mortgage finance, even though women hold approximately half of building society balances. Within a marriage she may be the token second name on a 'joint' mortgage application, though a significant minority of properties are still registered solely in the male householder's name. She has no impact on the size of loan granted if she does not work, and even if she is a wage earner she normally earns the second (lower) income on which a maximum 1 × the annual salary is advanced. Only when a woman is a single parent does she normally fall within the ambit of the public-sector housing authorities – half of divorced and separated women live in local-authority rented accommodation, while only a quarter are able to continue to live in a house with a mortgage, compared with a third of divorced or separated men (Equal Opportunities Commission, 1987, 44).

In fairness, the last two decades have seen *some* opening up of housing oppor-tunities for women. Many local authorities have become more sensitive to the needs of those enduring insufferable domestic circumstances, recognizing their responsibilities for rehousing under the 1977 Housing (Homeless Persons) Act and, in a minority of cases, providing refuge accommodation for battered women. Much of this accommodation is of very poor quality, however, and it is heavily concentrated in areas where housing is 'difficult to let'. These develop-ments have also taken place against a background of financial cutbacks and sustained attacks on public-sector housing, and the little that has been done remains under constant threat (see Chapter 2).

At the other end of the spectrum, building societies have recognized the opportunities afforded by the growing number of professional women with sound careers and regular incrementally-increasing salaries, particularly in regions like the South East of England. As a result one of the 'big three' building societies increased its lending to women from 8.2 per cent of all advances in 1975 to 15.9 per cent in 1985 and 19.3 per cent in 1988 (Nationwide Building Society, 1986; Nationwide Anglia Building Society, 1988), though nationally advances to 'one female' accounted for only 10 per cent of the total in 1986 (Equal Opportunities Commission, 1987). Even these women borrowers were dealing in an inferior market to their male counterparts; according to the Nationwide Anglia Building Society (1988, 1–11) they were more likely to buy:

(a)   older property (34 per cent of properties bought by female main borrowers were pre-1919, compared with 24 per cent of those purchased by male main borrowers)

(b)   flats and terraced property (36 per cent compared with 31 per cent bought terraced houses, 26 per cent compared with 14 per cent bought flats; only 9 per cent of female main borrowers bought detached houses, compared with 19 per cent of males)

(c)   property with fewer amenities (29 per cent had no central heating, compared with 23 per cent of those properties bought by male main borrowers; only 32 per cent had a garage, compared with 47 per cent)

(d)   cheaper property (an average price of £44,135 compared with £50,179), despite the fact that female borrowers were over-concentrated in the (expensive) South East

Although 58 per cent of female borrowers were in white-collar employment (half of them being classified as professional / managerial) they had significantly lower income levels – an average of £12,704 per annum compared with £15,980 per annum for male main borrowers in the financial year 1987/8 – and their salary status, if not their job title, was most significant in limiting them to inferior housing resources. However it should be remembered that even this unequal access to a valued resource is available to only a minority of women, in this case to those employed full-time, with average earnings of over £240 per week, and heavily concentrated towards the upper end of the job spectrum.

4.1.3 Women and Power

The interrelated interests of men, capital and the state have combined to restrict the influence and power of women, particularly in the wider domain, and this inference provides the framework for much of the substance of this chapter. Although there is no evidence for gender-specific differences in academic ability – if anything, females tend to do better up to the age of 16, accounting for 51.4 per cent of 'O' level and 55.4 per cent of CSE passes in 1985 (Equal Opportunities Commission, 1987, 13) – women still get considerably less out of post-compulsory education. The education system is dominated by men and male values – although women represented more than 78 per cent of primary school staff, they accounted for just 19 per cent of head teachers in 1985; in secondary schools there were only 750 female heads, compared with over 4000 headmasters, and almost 70 per cent of women teachers were concentrated on the bottom two scales. Such a pattern is repeated in higher education. In 1980 36 per cent of British geography departments had no women on their staff, and a further 22 per cent had only one (Women and Geography Study Group of the Institute of British Geographers, 1984, 125); furthermore only 7.9 per cent of new appointments to university departments in the 1970s were of women, compared with 11.6 per cent in the 1950s (Johnston and Brack, 1983).

In politics and public life women are notable by their absence, and there is little evidence to suggest that the limited progress being made in some areas of the housing and labour markets is being sustained. After the 1983 General Election there were just 23 women Members of Parliament (out of 650), one fewer than in 1945. At no time has there ever been more than two women in the Cabinet, and Mrs Thatcher has presided over an all-male team for her entire period in office. In local government the position is a little brighter, but women still account for only between one-fifth and one-quarter of councillors. Of those 45,000 people appointed to public bodies by the government, women account for 19 per cent, and they tend to be found on local and regional, rather than national bodies. The proportion of women chairs is also much lower than the membership (Equal Opportunities Commission, 1987, 56). Nor are those groups who one would expect to see campaigning for equal rights at the forefront. Despite some improvements in the 1980s, 'In no trade union is the proportion of women on the national executive committee, the Trades Union Congress delegation or among full time officials anywhere near the proportion of women among the members' (Equal Opportunities Commission, 1987, 58); for example, women constitute 73 per cent of the membership of the National Union of Teachers, but just 21 per cent of the Executive and 7 per cent of the full-time officials. It may be the case that women find it more difficult to devote time to the voluntary activity which often initially underpins a career in politics or public life, but this simply helps to highlight the extent to which women are confined, both in practice and by the expectations of others, in the domestic domain.

4.2 The Sexual Division of Space

4.2.1 Sexual Segregation

In sum, the findings reported above help to establish the extent of women's social and economic subordination. This subordination extends directly into the spatial realm. Spatial unfreedom is the very embodiment of female subordination. Just as there exists a sexual division of economic and social roles, so there also exists a sexual division of space (McDowell, 1983). The hoary old slogan, 'A Woman's Place' is more than an abstract truism. Even in a supposedly mobile society it has a literal truth in that the two sexes have differential access to space, with women enjoying far less freedom of spatial choice.

Since geographers tend conventionally to interpret the term segregation as referring to *residential* separation, we should perhaps emphasize here that the sexual division of space refers primarily, though not exclusively, to *activity* segregation. Given that the two sexes continue to cohabit and form families together, then by definition the residential location of the majority of women will be identical to that of their male counterparts. Indeed the residential position of most married women will actually be determined by that of their husbands. Hence, for the most part, the sexual division of space is a question of the different and unequal ways in which the two sexes use space – or are permitted to use it. Before returning to this question, however, we should recognize that residential segregation by gender is by no means unimportant and is likely to become increasingly important in the future with the growth in the population of single women – unmarried, separated, divorced and widowed. There is now a significant proportion of women whose social and spatial position in the housing market is not determined by that of a spouse, and this proportion continues to grow (Land, 1976). For unmarried, childless car-owning women in high-status, well-paid occupations this may well offer a genuine freedom of residential and locational choice. For others, however, it may simply mean a losing battle with a male-dominated housing allocation system which recognizes no housing needs beyond those of the conventional (small) nuclear family. McDowell (1983, 145) refers to ' . . . the constraints on access to decent housing for those minorities that are neither deemed in need by local authorities nor creditworthy by financial institutions'. Whatever the conscious motives of the gatekeepers, in practice their actions amount to 'discrimination against unsupported wives, single parents, widows and unmarried single women'. Such women tend to be over-represented in the various forms of housing and types of neighbourhood described as 'submerged' in Chapter 2. Moreover the financial and emotional costs exacted by such environments are unduly high for women because of their extreme dependence upon and isolation within their immediate home and locality (McDowell, 1983, 142).

Furthermore, irrespective of her marital, parental or residential status, there exists a complex array of barriers to a woman's everyday use of space. There are, for example, customary prohibitions on where she may or may not go: elite 'men only' clubs and, at the other end of the social scale, pubs and bars in traditional working-class communities where unaccompanied females are still frowned upon. A woman's place is definitely not the tap room of the Miners' Arms. Parallel to this there are many urban neighbourhoods which, because of their perceived physical threat, are effectively no-go areas, especially at night. For practical purposes it does not matter whether their reputation for violence, mugging, rape and prostitution is based on fact or fiction (see Smith, 1987). All told, these prohibited spaces add up to a very considerable expanse of forbidden territory.

4.2.2 Activity Segregation: The Domestic Sphere

One aspect of segregation which has been singled out for attention by feminist geographers is women's continued confinement within the domestic sphere of life. From the very outset feminist geography has been concerned with' . . . the social and spatial limitations that domestic labour and child-care set for women, both individually and collectively' (Foord and Gregson, 1986, 189). Given the recent steady increase in the proportion of females in the labour force, this might seem a somewhat surprising preoccupation. Against this, however, must be weighed the large numbers of women who work only part-time and the far larger number who are still outside the formal labour market. As McDowell (1983, 142) confirms, 'A growing number of women now combine work in the home with participation in the labour market but it is still the widespread expectation and practice that domestic labour is "women's work" '. Seen in this light, the trend towards female economic independence is clearly one of slow, uneven and personally costly progress and does not represent a genuine re-shaping of the male–female relationship. To a quite startling degree women in contemporary society are still locked within what Crompton and Mann (1986, 8) call the 'domestic world of the family, of production for use rather than production for exchange and of course, of the reproduction of human beings'. To emerge from this, as growing numbers are now doing, and to enter the 'mainstream world' of paid work, they must overcome numerous obstacles including a continuing responsibility for domestic matters.

Thus there is for many women an acute tension between the worlds of work (as formally defined) and home. For feminist theory, this tension between what Stacey (1981; 1986) has called the 'public' and 'private' domains is now recognized as vital in understanding the condition of women in contemporary society. Stacey's notion of the public domain includes not only the world of paid employment but also that pertaining to the affairs of state. This is contrasted with the private domain of home, family and domesticity. We might perhaps quarrel with the labels public and private, often associated with state ownership versus private enterprise; we might also note that the concept of the private realm comes close to replicating Tonnies's (1887) classic notion of *gemeinschaft*: yet, whatever the nomenclature, it requires no great subtlety of mind to appreciate that these two opposed categories coincide almost precisely with the categories male and female. Even where the categories do not exactly match in actuality, they most certainly do in their symbolism – one a world of hard-nosed hard-edged cut-throat aggression, the other a world of knitting patterns, fluffy toys and big decisions about which brand of washing powder . . .

While the last is intended as a caricature, it nevertheless highlights serious questions, not least about the peripherality of female space. Here we should recognize that the domestic sphere has been progressively downgraded in a cultural sense and devalued in a material sense. In contemporary society it is rarely regarded as of primary importance in its own right but rather as an ancillary back-up system for the more centrally important sphere of work and public decision-making. It is seen as in every sense dependent, reproducing, nourishing and sheltering those 'gainfully employed' functionaries who perform the 'real work' of society. The womb-like imagery is inescapable. Naturally enough the material rewards of those who operate the domestic system are commensurate with this peripheral and dependent status. Indeed, the housewife is not rewarded at all in any formal contractual sense of the term: she is the original voluntary worker. Challenging this, Walby (1986, 34) insists that housework must be acknowledged as' . . . a distinctive form of work. It is hard work, and the fact that it does not receive a wage should not be held to disqualify it from the status of work'. Yet it is so disqualified according to cultural norms. Walby continues,

> More significant is the distinctive social relations under which the work is performed by the woman. It is neither exchanged in a calculated bargain for a wage which varies in proportion to the effort expended, nor with an employer who may be exchanged easily; rather it is indirect, although nonetheless present (since a wife who refused to perform domestic services of various sorts is liable to be divorced), and changing the employer (the husband) is much more difficult than for a wage labourer (1986, 34–5).

We shall return to Walby's views on patriarchy shortly (4.3.2).

4.2.3 The Evolution of Public and Domestic Spheres

Popular wisdom tends to insist that the sexual division of labour is somehow natural, that women's confinement within an invisible and second-class sphere is biologically determined, an inevitable outcome of their reproductive role, their 'passive nature' and their 'congenital obsession with home-making'. Women are simply not built for participation in a competitive world, still less for leading others in the great struggle for survival. Despite the pervasiveness of these stereotypes, the long view of human history tells us that female dependency is neither eternal nor immutable. On the contrary, it is of rather recent invention. We use the word 'invention' advisedly to underline the effects of *culture* rather than biology in determining 'Women's Place'; and to make the point that that place has changed decisively over the course of history with changes in technology and social organization. Ironically enough, it seems that the closer human society is to Nature, the less the likelihood of any sexual division of labour – or indeed any other form of inequality. As Hayford (1974, 137) reminds us,

> At the smallest scale of human organized production there is no real distinction between public and private activities; the household and all its crucial relations are contiguous with the community as a whole. While the particular work of individuals may be differentiated to some degree, most adults could do all the different kinds of work required for the support of the group. . . . Sex roles develop in terms of their complementarity and often are not mutually exclusive in the nature of their functions.

Superficially these points might be dismissed as mere echoes of certain anthropological and archaeological truisms about women in ancient society. But the point to grasp here is that until the rise of feminist consciousness these facts were rarely deemed to be pertinent. Far from offering a challenge to the popular stereotype of women, social science has commonly succeeded in reproducing it. By assuming the status of women to be non-problematic, scholarship simply confirmed their marginalization and in effect lent its own seal of approval to sex discrimination. For example, until very recently gender has never been considered as a variable in its own right in studies of social stratification and inequality. According to Allen (1982, 137), the sexual division of labour has 'been accorded the status of a natural order . . . rather than a socially constructed set of relations' (see also Smith, 1978). Noting this 'intellectual sexism', Crompton and Mann (1986, 7–8) ask, 'Is the neglect of gender by stratification theorists therefore a reflection of the conscious or unconscious action (or inaction) of "malestream" sociology?' They proceed to argue that the study of society has in practice been a study of only part of that society, the public domain in which women appear only in footnotes and asides. Enclosed within the domestic sphere, the vast majority of women have remained invisible or visible only by shadowy inference (Tivers, 1985). These of course are the very accusations which are now directed at the social geographer.

However, some writers in the classical Marxist tradition have projected women's position as historically, rather than biologically, created. One particularly interesting example of the genre is Dange (1986), who combines the theoretical insights of Engels with the writings of ancient Indian scholars to produce a picture of 'primitive communism' in pre-feudal India. It seems that in the primitive commune (the predominant social unit of the period) all the now taken-for-granted institutions which demarcate 'A Woman's Place' were completely absent. Private property, class divisions, the sexual division of labour and private monogamous marriage were all unknown. Economically the keynote seems to have been social and sexual equality with 'collective labour and consumption . . . both men and women participated in the *Satra Yajna* or labour' (Dange, 1986, 44 and 46).

Relationships of this kind seem to have been common in essence to most hunting, gathering and rudimentary agricultural societies. Women's biological role as Mother does not appear to have seriously impeded her equal contribution as worker, both producer and reproducer. However with technological advance came a specialized division of labour which in turn paved the way for a growing division of male and female economic roles, with the former assuming an ever more politically dominant status. This 'enslavement of women', as Dange calls it, reached its apogee in feudal societies, where woman's status often approximated to that of a chattel.

Not unexpectedly, it was the Industrial Revolution and the formative period immediately preceding it which finally laid the foundations of today's sexual segregation. In the British case, Rowbotham (1973) sees the period from the seventeenth century onwards as the one in which the segregation of women's activities finally became institutionalized, part of the normal unquestioned social fabric. This took place on two parallel and related fronts:

(a)   the separation of public and domestic spheres and the enclosure of women within the latter
(b)   their exclusion from a growing range of occupations and industrial sectors.

The increasing seclusion of women within domesticity was closely associated with the rise of a middle class, for whom the leisured wife or daughter was a mark of status and among whom '. . . the idea gained currency that men should be able to support their wives from their wage' (Rowbotham, 1973, 5). Thus woman's changing role was a product of increasing economic productivity, a new class structure and a new morality, all of which ensured that the wives of such men as yeoman-farmers, craftsmen or tradesmen came to be gradually removed from active participation in the family enterprise (see also Davidoff, 1986). In fact, as the new bourgeoisie grew richer, their women even came to be relieved of their domestic duties altogether. As Galbraith (1977, 61) pithily remarks, 'The rich man might work himself. But he gained much distinction from the conspicuous idleness of his women.' Needless to say, this sexual segregation was regionally and socially uneven. In rural areas it was slower in the upland periphery of western and northern Britain and did not apply at all to the landless and working classes. For the rural poor it was a case of the 'double load of women's work at home and outside' (Rowbotham, 1973, 26). 'Capitalism had different consequences for them. Far from being excluded from production their life was one of ceaseless labour' (Rowbotham, 1973, 24).

On the matter of regional variations, McDowell and Massey (1984) have compared the vastly differing impact of emerging sexual divisions in four British localities – a coalfield location, a textile area, an inner London 'rag trade' centre and a remote area of rural Norfolk. The most extreme version of the 'bread-winner–homemaker' split occurred in the first of these, County Durham. Here an overwhelming predominance of mining employment coupled with a virtual absence of paid work deemed suitable for women meant that by the mid-nineteenth century

> the separation of men's and women's lives was virtually total. . . . Virtually all the men earned their livelihoods in the mines and the mines were an almost exclusively male preserve. . . . For miners' wives almost without exception and for many of their daughters unpaid work in the home was the only and time-consuming occupation (McDowell and Massey, 1984, 129–30).

As widely confirmed by a host of local studies, this near absolute activity segregation was replicated throughout almost all the coal-mining regions (Frankenburg, 1976; Massey, 1984; Williams, 1984). So too was sexist oppression, with male dominance in the household taking on an overt, exploitative and even brutal form (McDowell and Massey, 1984, 130–1).

Quite clearly, local and regional disparities in gender segregation are principally determined by the spatial division of labour (see Chapter 1). The changes described by Rowbotham proceeded furthest and fastest in areas based economically on a narrow range of industries where women workers came to be outlawed, both by male workers and by law (Rowbotham, 1973; Hall, 1982). Other regions with a different employment mix present a contrasting picture. Textile-based Lancashire was the archetypal example of a local economy based as much on female labour power as male. For McDowell and Massey (1984) this obviously granted a much greater degree of working-class female independence than was traditionally the norm in Britain. But they also show that within the work sphere certain occupations came to be a male preserve and, almost automatically therefore, to be classed as 'skilled'. Furthermore, their high rate of workplace activity did not relieve Lancashire women of their domestic responsibilities: like other working women elsewhere they were forced to become dual-purpose workers.

4.2.4 The Devaluation of the Domestic Sphere

Despite these highly significant regional and indeed class variations, we may nevertheless conclude that the rise of industrial capitalism had the general effect of sharply separating the sphere of work from that of the home and family, and of increasingly confining women within the latter. Equally significantly, this segregation of women in the domestic sphere was accompanied by a gradual devaluation of that sphere, as many of its traditional functions were removed or undermined (Hayford, 1974, 142). With the transfer of almost all manufacture to the new capitalist market sector, the productive *raison d'être* of the family virtually ceased to exist. Less obviously, there was a parallel, though less complete, decline in the family as a unit of consumption: its role in supplying its members with their daily needs became less and less necessary as more of these needs came also to be supplied by the market place. A couple of examples – the supplanting of homespun clothing by mass-produced articles, the rise of convenience foods – will suffice to highlight Mandel's (1975) point that the needs of the family are increasingly met by capitalistically produced commodities and capitalistically organized goods and services.

One of the most seminal discussions of this historic trend is provided by Braverman (1974; see Chapter 1), who introduces the term 'Universal Market' to describe the way in which formerly self-sufficient areas of private and family life have been displaced. He writes of the transformation of 'all society into a gigantic market place' (1974, 271) in which family, friends and neighbours are ousted by profit-seeking capitalists, not only for material goods but also for recreation, amusement, emotional needs and the care of children, the sick and the elderly. In the twentieth century, the State has also become a force to reckon with, with its expanding provision of welfare, medical, educational and other services. Not only has the family been commodified, it has also been socialized (West, 1980, 177–9).

It goes almost without saying that neither commodification nor socialization has completely eliminated the purpose of the woman-centred family and, as we shall see, the family continues to fulfil a major purpose within capitalism. Yet its value is now largely invisible. Ostensibly it appears as a shrunken residual of its former self and hence women's second-class status stems, in part at least, from the fact that they preside over a much diminished realm.

4.3 Female Subordination

4.3.1 Women and Class

If we are to adequately theorize women's position in geographical space, we must ask certain questions about their location in the class structure. Do they indeed occupy a distinctive position? Undoubtedly this is the very crux of the confrontation between feminism and 'malestream' social theory (see Gamarnikov *et al.*, 1983; Crompton and Mann, 1986). The latter, as represented in almost all the class analysis literature published before about 1970 (and much since), has consistently operated on the premise (usually by default rather than conscious design) that gender is not a salient dimension of class structure. This premise breaks down into three overlapping assumptions as outlined by Allen (1982, 139):

(a) that the family is the unit of class . . . (b) that women derive their class position (and their status) from their male kin, (c) that women are dependent in critical respects on men.

Even on the empirical plane these propositions seem shaky, given the proven occupational segregation of women and the differential earnings and status that accrue to 'men's' and 'women's' work. Theoretically, too, it begs certain vital questions, resting as it does upon an inadequate and very dismissive evaluation of women's economic contribution within the domestic realm.

What of housewives? Walby (1986, 35) is among those who argue plausibly for housewives as constituting a class in their own right: 'I would argue that house-wives (both full-time and part-time) are a class exploited by their husbands who also constitute a class'. Just as in the market place capital expropriates surplus value from wage labour, so on the domestic level husbands expropriate surplus value from their unpaid wives: ' . . . the housewive is involved in an unequal exchange relationship in which she receives maintenance for her labour' (Walby, 1986, 34). Though this may appear as an unjust indictment of those males who devote the bulk of their earnings to the upkeep of their family, it is nevertheless logically correct. What Walby is stressing here is the *relational* aspect of class, in this case an unequal power relationship in which the woman is not only dependent but also has no sort of contractual guarantee. Because of this the model applies equally to affluent middle-class and to impoverished working-class women. A common dependency cuts across the normal boundaries of class. Having made this point, Walby adds that only husbands and wives constitute classes in this sense, not men and women as such.

Since the discussion of women as a class is still more concerned with raising questions than supplying answers, we can hardly be definitive here. The only certainty is that we can no longer be satisfied with the stereotype of the housewife hanging on to the class structure by her husband's coat tails. Women occupy a distinct position, one which many would argue has been created and maintained by *patriarchy*.

4.3.2 Patriarchy

. . . a universal system of political, economic, ideological and, above all, psycho-logical structures through which men subordinate women (Vogel, 1981, 208).

In stressing the exploitative aspects of an inter-sexual relationship grounded in unequal power, Walby is employing one of the central concepts of feminist theory, patriarchy (Millett, 1969). A dictionary would describe patriarchy as the rule or authority of a father – or perhaps of a surrogate father figure, such as a tribal chief – and the very word 'patriarch' irresistibly conjures up cliched visions of biblical figures like Moses, Abraham or even the Old Testament God himself. Literally, then, patriarchy applies to situations where the male head of a dynasty exerts absolute rule over his wives and descendants, demanding and receiving total obedience and entitled to resort to sanctions or even physical force against acts of insubordination.

Extending this concept beyond these narrow confines, feminist theory has come to apply it to societies where the overt and aggressive wielding of male power is formally disapproved of but where a system of structured gender inequality nevertheless continues to operate in less explicit ways. Accordingly Lown (1983, 29) argues,

Rather than conceptualize patriarchy in terms of an outmoded familial form, my argument is based on the notion of unequal power relations of gender and age forming a central axis of historical and social change.

Judging simply by the empirical evidence (see 4.1), there can be little doubt that modern Western society answers Lown's conditions. As she says, patriarchal relations are not 'just a facet of one particular historical formation', but comprise 'a pivotal organizing principle of society' (1983, 29).

Not surprisingly, this all-inclusive interpretation has been frequently castigated as lax and ahistorical (Rubin, 1975; Cockburn, 1981), but Lown (1983, 31) counters by explaining that though the form and style of patriarchy is certainly subject to profound variations in time and space, its content and purpose remain essentially constant. Changes in social manners (and even the existence of equal opportunities legislation) simply act to conceal realities, so that while women in Western society are no longer effectively enslaved, they are still inhibited and enclosed in countless informal ways.

Patriarchy, then, is the systematic subordination of women by the exercise of male power legitimized by custom, tradition and myths about women's innate biological capacities. Its principal effect, some would say its very purpose, is to transfer material and social rewards from women to men. For Foord and Gregson (1986), patriarchy is very much this kind of dialectical arrangement, one in which women's disadvantage is both cause and effect of men's advantage: men benefit because women (housewives, mothers and low-paid workers) bear many of their social costs, create many of their social benefits and demand very little recompense for these services. Because of the 'personalized service' which they receive from their wives, 'it follows that men have a material interest in women's continued oppression' (Hartmann, 1981, 6).

Though Foord and Gregson (1986) commend patriarchy as a potentially fruitful organizing principle for feminist geographers, they give little practical guidance as to how it might be applied to the issue of women in geographical space. Perhaps the answer to this lies in our original proposition that feminist geography is ultimately concerned with women as oppressed by man-made space. By insisting that geographical space is itself an embodiment of the patriarchical relationship, we help to make sense of those locational, environmental and architectural forms – high-rise flats, peripheral estates, under-serviced suburbs – which are especially hostile to women's needs and which often extract extra and unnecessary costs from them. Putting it baldly, we might say that spatial arrangements tend to work for men but against women. This is no accident but the logical outcome of male power and female powerlessness – all the crucial decisions about the built structure of cities and regions were and still are taken by males: from the Victorian mill-owner building cottages for his workers to the present-day local authority housing department, from the old-style landlord to the modern property developer, from the nineteenth-century Borough Engineer to the latter-day professionals engaged in the production of space, virtually all were and are male. And they have constructed man-oriented geographic space. Even where women have been included in their calculations, this has been women as seen through man's eyes, women's needs as defined by men not by themselves. This is particularly evident in the geographical separation of home (the domestic sphere) and work (the public sphere), a process which, whether conscious or unconscious on the part of male space builders, is entirely consistent with the need to slot woman into her correct place. By accentuating her remoteness from the public sphere, it both reflects and perpetuates her second-class status. By inhibiting her labour-force participation it keeps her at home 'where she belongs' serving the needs of her husband and children. This segregation is one of the means by which patriarchy is implemented and institutionalized.

4.4 Marxist Feminism

4.4.1 Capital and Patriarchy

One notably penetrating analysis of the patriarchal qualities of urban space is that by McDowell (1983). In this paper she demonstrates the way in which a variety of residential forms ranging from planned neighbourhoods through council estates to owner-occupied suburbs have been predicated on patriarchal ideologies – 'the anti-urban ideal and the ideal of community', 'state housing and the domestic ideal', 'suburban privatization and the glorification of domesticity' – all designed to legitimize women's place in the social and spatial order. Interestingly, however, McDowell quite explicitly rejects patriarchy as a primary determinant of women's position.

> Far from seeing gender relations in general or patriarchy in particular as a separate structure, I want to argue that women's oppression in capitalist societies needs to be explained by a class analysis at the *theoretical* as well as at the empirical level. Unlike Foord and Gregson I do not accept that child-bearing and heterosexuality can be theorized separately from an analysis of class relations in capitalist societies.

In saying this McDowell is engaging in one of the central debates in feminist theory, that surrounding the relationship between patriarchy and capitalism. Is female subordination primarily for the benefit of men *per se* or is it ultimately for the benefit of capital? In most essentials this debate parallels that within the field of ethnic relations about whether racism or capital is determinate in the oppression of black people. Within feminism this debate is crystallized in the collection of papers edited by Sargent (1981), whose theme 'the unhappy marriage of Marxism and feminism' accurately expresses a widespread feminist dissatisfaction with the Marxist method of reducing all forms of exploitation and oppression to the class struggle and paying little or no attention to patriarchy as a determining force. Thus Hartmann in the leading paper complains that Marxist writers ' . . . consistently subsume women's relation to men under workers' relation to capital' (1981, 4).

Hartmann's chief thrust is that capitalism and patriarchy constitute a kind of dual system – 'in capitalist societies a healthy and strong partnership exists between patriarchy and capital' (1981, 19) – with each deriving substantial and crucial advantages from the existence of the other. But there are contradictions in the relationship. On the one hand, capital accumulation benefits from cheap female labour in factories and offices. Yet on the other hand, 'the vast majority of men want their women at home to personally service them' (1981, 19). In this sense, capital is denying male workers their customary patriarchal 'rights' and therefore the interests of patriarchy and capital are opposed in this instance.

Despite these objections Marxist feminists would insist on a 'class-based explanation of women's subordination' (McDowell, 1986; see also West, 1980; Vogel, 1981; 1983) in which patriarchy works not only to men's immediate advantage but also to capital's ultimate advantage. In such an analysis patriarchy is given *relative* autonomy but is not seen as being determinate in the last instance. As we have stressed, patriarchy is by no means a unique creation of capitalism but it does assume a specific form and function within capitalist social relations. Under capitalism, patriarchy does not spring independently out of culture but has a distinct material purpose: or, as Vogel (1981, 210) puts it, '. . . behind the serious social, psychological and ideological phenomena of women's oppression lies a material root'. Its material purpose is that of underwriting the process of capital accumulation. It effects this by:

(a)   Ensuring a supply of female workers as part of the reserve army of labour
(b)   allotting women a key role in the reproduction of labour power.

4.4.2 Women in the Reserve Army

In Chapter 1 we discussed capital's reliance upon what Marx called the 'reserve army of labour'.

> This surplus population . . . [is] . . . the lever of capital accumulation, nay, a condition of existence of the capitalist mode of production. It forms a disposable reserve army . . . a mass of human material always ready for exploitation (Marx, 1973, 592).

Following Harvey (1982) we stressed the way in which expanding industrial capital has always drawn on migrant labour from pre-capitalist external space to boost the supply of labour and thus drive its price down. Migrant labour is also eminently exploitable in itself, being industrially inexperienced, politically unorganized, socially insecure and culturally alienated from local workers (see also Chapter 5).

There are close and obvious parallels between the position of women in the labour market and that of migrants. Women workers too are often 'super-exploited' in that they tend to earn low wages, commonly below subsistence. According to Braverman (1974), this is because they too are part of capital's reserve army, recruited not necessarily from underdeveloped regions but locally, from the domestic labour reserves of home and family. Noting the increasing rate of female labour-force participation, Braverman argues that women have now become the prime supplementary reservoir of labour:

> Women form the ideal reservoir of labour for the new mass occupations. The barrier which confines women to much lower pay scales is reinforced by the vast numbers in which they are available to capital (1974, 385).

Even though Braverman based his analysis on trends in the United States' labour market in the 1960s, it is widely applicable throughout the advanced capitalist world and may indeed hold an even greater relevance for the 1980s. He emphasizes that women's activity rates, though rising, are still comparatively low and that this represents a form of concealed unemployment on a massive scale. In most advanced capitalist states official unemployment figures include only those who declare themselves to be currently seeking a job but do not include 'discouraged non-seekers' such as the great regiment of married women who, despairing of obtaining any job which pays them a living wage, fall back into full-time domesticity (Braverman, 1974, 399; see also Beechey's 1982 critique).

From the employer's point of view the advantage of female labour when it does enter the market is that it can command far less pay than male labour. In much of the western world, pressure for equal pay has redressed the balance to a degree but, as established earlier, it is still the case that women are over-concentrated in low-wage sectors, paid less for the same occupation and over-represented as part-time workers. Many can be said to receive less than subsistence in that their earnings fall below 'the average normal level of the working class' (Braverman, 1974, 398). In Marxist terms, this constitutes super-exploitation.

There is, as Braverman observes, a 'barrier' to female job mobility, one which places them in a marginal position in relation to the rest of waged labour. Although Braverman does not spell it out, the barrier is patriarchy, that system of structures and ideologies which fixes a women's place in society. Springing originally from the early middle-class morality noted by Rowbotham (1973), various notions about the nature of femininity and the right and proper role for women have developed from a status of mere assertion to one of sacred truth. Thus by the latter half of the last century it had become no less than a social norm that women have no genuine claim to participate in the world of paid work, except in a limited capacity or in special circumstances. If a women's place is in the home, then a (married) woman in the workplace is an abnormality, perhaps even a deviant individual.

Exceptions can be made for women prior to marriage or for jobs like typing, cleaning, nursing or infant teaching which demand a 'woman's touch', but by and large women are defined as part of a household in which it is a male bread-winner's duty to provide for. As such she has no 'need' to work. By doing so she may even be taking the bread from the mouths of another family.

As is demonstrable in instances like the Lancashire cotton industry, the female work role is certainly open to flexible interpretation to suit the needs of particular fractions of capital at particular times (McDowell and Massey, 1984). In general, however, they have always had to struggle against an ideology which states that women in the workplace are there only on sufferance. Women's place at work has become delegitimized, often so thoroughly that women themselves have been forced to accept this view of their 'proper' estate. Consciousness-raising among women themselves has always been one of the crucial tasks of feminist politics (Rowbotham, 1973). And for all the advances achieved by the feminist struggle, patriarchal dogma dies hard. It continues to play a key part in legitimizing capital's use of women workers and in creating a form of labour which can be taken into employment by capital on its own terms.

Feminist geographers and regional analysts have not been slow to recognize the part played by patriarchy in the capitalist restructuring of regional and local economies. Contributions made by Massey (1984) and McDowell and Massey (1984) have paid particular attention to the way in which traditional heavy industrial economies in South Wales and Durham have been converted into permanent pools of high unemployment, pools into which light manufacturing and service industries are able to dip at will. This has served to 'decompose' the old working class, based as it was on powerful male-dominated trade unionism. It has also served to open up the female reserve labour army of these areas, that multitude of women so long locked away in the service of male breadwinners. This labour not only 'comes cheap', as Massey (1984) observes, but it is also assumed to possess the supreme virtues (from capital's point of view) of passivity, unawareness of rights and a 'natural' tolerance of mind-numbing repetitive work.

Writing on the electronics industry, a leading newcomer to the coalfields,

Massey (1984) describes the geography of a rigidly hierarchized social system of production, in which the high status functions of research and development are predominantly concentrated in the South East of England with generally only the lowest routine activities being hived off to the periphery. This regional division of labour runs parallel to a gender division in which it is taken for granted that males predominate in management, male skilled workers perform most of the intermediate technical production, and women fill the mundane semi- and unskilled jobs at the bottom of the hierarchy. Their position there is justified on grounds which can only be described as blatantly sexist.

> At the bottom end of the ladder, the assembly workers are equally assumed to be female. This is usually argued to be a result of the requirements of production because of women's supposedly greater 'dexterity'. . . . . If dexterity really is the issue it seems surprising that not more women get to do the detailed fiddly job of brain surgery'' (Massey, 1984a, 140–1).

Further evidence on the part played by the female reserve in spatial restructuring can be gleaned from the Women and Geography Study Group of the Institute of British Geographers (1984), Murgatroyd et al. (1985) and, for a critical commentary, Bowlby et al. (1986).

4.4.3 The Domestic Mode of Reproduction

In Chapters 2 and 5 we give considerable prominence to the concept of reproduction. We note that its relationship to capital accumulation is both an essential and a conflicting one. The reproduction of labour power is a necessity to capital but a costly one, in that workers' wages (their subsistence) represent a proportion of value which capital must sacrifice. The higher the subsistence the lower the surplus value. In this connection several writers, beginning with Morton (1971), have noted that women's unpaid labour in the family helps to relieve the tension between accumulation and reproduction. By providing free childcare, cooking, cleaning, shopping, and other services, wives are a necessary link in the chain of reproduction and materially reduce the costs to capital of reproducing their husbands' labour power (Vogel, 1983). If a male worker's wage had to cover the cost of purchasing all catering, nursery and sexual services on the open market as commodities, capital's surplus value would be seriously eroded. The classical Labour Theory of Value states that capital is compelled to pay at least a bare subsistence wage. Yet were it possible to cost the unpaid labour of wives, it would become clear that historically, capital has always reproduced its labour force at something below subsistence wages – given, of course, that few husbands pay their wives the market rates for services rendered. (See Chapter 5 for parallel arguments on ethnicity as a subsidy to capital).

In this way the process of advanced capitalist accumulation is subsidized by a non-commodity mode of reproduction, with the family as its basic unit and the wife as the lynch-pin of the family. At an early state in the feminist debate on domestic labour it was suggested that the domestic sphere actually constituted a kind of pre-capitalist survival, nesting within the advance capitalist social formation but external to the capitalist economy (Benston, 1969). Subsequently, however, writers like Dalla Costa (1972) have argued conclusively that, as reproducers of labour power, housewives are anything but detached from that economy. Submerged, marginal, second-class they may be, but by no means external.

Even so there is a definite case for arguing that patriarchal relations within the family are intrinsically pre-capitalist, or at least non-capitalist, in character, based as they are on personal marital commitments rather than the cash nexus. As West (1980, 176) explains,

> . . . the claim of the full-time housewife on the wage packet . . . can only result from a personal bargain struck with her husband. . . . The obligation on husbands to maintain wives is unenforcable until divorce, and even then practically so. Indeed, even with rising wages, it is not uncommon for wives to receive a less than proportional increase in housekeeping, if any at all, out of the larger wage packet.

Thus does patriarchy mesh with class expropriation. By obliging wives to service their husbands on a non-capitalist basis, patriarchy helps to ensure that capital need not meet the full costs of reproducing its male labour power. And women as mothers also help to defray the cost of reproducing the next generation. (For further discussion see also Smith, 1978; Middleton, 1983.)

Needless to say, the domestic mode of reproduction has not always existed in its present shape nor did it spring into being instantaneously. Historically it evolved from the eighteenth century onwards as part of the evolution of industrial capitalism, which required large masses of propertyless labour. To this end the domestic sphere came to be gradually segregated from the work sphere and increasingly reserved for the purpose of reproducing the latter's labour power. But, as is invariably the case when analysing capitalist development, we find a complex variety of conflicting strands. Rowbotham (1973) shows that in the early decades of the Industrial Revolution in Britain, the domestic mode of reproduction was placed under great strain because of the value employers placed on women as cheap labour. The increasing employment of women in factories and mines led to inevitable neglect of children and to frequently expressed fears (by political and religious leaders) about the imminent breakdown of family life (McDowell and Massey, 1984). In effect the state itself had to intervene to protect the domestic mode by legislation to restrict women's working hours and the occupations they could enter (Hall, 1982).

This is only one of the many possible and actual contradictions between capital and patriarchy. At various points in this chapter we have touched upon others. For example Braverman's (1974) 'Universal Market' opens up new market opportunities for capital accumulation by commodifying certain family needs. Large profit-making industries supplying such wants as processed foods, household appliances, consumer goods and services owe their entire existence to this commodification of the family. But at the same time the commodification process acts to undermine the family as a reserve of labour cheaply reproduced on the back of the unpaid housewife.

Similar contradictions are thrown up by the socialization of the family. Here the state provision of schools and hospitals has relieved women of many of their caring duties, thus releasing them to enter the labour market. By doing this the state is directly freeing a portion of the reserve army for use by capitalist employers. But in so doing it is also threatening the labour-reproducing capacity of the family by removing its very source of unpaid services.

The issues raised here have clear general relevance to geographers in challenging certain widely held assumptions about the workings of society. They also have direct bearing on many of the established research preoccupations of social geographers. Of the many potential applications, one which immediately springs to mind is the field of neighbourhood and community studies, where students from many disciplines have puzzled over the changing role of these small-scale institutions in a large-scale, ever more centralized, society. There is a conspicuous link between the concepts of family and community: the former is frequently seen as the building block of the latter and both embody a form of relationship built around feeling and emotion rather than the cash nexus or legal contract. As Bowlby *et al.* (1986, 329) put it, 'community and home are the arenas in which people try to make space, to free themselves from relations of capitalist production'. In this respect, they symbolize a kind of human association which is the very antithesis of capital but which exists symbiotically within the capitalist social formation.

The pivotal position of women in both family and community is testified by many of the classic community studies (the most widely cited being Young and Wilmott, 1957), but this is a position which needs to be re-evaluated as part of a wider reassessment of such concepts as neighbourhood and community. In the light of the arguments raised in the present chapter, it is evident that many cherished beliefs about the virtues of community should be carefully scrutinized. In the final chapters of this book, we raise the possibility that community showers its benefits not so much upon its own members as upon outside interests, notably the powerful duo of capital and state. Given the central role of women within the communal living space, it is likely that they are as much oppressed in their roles as neighbour, citizen, voluntary worker and local resident as in their roles as wife or worker.

Recommended Reading

An obvious introduction to the study of *Geography and Gender* is the text of the same name produced by the Women and Geography Study Group of the Institute of British Geographers (Hutchinson 1984). This book successfully attempts to outline basic theoretical concepts and reinforce them with issue-based analyses of women's access to space, facilities and employment, and it remains an accessible first source. An understanding of the extent to which such an analysis has been absent from the discipline in the past can be gained from Janice Monk and Susan Hanson's (1982) review of sexist bias in the content, methods and purposes of geographical research, 'On Not Excluding Half of the Human in Human Geography' (*Professional Geographer*, 34, 11–23). Two further texts have recently emerged to begin to redress this imbalance. Momsen and Townsend's edited collection, *Geography of Gender* (Hutchinson 1987), offers a wealth of comparative material in a non-Western context, whilst the newly-published *Women in Cities*, edited by Jo Little, Linda Peake and Pat Richardson (Macmillan 1988) is largely focused on the developed world.

In producing this chapter we have made use of Linda McDowell's 1983 article 'Towards an Understanding of the Gender Division of Urban Space' (*Environment and Planning D*, 1, 59–72) and her jointly authored chapter (with Doreen Massey), 'A Woman's Place', in Massey and Allen's edited collection *Geography Matters* (Cambridge University Press 1984). As Massey indicates in her foreward to *Geography and Gender*, 'Feminism is clearly on the agenda in geography' (1984, 11) and, hesitantly, many contemporary analyses of resource allocation identify gender as a key variable when outlining variations in life quality and opportunity. Nowhere is this done more effectively than in Seager and Olson's fascinating, and at times shocking, international atlas, *Women in the World* (Pluto Press 1986); its concerns are global, reaching far beyond the immediate socio-spatial confines of our chapter, and its evidence is a little uneven and at times rather chaotically presented, but its findings are of great significance. It provides an excellent context for informing the theoretical debate.

Considerable theoretical inspiration can be drawn from outside geography. For a very basic and straightfoward introduction Michelle Stanworth's chapter on 'Gender Divisions in Society' in Bilton *et al.*'s *Introductory Sociology* (Macmillan 1987, 148–95) is useful, as is the same author's more specialized consideration of *Gender and Schooling* (Hutchinson 1983). The debate on *Gender and Stratification* is fully explored in Crompton and Mann's edited collection of the same name (Polity 1986); in particular the chapter by Walby is worthy of special attention. Similarly, the Marxism: Feminism debate referred to in 4.4 above, is extensively reviewed in Sargent's edited critique, *The Unhappy Marriage of Marxism and Feminism* (Pluto 1981). Readily accessible are the pioneering writings of Dale Spender and Ann Oakley (among others); both are read widely, and deservedly so.

5
Racial and Ethnic Minorities

The second chapter on social groups focuses on those who can be identified by skin colour and/or cultural traits, racial and ethnic minorities. In the United States, where the cities of Detroit, Philadelphia, Cleveland, St Louis and Baltimore already have a black voting majority (Morrill and Donaldson, 1972), the presence of racially distinct groups has long attracted socio-geographic attention (e.g. Hartshorne, 1938). Geographical research on ethnic minorities in Britain is, for the most part, a far more recent phenomenon, reflecting New Commonwealth migration since the 1950s. Studies of black residence in British cities have tended to lean heavily on their American antecedents for theoretical constructs and empirical comparisons, and this chapter attempts to integrate research from both sides of the Atlantic. The chapter commences by defining the terms used (and misused) in studies of racial and ethnic minorities before focusing on spatial patterns of concentration and segregation. The dominant assimilationist theory in British ethnic geography is critically reviewed, and managerialist and structuralist alternatives are assessed. The chapter ends by reviewing the status of minority groups within a wider social and economic framework, focusing on black accommodation and resistance.

5.1 Ethnicity, Race and Minority Group as Analytical Categories

The terms 'race', 'ethnicity' and 'minority' are among the most misused in modern English vocabulary. Misapplied as slogans, loosely defined and shuffled around as if they were interchangeable, they bring confusion to students, decision-makers and the general public alike. Since they came to the forefront of British life three decades ago, their usage has remained constantly blurred. For example, the titles of public bodies administering to the special needs of black minorities often display the term 'ethnic' (or even 'community') as their key word in preference to 'race': this is despite the patent truth that it is the ascribed stigma of race which more than any other factor excludes black people from full participation in British society (Smith, 1976; Scarman, 1981, para. 6.35.; Banton, 1983; Cashmore and Troyna, 1983; Brown, 1984). The readiness to adopt 'ethnic' phraseology undoubtedly springs from its soothing anodyne qualities. 'Ethnic' is the kind of word unlikely to provoke anyone, possibly because, to the majority of the public, its meaning is obscure. Yet, as van den Berghe (1978, xiv) reminds us, to take refuge in euphemisms is to sacrifice rigour and gain nothing in return: 'abandoning a useful analytical category . . . does not exorcise the evil of racism. To label race ''ethnicity'' does not make it so.'

In seeking clarity of definition, many scholars in the minority-relations field have taken Wirth (1945) as their starting point. The following short extract from Wirth's writings is frequently cited or quoted, in which he defines a minority as

> a group of people who, because of their physical or cultural characteristics, are singled out . . . for differential and unequal treatment and who therefore regard themselves as objects of collective discrimination (cited in van Amserfoort, 1978, 218).

Inevitably this statement has received much critical scrutiny, not all of it favourable (Rex, 1970; Cohen, 1974), but it does help us to clarify certain basic concepts.

Minority status is commonly a state of exclusion, of non-membership or at best junior membership. Certain physically identifiable groups are disqualified from a satisfactory share of society's resources and hence, whatever their official legal status, they may be regarded as having been denied equal rights as citizens. According to Wirth, the two properties of ethnicity and race are linked to minority status in that they form two of the principal criteria for exclusion. Minorities are disqualified because they are perceived as alien, as deviating from the majority population's definitions of social normality. *Ethnic* minorities are groups who are culturally differentiated from the majority population. Most frequently, the ethnic label is attached to groups incorporated by international migration into a new nation. Whether or not they immediately relinquish their original nationality, such groups tend to retain a separate identity long after initial migration. They may be differentiated from the receiving society by language and religion and are almost certainly distinct in custom and tradition. Their visibility in the eyes of the majority is heightened by their cohesiveness as communities. The recent immigrant's primary loyalty is to his co-immigrants from the same homeland or region, with the immigrant community promoting and preserving cultural traditions which are a source of psychological security to its members but which are alien to most of the citizens of their adopted country.

Racial minorities are groups categorized and set apart by the majority on the basis of phenotypical features: i.e. they are identified as outsiders because certain of their genetically inherited physical features do not conform to the generally accepted norm. The key divide is often skin colour: from the early days of slavery and colonialism white society has deemed non-whiteness as indicating intellectual and cultural inferiority. Although modern scientific analysis has confined doctrines of racial inferiority to the academic dustbin (Walvin, 1973), racism nevertheless persists as a social dynamic, however perverted. The key difference between ethnic and racial characteristics is, of course, related to their permanence. Ethnic minorities, such as Irish and East European migrants to the British mainland, have progressively become assimilated into the social and spatial structure of British society. To be sure, the relics of greater cultural distinctiveness remain in the religious and cultural establishments of the inner city, but such groups are now largely integrated in both Britain (Jackson, 1963) and the United States (Lieberson, 1963; Hawley, 1981; see, however, Kantrowitz, 1981 for an alternative interpretation). In contrast racial features, and specifically skin colour, are characterized by their permanence. However strongly a black minority may desire integration into main-stream society a visual badge renders them distinct and therefore vulnerable to the prejudices of the white majority. While an ethnic minority can progressively change its behavioural characteristics, no such option is available to a racially distinct group.

5.2 The Spatial Dimension of Race and Ethnicity

5.2.1 Initial British Studies

Until the late 1950s British domestic society had been characterized by a high degree of racial homogeneity. The nineteenth century had of course seen the incorporation of many foreign-born migrant groups into Britain; indeed, this was virtually inevitable given the insatiable labour demands of an expanding industrial economy. The two decades after 1945 represent a more recent expansionary phase (related to post-war recovery), which also sucked in immigrant labour. However the historical uniqueness of the post-war period is that immigrant workers originated predominantly from the 'New Commonwealth' regions of the Caribbean and the Indian sub-continent. Prior to this, settled black communities were almost unknown outside parts of London and a handful of other port cities, of which Liverpool and Cardiff were perhaps the most noteworthy (Fryer, 1984).

By the 1960s, academic curiosity had been whetted, not simply by the novelty of a visible black presence in a hitherto 'raceless' or 'colourless' society but by the intense political relevance of the twin issues of race and immigration (Hiro, 1973, 46–9). Among the early anthropological, sociological and general studies, there were several which included a discussion of spatial/residential aspects of black settlement in British cities (Collison and Mogey, 1959; Glass, 1960; Davison, 1962; Patterson, 1963). The awakening of interest on the part of professional geographers post-dated these studies and was heralded by notable contributions from Peach (1968), Doherty (1969), Jones (1970) and Dalton and Seaman (1973) among others. For any reader who cared to build up a composite picture from these writers, it became clear that the geography of British blacks was as distinctive as every other aspect of their incorporation into a new homeland.

The common theme emerging from the above studies was spatial concentration. By the 1960s the distribution of black immigrants had been shown to be highly skewed, a pattern visible at differing levels of geographical scale.

At the regional and inter-urban scales it was clear that black immigrants were heavily concentrated at the top of the urban hierarchy and in five of the eleven standard regions of the United Kingdom. They were (and are):

(a) over-represented in the large conurbations of Greater London, the West Midlands, West Yorkshire and East Lancashire, and in the two major East Midlands cities of Leicester and, to a lesser extent, Nottingham.
(b) under-represented in conurbations and large cities in other regions such as South Wales, Central Scotland and the North
(c) heavily under-represented in smaller settlements irrespective of region (Jones, 1978; Figure 5.1) Peach (1968) leaves us in no doubt that, prior to immigration control (from 1962 onwards), the forces creating this pattern were economic in nature. The total volume of immigration was a function of national labour demand, while its direction within Britain was governed by regional and local variations in such demand.

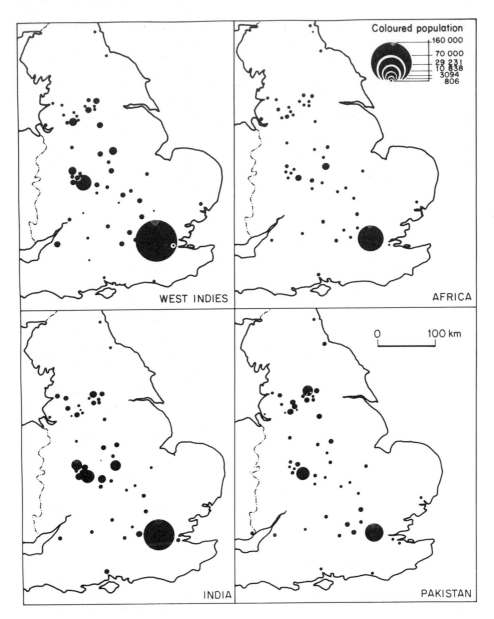

Figure 5.1 Distribution of total coloured population in England and Wales, 1971.
  Source: Jones P.N. 1978, 527.
  Note: The figure shows the distribution of the four main groups of black immigrants to England and Wales since 1950. As the maps show, the two principal cities, London and Birmingham, are significantly represented, though the rapidly growing Pakistani population is more widely distributed with significant concentrations in the textile towns of East Lancashire and West Yorkshire. Equally meaningful are the areas which display an absence of black immigrants; these include all rural areas, the North-East, East Anglia and the South West, and the prosperous commuter belt of the South-East.

At the urban scale, blacks were noted to be restricted to a relatively small fraction of the total urban space. Already a characteristic black settlement pattern had emerged, with households clustered within a small number of local neighbourhoods and absent (or present in only token numbers) throughout the rest of the city. Clusters were typically located in the innermost portions of residential space and absent from suburban areas. Though contemporary authors directed much of their attention to London (Patterson, 1963; Doherty, 1969) and Birmingham (Jones, 1967; Rex and Moore, 1967), retrospective work on other cities e.g. Coventry (Winchester, 1974), Nottingham (Husain, 1975), Huddersfield (Jones and McEvoy, 1978) and Bradford (Cater and Jones, 1979), has helped to confirm their findings as generally valid in the 1960s. The black inner city had begun to form part of the stock political vocabulary.

5.2.2 American Antecedents

In searching for processes behind patterns and for a theoretical and comparative context for their work, the early British 'ethnic' geographers were unavoidably drawn towards the United States experience. At that time, urban geography as a whole had recently discovered its debt to the Chicago School of human ecology, according to Peach (1975a, 1–2) 'the fountainhead from which all else flows'.

The study of minority-group spatial relationships had been pioneered by Robert Park (1926) as part of his overall concern with the interaction of social process and spatial form. Park's work was a clear and direct response to the American city of his day. Several decades of industrialization and urban growth coupled with mass migration from a score of European countries and the beginnings of an urban migration by America's own rural blacks had combined to produce an urban population of complex plurality, divided along countless lines of national origin, language, religion and race. Social divides were fairly faithfully reproduced by physical divides. Each group was segregated into ethnic and racial sub-territories so that larger cities like New York, Boston and Chicago came to resemble collections of minature nation-states, enclaves known in popular parlance by such epithets as Little Sicily, Greektown and the Black Ghetto.

All of this gave clear substance to Park's cardinal precept that *spatial location is the tangible and concrete counterpart of social location*. Spatially, newcomers to the city tended to be segregated and located in deprived areas, both of which was indicative of minority status. *Residential segregation* can thus be seen as the spatial expression of outsider status. The term refers to the physical separation of minority and majority residents, usually taking the form of a concentration of minority residents in restricted urban sub-areas (ethnic colonies or enclaves) where majority members are largely absent. Those whose culture or race placed them outside the host society's norms lived in locations isolated from contact with mainstream society. *Residential deprivation* was (and is) an expression of subordinate status. Migrant enclaves in the city of Park's time were distinguished by their qualitative inferiority to the rest of the city, especially with regard to housing. Located mainly in the inner city, they included much of the oldest, most decayed and most densely built-up housing stock, residual dwellings unwanted by established members of society. By this time, suburbanization was already operating as a major urban dynamic, relocating the established population on the new urban perimeter and creating vacant inner-city space for fresh waves of incomers.

This pattern of inner-city segregation of migrants is a recurrent feature constantly reproduced throughout the urban industrial world in the present century. It may be said to result from two mutually reinforcing sets of disadvantages which confront migrants on entry to a new society. The first of these is host hostility, with the new arrivals being perceived as intruders whose share of scarce social resources is disputed by the established population, who may use individual and collective means to limit access to these resources. The second is migrant disability when, as in the case of the American city, migrants originate largely from rural and/or overseas territories. Although such disabilities can be over-emphasized, migrant groups may bring only pre-industrial skills to the industrial city and be therefore unable to compete effectively for urban resources. This disadvantage is, for first generation migrants, frequently compounded by language and literacy barriers. Even if host hostility were absent, an 'unskilled' and poorly educated new migrant would still be obliged to enter both the job market and the housing market at the bottom.

5.3 The Doctrine of Assimilation

5.3.1 Social and Spatial Mobility in the American City

For any liberal thinker the situation described above – segregation and relative deprivation – was profoundly unacceptable and in flagrant breach of the constitutionally-enshrined American ideal, and much subsequent scholarly effort in the field of minority relations has consequently been dedicated to resolving the contradiction between the high-minded tenets of the 'American Creed' (Myrdal, 1944) and the less exalted practices to which minorities have been habitually exposed. The theory of assimilation represents the first systematic attempt to resolve this dilemma. Although assimilationist thought originated as popular ideology (Newman, 1973), it was given coherent form by academics, among whom Park and the Chicago sociologists played a prominent role. The central proposition put forward by the Chicago School was that minority status represents a transitional stage *en route* to eventual social acceptance and economic parity. Crippling though the disabilities of the migrant could be, they are nonetheless liable to change. Job skills, language and customs are learnable; old ties can be severed and replaced by new loyalties, the immigrant can readily acquire 'the language, manners, social ritual and outward form of his adopted country' (Park, 1926). If not first generation immigrants, then certainly their children and their children's children will become progressively 'Westernized', as Gans (1962) demonstrated in his study of Italian Bostonians: whereas the migrant generation itself continued to operate inside an Italian community, living, speaking and breathing Italian in a kind of Sicilian space–time capsule, their grandchildren were engaged in a lifestyle shaped almost entirely by membership of their American High School peer group.

Thus the assimilationist view of society presents inter-group conflict as a temporary maladjustment (on the part of minorities), eventually to be resolved by the disappearance of those traits which serve to label and stigmatize minority members and to deny them full access to the economic fruits of the society. The ethnic enclave is viewed as a transitional space within which immigrants may find protection from an alien society until such time as they are capable of affiliating to that society. Its main function is that of a social womb or chrysalis.

A womb is of course designed to be vacated when its occupant is capable of independent life, and so it is with the ethnic enclave. As ethnic group members acquire the fully formed industrial and cultural attributes necessary for survival in the external world, so they migrate to other more socially acceptable parts of the city. Changes in economic and social status tend to be reflected in changes of location. Upward social mobility and loss of subordinate status are matched by outward spatial mobility and desegregation. A number of research studies in the ecological tradition testify to the rapidity of ethnic desegregation in twentieth-century America, with the work of Duncan and Leiberson (1959) and Leiberson (1963) being especially influential. Their studies of white ethnic groups in selected cities use the Index of Dissimilarity[1] to demonstrate a consistent decline in segregation levels over time and to illustrate that this desegregation is correlated with occupational mobility, rising literacy and an increasing ability to speak fluent English. Residential desegregation is thus part of a wider process of assimilation into American life in which immigrants and their descendants become closer to American norms in the spatial as well as the behavioural sense.

5.3.2 Early British Applications

By the early 1970s the spatial and residential characteristics of Britain's black population were coming under increasing scrutiny by geographers, prompted by the heightening controversy surrounding the issue of race relations and armed with the results of the 1971 Census. Although many of the studies are not notable for an eagerness to place residential segregation in a theoretical context, it is possible to identify a definite leaning towards the assimiliationist perspective. From a wealth of detail, neatly summarized by Peach (1975b), we can arrive at two broad conclusions about the nature of black segregation up to and including the early 1970s:

1 Segregation between black and white was not extreme. A large proportion of New Commonwealth immigrants and their families were living in racially-mixed areas, with comparatively few in heavily black-dominated neighbourhoods. Most significantly, analyses of the residential segregation of black minorities found that nowhere in Britain did levels reach the heights recorded for Negroes in the United States (Taeuber and Taeuber, 1965). The implication was that inter-racial social distance was far less in Britain than in the United States.
2 In the case of West Indians, a certain degree of desegregation had occurred between the two censuses of 1961 and 1971. Hence, if desegregation is indeed one of the active components of assimilation, then West Indians were already advancing along the path towards full membership of British society.

---

1 The Index of Dissimilarity and the Index of Segregation are measures of the extent to which the residential distribution of two separate sub-groups differ from each other within a given area and at a given scale level. It is thus possible to compare the distribution of, for example, the Asian-origin population with the remainder of the city's population at Enumeration District level and to calculate an index on a scale ranging from 0 (spatial assimilation: an even distribution) to 100 (total spatial segregation). This index is then comparable with indices calculated for differing ethnic groups in differing cities at differing points in time at an equivalent spatial scale (Duncan and Duncan, 1955; Timms, 1965).

5.3.3 Geographical Myth versus Historical Reality

The situation depicted by the majority of these geographical studies, a degree of racial intermixing at the neighbourhood level, increasing entry of black families into council housing, and gradually diminishing segregation levels, may have hinted at tension and racially unfair competition but failed to convey the true spirit of the times. Outside the confines of geographical academia the world was a far less peaceful place. Although the actions of the Klu Klux Klan, race riots, and the Civil Rights and Black Power movements on the American side of the Atlantic provoked the greatest attention, the steady growth of the British black population was accompanied by the equally steady growth of racial hostility. Often this was covert rather than overt, though on occasion it manifested itself violently in the form of civil disturbances (e.g. Nottingham and Notting Hill in 1958) and in attacks on individual black people and property (e.g. 'Paki-bashing', generally by young right-wing extremists since the late 1960s). On the political front it surfaced in a series of legislative measures specifically designed to curb black immigration (Runnymede Trust *et al.*, 1980, 30–54); in the re-emergence of fascist parties preaching discredited doctrines of racial superiority (Billig, 1978); and in the rise to public prominence of parliamentarians dedicated to immigration control and repatriation (Foot, 1969).

Less dramatic but in the long term far more damaging was the accumulating evidence that blacks were systematically disadvantaged in the competition for social and economic resources, especially in the two key markets of jobs and housing. A series of surveys and research reports (Daniel, 1968; Deakin, 1970; Smith, 1976; and, most recently, Brown, 1984) established beyond reasonable doubt that blacks were over-represented in low-paid, unskilled, menial, arduous and otherwise undesirable occupations and industries, despite in many cases being over-qualified for their allotted jobs. They were also more likely to be unemployed. In the housing market they were over-represented in old, unfit or overcrowded housing in undesirable inner-urban areas.

These sources also left little doubt that race itself was the prime criterion on which unequal competition was based. Consciously or unconsciously, blatantly or (with the passage of anti-discrimination legislation) more subtly, on the individual and institutional levels, the 'host' population was exercising its power to restrict the roles of a sub-population whose alien racial status was visibly and permanently apparent. Not for nothing did black spokespersons complain about a built-in racial bias in British systems of resource allocation (Sivanandan, 1976).

Given these realities, it is evident that assimilation is a less than workable model when applied to inter-racial, as opposed to inter-ethnic, relationships. This has long been recognized in the United States, where Lieberson's authoritative study in 1963 had found that the process of spatial desegregation operated only in the case of white ethnic groups of immigrant ancestry. Black segregation from whites was found to be constant and at a very high level of intensity, a pattern confirmed by the Taeubers' 1965 study of 207 American cities and, more recently, by Van Valey *et al.* (1977). This evidence reinforces van den Berghe's assertion that 'a racial phenotypical definition of group member-ship is far more stigmatizing than an ethnic definition and typically gives rise to far more rigid social hierarchies' (1978, xiv). Moreover, the fundamental pro-perty of race as a minority criterion is that it is ascribed not chosen and cannot therefore be cast off in the way that one converts to a new religion or swears allegiance to a new nation. Blacks do not have the option of renouncing that which is used to set them apart by the majority.

By no means all interested geographers were unaware of these difficulties. Several were beginning to raise questions about the desirability and practicability of black spatial dispersal as a policy goal (Lee, 1977) and about the manner in which measures of segregation and concentration should be interpreted. Even so, there seemed little urgent concern with the glaring discrepancy between the spatial evidence and the apparent real-life experiences of black Britons. If racism is a decisive force at the economic and political levels, why is it less explicit in spatial terms? Is the link between society and space a mirage?

Although the majority of studies of the segregation of black residents in the United Kingdom had emphasized the relative lack of racial separation, with dissimilarity indices 20, 30 or even 40 points below those calculated for American cities (Peach, 1975b; Jones, 1976; 1979), it may be, however, that black Britons were (and are) far more sharply divided from their white neighbours than has been previously recognized. Jones and McEvoy (1978) have argued that many British analyses of segregation have been undertaken at too coarse a scale level to adequately reflect the comparatively small pockets of black residence in many British cities. More recently Jones (1983/4) has claimed that studies in the United Kingdom, by referring to present-day segregation levels in the United States as a source for comparison, have unwittingly overlooked the evolutionary stage of black residence in British cities. Black areas in British cities should be compared with the emergent Negro ghettos in North America in the inter-war period and, according to Jones (1983/4), striking parallels can be found. To appreciate this point we need first to know something of the historical development of the American black ghetto.

5.4 Racial Assimilation – A Non-Option?

5.4.1 The Black Ghetto in the United States

In Northern cities in the United States the characteristic black settlement form is the ghetto, a territory rigidly segregated from white neighbourhoods and a permanent base rather than a springboard for the dispersal of its inhabitants (Morrill, 1965; Rose, 1969; 1970; 1972). The term ghetto, which originally applied to the Jewish quarters of pre-Industrial European towns and carried with it an implication of enforced segregation, creates intense emotion. Although no more than a variation on the minority enclave theme, it forms a quantitatively and qualitatively distinct category, implying severity, persistence, greater territorial extent and, most importantly, enforcement.

In most large American cities, the residential space occupied exclusively or almost exclusively by blacks is so extensive and populous as to justify the label 'city-within-a-city'. As the urban black population has grown, so the ghetto area has continually expanded. Before the First World War, the tiny black minorities of the northern industrial cities were typically contained within very small scattered clusters (Weaver, 1948; Drake and Cayton, 1962; Spear, 1967), but by the 1920s these had been replaced by large consolidated enclaves (Spear, 1967) which in turn have been intensified, enlarged and amalgamated by subsequent population increase. While it does not 'cause' segregation, demographic change is evidently an important secondary variable and a key factor to bear in mind in

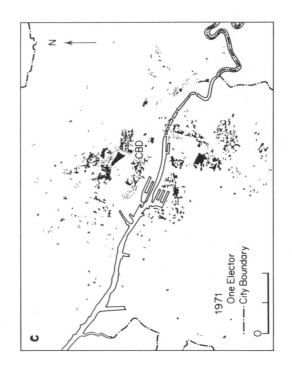

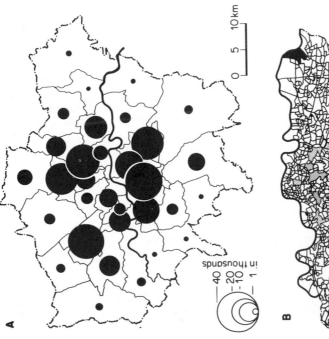

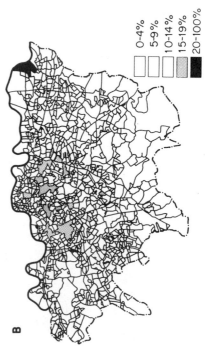

C

1971
· One Elector
—··—·· City Boundary

CBD

N

A

in thousands
40
20
10
1

0   5   10 km

B

0-4%
5-9%
10-14%
15-19%
20-100%

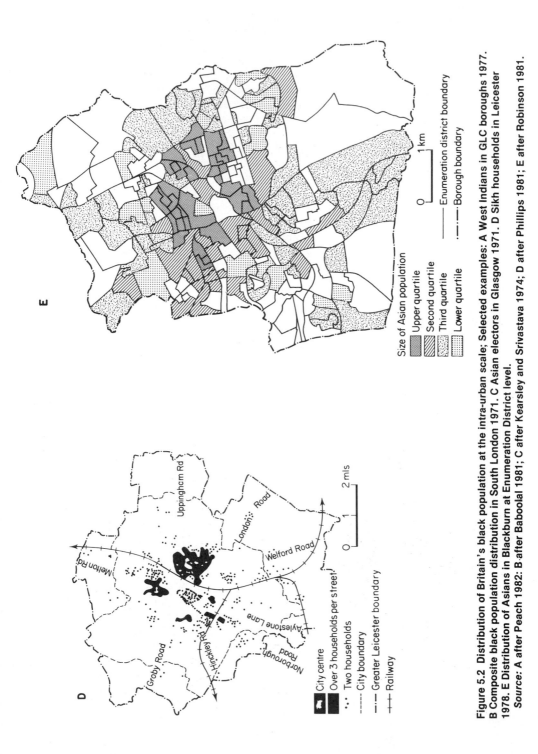

Figure 5.2 Distribution of Britain's black population at the intra-urban scale; Selected examples: A West Indians in GLC boroughs 1977. B Composite black population distribution in South London 1971. C Asian electors in Glasgow 1971. D Sikh households in Leicester 1978. E Distribution of Asians in Blackburn at Enumeration District level.

Source: A after Peach 1982; B after Baboolal 1981; C after Kearsley and Srivastava 1974; D after Phillips 1981; E after Robinson 1981.

Size of Asian population

- Upper quartile
- Second quartile
- Third quartile
- Lower quartile

—·— Enumeration district boundary
—··— Borough boundary

0   1 km

D

- City centre
- Over 3 households per street
- ·.· Two households
- —·— City boundary
- — Greater Leicester boundary
- +++ Railway

Uppingham Rd
Melton Rd
Graby Road
Hinckley Rd
Narborough Road
Aylestone Lane
Welford Road
London Road

0   1   2 mls

making transatlantic comparisons. It may well be that the recorded contrasts between British and American residential structure are due more to demographic immaturity than to basic cultural or behavioural differences in the two societies. The greatest contention over the use of the term ghetto lies, however, in its implication of enforcement. In the United States segregation, in the rural South and the urban North, was legally enforceable until a Supreme Court ruling in 1948. The northward migration of the early twentieth century found no promised land, the plantation being replaced by the ghetto (Meier and Rudwick, 1966). Even the liberalizing of legislation had little if any effect – *de facto* segregation is no less real than its *de jure* version. In the words of a character from Arthur Haley's *Roots*, 'there's one law that's in the book and another that folks lives by', and we should recognize that urban American segregation has been imposed on blacks through covert and informal means rather than official sanctions (Drake and Cayton, 1962; Ossofsky, 1963). While it is certainly true that many of the northern municipalities initially imposed residential zoning on their cities, this was declared unconstitutional as early as 1917 (Meier and Rudwick, 1966). However individually, collectively and at the institutional level whites developed a range of discriminatory practices within the letter of the law, a fact to be borne in mind in view of the claim that ghettos do not exist in Britain since we do not formally legislate them into being (Burney, 1967; Jones 1976; Scarman, 1981).

In outline, then, American racial residential geography is characterized by the *de facto* barring of the vast majority of blacks from the bulk of urban territory and their confinement in white-abandoned areas whose core is made up of deteriorated slum property. Not unexpectedly, poverty, overcrowding, poor living conditions and a decayed environment are reflected in disturbingly high levels of morbidity and mortality and a depressingly extensive range of related disorders (de Vise, 1968).

5.4.2 The Ghetto as Colony

With the realization that the ghetto results from a one-sided and seemingly perpetual black–white conflict, the need for intellectual readjustment became urgent. In America, radical academic thinking was much influenced by the black nationalist concept of the ghetto as 'internal colony'. The theory of internal colonialism was articulated by writers such as Blauner (1972) and Tabb (1975), who argued that racism as a product of centuries of colonization by which the non-white peoples of the world were politically subjugated and economically exploited by the white nations. The doctrine of racial inferiority served to justify the barbaric instruments (slavery in particular) used to subordinate colonial peoples.

However, colonial exploitation is not confined to colonial territories, since Third World peoples have also been incorporated by migration and slavery into metropolitan countries, where their exploitation as cheap labour continues to operate and to be legitimized by racist ideology. In this context, the ghetto may be interpreted as the 'Third World within', a space existing in the same economic relationship to the surrounding white city as does the colony to the metropolis (Blaut, 1975; Tabb, 1975).

It is perhaps surprising that the colonial model, with its close links to core–periphery theory, has stimulated little reaction from geographers. Indeed, in British race relations the model appears to have exerted little influence on practitioners of whatever discipline. Although a handful of sociological works have utilized concepts such as 'colonial underclass' (Rex and Tomlinson, 1979), the general preference has been for approaches derived from classical Marxism or Weberianism. Marxists in particular have been highly critical of the colonial model, despite the latter's explicit reliance on Marxist concepts such as exploitation, surplus expropriation and the international division of labour. The decisive

Table 5.1 Ward level dissimilarity indices: West Indians and Asians in British cities

City	Author and data	Groups compared and data source	Index
London	Lee (1977)	West Indians:Total population (1961 C)	55.2
London	Lee (1977)	West Indians:Total population (1971 C)	49.2
London	Lee (1977)	Asians:Total population (1961 C)	30.2
London	Lee (1977)	Asians:Total population (1971 C)	33.8
London	Peach (1975b)	Caribbean:England & Wales (1971 C)	50.9
London	Peach (1975b)	India:England &Wales (1971 C)	38.2
London	Peach (1975b)	Pakistan:England & Wales (1971 C)	49.0
Birmingham	Jones (1967)	West Indians:Rest of Population (1961 C)	55.0
Birmingham	Jones (1967)	Indian & Pakistani:Rest of Population (1961 C)	45.0
Coventry	Winchester (1974)	West Indians:Total population (1971 C)	34.0
Coventry	Winchester (1974)	Asians:Total population (1971 C)	61.0
Blackburn	Robinson (1980)	Asians:Not specified (1977 ER)	55.4
Huddersfield	Jones & McEvoy (1978)	Asians:Total population (1971 ER)	55.0
Glasgow	McEvoy (1978)	Asian:Non-Asian population (1976 ER)	69.2
Bradford	Cater & Jones (1979)	Asian:Non-Asian population (1977 ER)	61.1
Bradford	Cater (1984)	Asian:Non-Asian population (1982 ER)	57.8

Sources: As reference list except:
Cater J. (1984), Immigration, Segregation and Retail and Service Business in the Northern City, unpublished Ph.D. thesis, CNAA.
McEvoy D. (1978), The Segregation of Asian Immigrants in Glasgow, *Scottish Geographical Magazine* 94, pp. 180–183.
Notes: C – Census; ER – Electoral Register.
Segregation levels tend to be lower in London than elsewhere, and higher for West Indians than Asians. However evidence from other cities suggests that, while West Indian segregation has fallen with dispersal into public sector housing, Asian segregation levels have stabilised and are at a higher point. This is particularly true in the Northern industrial towns with their plentiful supply of inexpensive, often low amenity, terraced housing. Given the increasing indices at finer spatial scales (Table VII.2.) and the evidence from the P* Index (VII.4.c.), we have reason to conclude that Asian segregation in particular is relatively intense in many cases; it also appears to be persistent.

split between Marxist and colonialist approaches is the primacy which the latter accords to race rather than class and its insistence on black separatism as the logical means of black liberation (Banton, 1972). For most Marxists, blacks are inseparable from the rest of the proletariat: their struggle is an integral part of the class struggle as a whole.

5.4.3 British Parallels? The Inner City Revisited

What evidence is there of ghetto formation in the United Kingdom? At first sight the answer must be very little. A wide range of geographical studies in the early and mid-1970s, using dissimilarity measures, found ward level indices ranging from 34.0 to 55.2 for West Indians and 30.2 to 69.2 for Asians or Asian sub-groups (Table 5.1). Clearly the first criteria of ghetto definition, intense spatial concentration, appeared unfulfilled.

A closer study suggests, however, that we perhaps should not be so dismissive. Although, as Peach (1975b) has illustrated, West Indian segregation levels are declining, largely consequent on the group's movement into the widely dispersed public-sector housing stock, the extent of Asian segregation, as measured by dissimilarity indices, seems relatively constant. It may also be that the most commonly used scale of analysis, the ward, offers too coarse a spatial net to pick up the detailed locational pattern of what is a relatively small, though growing, urban minority. Studies using electoral registers, with data that can be aggregated at any spatial scale, have demonstrated dissimilarity indices as high as 85.3 at the street level, 78.3 at the 250m by 250m grid level and 73.3 at the Enumeration District level (Table 5.2). Such indices are comparable with those calculated at similar scale levels for blacks in American cities in the 1920s and 1930s and are within a few percentage points of the levels identified by Taeuber and Taeuber (1965) and Van Valey et al. (1977) in the United States.

The dissimilarist measures have an additional limitation which may have also contributed to an unwitting underestimation of the 'segregation experience' in British cities. Lieberson (1981) has recently pointed out that indices of dissimilarity and segregation are insensitive to the ethnic composition of the total population. This is most easily understood from an example. If we assume a city with a black: white dissimilarist index of 0, the residential distribution of blacks and whites at the given scale level is identical and spatial integration is apparently complete; 10 per cent of the city's population is black, and 10 per cent of each sub-area is black. In this situation the 'average' black will find that 90 per cent of his neighbours are white; clearly the experience of integration means different things to different races. Let us now suppose that the proportion of blacks in the city rises to 50 per cent and (miraculously) the value of the index of dissimilarity remains at zero. Each sub-area now contains a population divided 50 : 50 between blacks and whites and, while there appears to be no change in the totally integrated pattern, the chances of inter-racial contact are dramatically altered for both groups. The average black resident has become markedly more isolated from contact with whites while, in contrast, whites have become more exposed to blacks.

In the context of a rapidly growing black minority this probability measure, known as P*, is particularly important. Unless rapid desegregation occurs, the experience of the average black resident will be of an increasingly polarized

Table 5.2 Dissimilarity indices at varying spatial scales

City	Author and data	Groups compared and data source	Index
a. Kilometre square level			
Huddersfield	Jones & McEvoy (1978)	Asians:Total population (1971 ER)	66.6
b. Tract or polling district level			
Birmingham	Woods (1975)	West Indians:Total population (1971 C)	56.0
Birmingham	Woods (1975)	Asians:Total population (1971 C)	69.0
Oxford	Peach, Winchester & Woods (1975)	West Indians:Total population (1971 C)	43.0
Oxford	Peach, Winchester & Woods (1975)	Asians:Total population (1971 C)	43.0
Glasgow	McEvoy (1978)	Asian:Non-Asian population (1976 ER)	73.0
c. 500 metre^2 grid			
Huddersfield	Jones & McEvoy (1978)	Asian:Total population (1971 ER)	70.4
d. Enumeration district			
Nottingham	Husain (1975)	New Commonwealth:Total population (1971 C)	54.7
Blackburn	Robinson (1981)	Asians:Indigenous population (1977 ER)	73.3
e. 250 metre^2 grid			
Bradford	Cater & Jones (1979)	Asian:Non-Asian population (1977 ER)	78.3
f. Street level segregation			
Blackburn	Robinson (1981)	Asians:Indigenous population (1977 ER)	75.3[a]
Huddersfield	Jones & McEvoy (1978)	Asians:Total population (1971 ER)	81.6
Bradford	Cater & Jones (1979)	Asian:Non-Asian population (1977 ER)	81.0

Source: As reference list except:
   Peach C., Winchester S. and Woods R. (1975), The Distribution of Coloured Immigrants in Britain, *Urban Affairs Annual Review* 9, pp. 395–419.
   Robinson V. (1981), The Development of South Asian Settlement in Britain and the Myth of Return, in Peach C. *et al*, eds., *Ethnic Segregation in Cities*, Croom Helm, London, pp. 149–169.
   Woods R. (1975), Dynamic Urban Social Structure: A Study of Intra-Urban Migration and the Development of Social Stress Areas in Birmingham, unpublished D. Phil. thesis, University of Oxford.

Notes: As Table 5.1. except:
   a – calculated for households; all others calculated for individuals

As the table shows, segregation levels tend to increase as we focus down on smaller units of the urban fabric, subject to variations in the city, the groups compared and the data source. In part this is due to a mathematical effect of reducing scale, but it does illustrate the comparatively high levels of segregation which have been generated in studies of British cities. The finer scale levels may also relate more closely to a person's 'neighbourhood' rather than the large heterogeneous wards of perhaps 15–20,000 people, and a smaller spatial mesh is also useful in isolating small minority populations 'lost' in a coarse- grained analysis.

existence. Clearly the dissimilarity indices demonstrate the persistence of Asian segregation in particular, and it would be reasonable to anticipate the increasing isolation of the minority as it increases in size. To date there has been relatively little testing of the P* index in British cities (Robinson, 1980; Baboolal, 1981; Jones, 1983/4), but Robinson's work on Blackburn has demonstrated that, at the enumeration district level, the statistical probability of Asian contact with Asian had increased from 28 per cent to 51 per cent between 1971 and 1977. As the author points out, the reality may be even more stark, since P* can only measure the probability of opportunities for interaction and the not the probability of interaction itself. Since ethnic avoidance appears to be the norm even when opportunities for interaction are available, the black minority may experience a distinctly isolated life.

It is clear then that the extent of ethnic concentration in British cities is far from slight. This alone does not justify the use of the term 'ghetto'. Although Britain does not share America's history of explicit racist legislation, and in Britain governments passed anti-discrimination laws in 1965, 1968 and 1976 (usually when tightening immigration control (!)), it is worth re-emphasizing the point that *de facto* segregation is as real as the *de jure* type (Cell, 1982). The concentration of black minorities also tends to occur in the least desirable locations within the urban environment, the least popular council housing, multi-occupied, privately rented property or, most commonly in the Asian case, voluntarily relinquished inner-city terraces. There is no doubt that such minorities receive a disproportionate share of the very worst of the housing stock. Taking one city as an example, Bradford in West Yorkshire, in 1981 (despite a decade of extensive housing improvement) the black population were more than three times as likely to live in overcrowded conditions, more than three times as likely to lack a bathroom and almost twice as likely to lack an inside toilet (City of Bradford Metropolitan District Council, 1984). This pattern is repeated nationally.

5.5 Ethnic Choice and Cultural Pluralism

5.5.1 Ethnic Choice

The occupation of ageing filtered dwellings in the run-down inner areas of British towns and cities has often been attributed to minority choice. Although social and economic factors are commonly accepted as an explanation of West Indian settlement patterns (Lee, 1977), Asians in particular are seen as a separate case. Reference is often made to the strength and diversity of Asian cultures, and to a desire of the Asian population to preserve cultural distinctiveness. The segregated ethnic enclave is thus seen as a protective space, the territorial base for the development of self-help institutions and even a separate sub-economy. The principal proponent of this view has been Badr Dahya (1972; 1973; 1974). Dahya claimed that Asians chose to live in back-to-back artisan's cottages and Victorian terraced houses since they are located centrally and afford easy access to the city centre, public transport and inner-city industry. Only in terms of native white evaluations does he see such housing as inferior. Dahya's views received considerable support in the social, anthropological and geographical literature. For example, Kearsley and Srivastava's (1974) interesting paper explains the evolving spatial concentration of Asians in parts of Glasgow as 'self-segregation'. More recently Robinson (1979, 7) has rationalized Asian occupance of low-amenity dwellings by reference to their apparent desire for 'non-participation in the morally-lax, status conscious British society'. He claimed that his findings in Blackburn 'not only support Dahya's thesis of return migration, but also suggest that the Asian element neither sees itself as part of, nor is concerned with, the larger white society' (Robinson, 1979, 38).

This enthusiastic acceptance exceeds even Dahya's position, since he states that 'the analysis offered above relates to the early stages of Pakistani settlement in England' (1974, 114). Much of his fieldwork was carried out between 1964 and 1966, and he acknowledges

> It is likely that at a later stage, the immigrants may begin to re-evaluate their position *vis-à-vis* Britain / Pakistan and according adopt new values and aspirations and seek recognition in British society or, alternatively, modify their present perspectives and yet, given the external constraints . . . may find their residential mobility blocked (1974, 114).

Although some immigrants undoubtedly retain strong cultural ties and a 'myth of return' to their homeland, such aspirations are unlikely to be fulfilled and goals, especially for subsequent generations, will continue progressively to change. To some extent this is recognized by Robinson (1986) in his most recent work.

5.5.2 Cultural Pluralism

Although the term 'cultural pluralism' is rarely employed in Britain (integration being the commonly preferred alternative), it nonetheless provides the theoretical underpinnings of the ethnic-choice model. According to Newman (1973), the basic concepts of cultural pluralism were hatched in the United States in the 1920s and represent an attempt to rewrite history from the minority viewpoint, in direct challenge to the assimilationist model of ethnicity. The answer lies in the attack functions of the ethnic community. The positive bonds of ethnicity, such as solidarity and bonds of loyalty within the group, can be mobilized to the collective advantage of the group in competition with other groups (Parkin, 1979; Boal, 1978). In America, this seems to have taken two main forms:

(a)   through the use of the democratic electoral process – the election by ethnic voters of ethnic politicians who then use their position to further the interests of their group (see Glazer and Moynihan's (1963) classic study of pluralism at work in New York)

(b)   through the rise of ethnic-owned businesses, supported by the financial backing, patronage and labour of fellow group members: this particular ladder has been climbed with spectacular success by the Jews (on ethnic business see Light, 1972; Cummings, 1980).

Whatever the mechanisms peculiar to each group, there is little doubt that white ethnics in America have generally achieved a high degree of occupational mobility. There is more controversy over the degree to which they have retained a separate identity, with one school arguing that ethnicity has become an even more vital force in American life since the 1960s (Glazer and Moynihan, 1975) and another dismissing it as being little more than lip service to ancestral memories (Gans, 1979).

As always, spatial change is the barometer of socio-economic change. In America, ethnic entry into the mainstream job market has been matched by entry into the mainstream housing market: groups such as the Irish, Poles and East European Jews are no longer confined to Burgess's (1925) submerged 'zone in transition' but have characteristically relocated in the middle and outer zones. But, and this is the crucial contrast with assimilation, they have not allowed themselves to be randomly dispersed into suburban homogeneity. Rather, as the work of Kantrowitz (1969; 1981), Guest and Weed (1976) and Jackson (1981) testifies, they have remained segregated, recreating their ethnic enclave in a new and superior location (Newman, 1973). This combination of economic success with persistent, though not especially acute, segregation captures the essence of cultural pluralism, the split-level lifestyle whereby ethnic group members compete freely and equally with other groups in the open market but conduct much of their private life within the ethnic community (van den Berghe, 1978).

5.5.3   Racial Pluralism: Another Non-Option?

Returning to the question of ethnic choice, it is evident that this can only be a reality for groups who have attained economic, social and political parity with the dominant majority. For example, if newly affluent Boston Italians decide to leave the inner city and reconstitute themselves as an Italian community in the suburbs, this is to exercise a genuine option, since members of this group possess sufficient financial and social standing to disperse into non-Italian areas if they so wish. This is hardly applicable to Asians, Africans or West Indians in Britain, economically and socially disadvantaged as they are. Indeed, it is rarely applicable to racially stigmatized groups anywhere. This realization dawned on America during the ghetto riots of the 1960s when, as the Kerner Commission noted, blacks had made little significant economic or residential advance during the entire post-war period (National Advisory Commission for Civil Disorders, 1967).

In every sense, the Kerner Report was an official acknowledgement that the 'ethnic route' was a non-option for blacks, a point also accepted in Glazer and Moynihan's revised assessment of black prospects in the second edition of *Beyond the Melting Pot* (1967). Prior to this, black political action had centred upon the Civil Rights Campaign, whose implicit purpose was to establish the same kind of social legitimacy for a racial minority as that enjoyed by white ethnic groups. The eruption of ghetto riots was triggered (though not caused) by frustration at what appeared to be a white American rejection of this approach. This rejection left just one apparent course of action, violent revolt.

5.6 Structuralist and Radical Alternatives

From the American experience it is clear that the pluralist and assimilationist theories relate to ethnic rather than racial groups, yet, as we have illustrated, they continue to be applied uncritically in assessing the futures of New Commonwealth origin residents in the United Kingdom. While British geography has undergone a significant radicalization in the past decade (Johnston, 1983), students of black segregation in the United Kingdom continue to claim that 'the balance of factors involved in the segregation of coloured minorities suggests that internal, or sub-cultural, factors are more powerful than the external factors, or structural constraints' (Jones, 1979, 162). This is despite the fact that, as the ensuing section illustrates, the black population is evidently disadvantaged in both the possession of, and access to, scarce and valued resources.

5.6.1 The Managerialist Perspective

Since Pahl's seminal work in the 1970s (Pahl, 1975), human geographers have focused considerable attention on the allocators of scarce urban resources, particularly in the housing market (Boddy, 1976; Gray, 1976; Williams, 1978). This interest in the institutions which determine individual life chances within capitalist society seemed particularly apposite for studies of black residential pattern, given the extent of racial discrimination identified in both official and academic surveys; for example the 1967 Political and Economic Planning Report, using 'situation tests' in which a black, a white immigrant and a British-born white applied for opportunities in the housing and labour markets, found that discrimination towards the black occurred in some 65 to 75 per cent of cases. The discrimination was specifically racial, since blacks were typically seven times more likely to be discriminated against than members of the white immigrant control group (Daniel, 1968). In recent years discrimination may have become more covert (Smith, 1976), but there is no doubting its entrenched existence (Scarman, 1981) and the fact that it is exacerbated in times of economic recession (Brown, 1984). However, geographical studies of black minorities in a managerialist framework are few and far between. Although Duncan (1977) and Cater (1981) have looked at access to, and the quality of, owner-occupied housing, and Skellington (1980) has focused on the allocation of council properties, most studies of racial segregation have continued to emphasize internal processes not external constraints. As Sims (1981, 123) states, 'geographers, for their part, have placed greater emphasis on the primacy of ethnic choice and the positive forces of ethnic association in determining the housing decisions of coloured immigrants.'

Although an important focus for research in the mid-1970s, it has become clear that, while managers do play an important role in filtering access to scarce resources, their role is subject to severe limitations. In part this has been countered by a focus on 'higher managers', switching attention from, for example, the local building society manager and seeing him merely as an implementer of his company's policy with very little personal autonomy (Pahl, 1979). Even this, however, fails to highlight sufficiently the macro-level factors; the state of the capitalist money market and the political composition of government are just two higher-level influences which have a direct, though less explicit, impact on the availability of resources. While the managerialist framework may offer some insight into the position of blacks in the housing and labour markets, it has little application to wider issues concerning the social, political and economic status of minority groups within society; for this we must turn to more overtly structural explanations.

5.6.2 A Marxist Perspective

Since Cox's seminal work in 1948, there has evolved a fairly coherent (though not necessarily consensual) Marxist perspective on race and racism. It is the more recent work in this tradition, most notably the growing literature on the class position of migrant labour (Castles and Kosack, 1973; Miles, 1982), which offers the most appropriate starting point for an analysis of black segregation in Britain.

Unlike the ideas of internal colonialism (5.4.2 above), a Marxist approach requires us to suspend any preconceived belief in the exclusion of blacks and immigrants 'outside and below the class structure' (Fig. 5.3). Since traditional Marxist analysis rests on the fundamental concept of capitalist society polarized into two primary classes, it has no room for additional categories such as a black underclass. On the contrary, as suppliers of labour power, the black relationship to the means of production is identical to that of the working class as a whole and they are subject to the same basic form of class exploitation. Blacks share the same class interest as any other section of the proletariat.

However, as we are already aware (Chapters 1 and 2), the working class is

Figure 5.3 Ethnic and racial status in society: possible outcomes

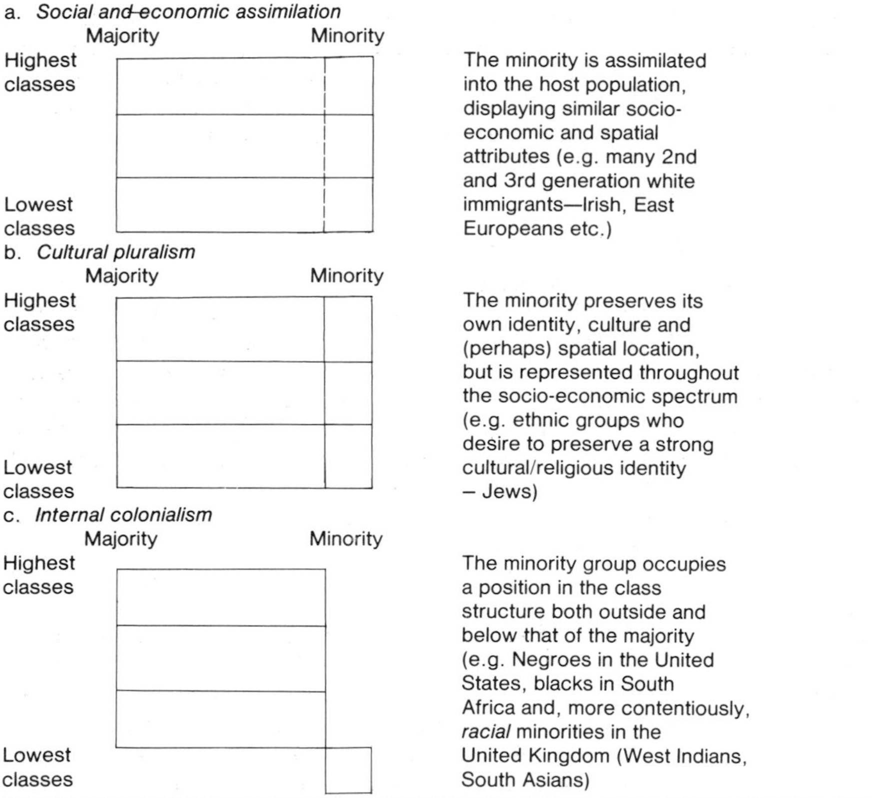
a. Social and economic assimilation
Majority
Minority
Highest classes
Lowest classes
The minority is assimilated into the host population, displaying similar socio-economic and spatial attributes (e.g. many 2nd and 3rd generation white immigrants—Irish, East Europeans etc.)
b. Cultural pluralism
Majority
Minority
Highest classes
Lowest classes
The minority preserves its own identity, culture and (perhaps) spatial location, but is represented throughout the socio-economic spectrum (e.g. ethnic groups who desire to preserve a strong cultural/religious identity – Jews)
c. Internal colonialism
Majority
Minority
Highest classes
Lowest classes
The minority group occupies a position in the class structure both outside and below that of the majority (e.g. Negroes in the United States, blacks in South Africa and, more contentiously, racial minorities in the United Kingdom (West Indians, South Asians)

The figure identifies in schematic form the main theories of the status of ethnic and racial minorities in society. The literature on British blacks has tended to adopt, in sequence, the first and second models. The third model has been extensively applied to black minorities in the United States and, in the view of the authors, could be profitably employed in the British case.

politically and ideologically fragmented into sectional interest groups, who fail to perceive themselves as members of a single solidary class. Consciousness of status, gender and other differentials prevents the proletariat asserting itself as a unified political force. With mass migration from the Third World, race has now emerged as an additional fissure in British class consciousness; or, as Miles (1982, 159) puts it, 'a new dimension is added to class struggle as a result of a racialized fraction within the working class'. Dormant popular stereotypes derived from a colonial past have been activated in the post-colonial present (Rex, 1970; Rex and Tomlinson, 1979), with black labour being perceived by the majority of the white working class as non-legitimate competitors to be resisted, subordinated or avoided.

Blackness, then, is assigned to a special category and its bearers to 'a class position with a distinctive set of economic functions in relation to capital' (Green, 1979, 36). In effect, blacks operate as a submerged sub-stratum within the proletariat, performing certain functions undesired by 'free' labour. Migrants entering Britain in the 1950s and 1960s were generally reserved for two principal economic tasks, to top up the reserve army of labour which enables wage rates to be depressed, and to act as replacement labour for unwanted jobs (Sivanandan, 1976). As in North America and continental Europe, the picture in Britain was one of an increasingly mobile indigenous labour force whose occupational expectations would have been unfulfilled in the absence of replacement labour for noxious, dangerous, menial and low-paid industries and occupations.

5.6.2.1 Social Reproduction and Black Segregation

Inevitably the class relations of blacks in the sphere of production are accurately reproduced in the housing market (Doherty, 1973). The ghetto-like pattern of many of Britain's black communities is entirely consistent with the low-status, unwanted economic roles wherein they are confined – urban reservations for the reserve army of labour. Even so, while this may be accurate as a figure of speech, it does not amount to more than a refinement of the orthodox view of the ghetto as inferior residential space. To make progress, we now need to take account of the role of the ghetto in the process of social reproduction.

A pure migrant-labour system on the lines operated by several continental European states confers the advantage of extreme flexibility. Not only may labour be imported 'ready for use', it may also be re-exported during times of high unemployment, so that the state is not obliged to support unnecessary workers in idleness (Castles and Kosack, 1973). In the case of Britain, however, it has never been possible for the state to operate a fully-fledged contract-labour system consistent with the needs of capital. Until recently, and despite the growing rigour of immigration legislation, the legal status of the Commonwealth citizen limited the restrictions which could be placed on the permanent settlement of immigrant workers or their dependents. Consequently New Commonwealth immigration has given rise to a settled and increasingly native-born black population, whose function as replacement labour is increasingly redundant (Chapter 1), but whose right to make demands on collective consumption is no less than that of other British citizens.

Yet it is far from correct to assume that the state is now saddled with the entire responsibility for maintaining surplus black labour. While it may be impossible to repatriate labour to its countries of origin, it is certainly possible to repatriate it to the ghetto. In practice, housing provision, the care of the unemployed and the elderly, and many other forms of social reproduction may be devolved from the state on to the black communities themselves, thus ensuring that black labour at least partially reproduces itself. Though permanent settlers in a literal sense, many blacks continue to make a migrant-like contribution to social reproduction: to the extent that they 'manage their own affairs' and 'solve their own problems' they relieve the state of the necessity of doing so. To argue this is of course to turn the conventional ethnic choice model (5.5 above) on its head. The much vaunted freedom enjoyed by Asians and, to a lesser degree, West Indians, to operate their own businesses and housing sub-markets can be seen as of ultimate benefit to the state, with blacks paying a price for the benefits which their ethnic community supposedly confers upon them. As Green (1979, 21) observes, 'blacks resident in Britain receive less from state welfare because they tend to live in decaying inner-city areas, use underfinanced school and hospital services and generally benefit less from council housing'.

As yet there have been few attempts to assess the impact of this self-reproduction. Jones (1983/4) has tried to measure the degree to which Pakistanis are under-represented in council housing, arguing that this represents considerable savings to the public purse. Similarly, it is possible to demonstrate how Asians are penalized, in the form of lost housing subsidy, lost tax relief and lost price appreciation, for their occupation of old inner-city terraced dwellings (Ward, 1982; Cater, 1984). These studies, which we have more fully reviewed elsewhere (Cater and Jones, 1987), lend support to the model of the ethnic community as an agent of self-reproduction, making it possible for blacks to survive in the absence of a full social wage. Even so, there remains enormous scope for research into unanswered questions, such as the role of black business as a 'sponge' for unemployment, and into the evidently differing positions of Asian and West Indian minorities and the various ethnicities which comprise these larger populations.

5.7 Accommodation and Resistance

Were we seeking a straightforward assessment of the political (as opposed to the material) value of ethnicity for the capitalist order, we would begin by emphasizing its role as a social stabilizer, an institution which encourages racial minorities to accommodate to the social order rather than to confront it. In as much as the ethnic community provides its members with what they consider to be (or are prepared to accept as) a viable alternative to mainstream society, then this will limit their demands on that society and hence minimize potentially damaging competition for scarce resources. As suggested elsewhere (Cater and Jones, 1979), ethnicity may actually be regarded as complementary to racism in that it allows, even encourages, migrants to accommodate to the subordinate roles which a racist society ascribes to them. As long as migrant groups like Asians continue to conduct their lives in a separate cultural/spatial capsule in which they are 'satisfied' with the ownership of second-class resources, then conflict with the native population remains dormant – in principle at least. Much of the literature lauding the benefits of ethnic 'autonomy' is in reality a celebration of a one-sided social peace which is entirely dependent upon an acquiescent black population.

5.7.1 Popular Racism

In practice even this distorted equilibrium has failed to materialize. It seems that however low the profile adopted by black communities, their presence has never been fully accepted. Racism is non-appeasable. In a history documenting the escalation of intra-class conflict between white and black workers, Sivanandan (1981/2) reminds us that as long ago as the early 1950s, when black immigration to Britain was still demographically insignificant, immigrants were already experiencing discrimination in housing and employment (see some of the early studies of housing segregation such as Glass, 1960; Patterson, 1963). Extremist anti-immigrant political parties also emerged at that time and racial attacks 'became a regular part of immigrant life in Britain' (Sivanandan, 1981/2, 116).

Despite the optimism which many students of residential segregation were continuing to express even into the 1970s (cf Lee, 1973; Peach, 1975b) the 1950s were, according to Sivanandan, merely the prelude to a rising tide of race conflict. Both covert *de facto* discrimination and open violence have steadily mounted. According to the Joint Campaign Against Racism, 'Racial attacks rose to 20,000 last year, with Asian families most frequently singled out for rough treatment. . . . Most Asian families have first-hand experience of some form of abuse' (*Observer*, 27 July 1986).

In any attempt to unravel the relationship between race and class the central problem is that the racist behaviour described above is to all intents and purposes a working-class phenomenon, and a very deep-rooted one at that: 'a submerged but powerful feature in British political culture' (Joshi and Carter, 1984, 53). As we have seen, the historical roots of racism are traceable to colonialism, with its racial division of labour in which the superexploitation of non-whites was justified by their alleged genetic/cultural inferiority. Black immigrants arrived in Britain during the final era of colonial withdrawal when 'all that was left of the colonial enterprise was the ideology of racial superiority' (Sivanandan, 1981/2, 116). But this negative ideology was all that most inner-city residents had as a means of relating to their new neighbours and it rapidly assumed a dominant role in working-class perceptions of the newcomers (Rex, 1973; Rex and Tomlinson, 1979; Miles and Phizacklea, 1981; 1984; Miles, 1982). As a journalist of the 1960s so percipiently put it, 'the last man in the queue is always more resentful of intruders and competitors than the man at its head' (*New Statesman*, 10 May 1968), and from the outset it has always been the most deprived and *marginalized* white workers who have provided the rank and file membership of racist political organizations (Billig, 1978; Gilroy, 1987).

Decolonization is only one aspect of an historical context which made racism an almost inevitable response for many sections of white labour. Immigrants entered an inner city already decaying economically and physically, in which established communities were being systematically dismantled by the urban renewal process, often leaving behind only the most disadvantaged sections of the population – the elderly and those failing to qualify for the suburban exodus (Rex and Moore, 1967; Rex and Tomlinson, 1979). For white residents and workers bewildered by the scale and pace of urban change, immigrants were and are seen as a cause rather than a symptom of this decline: 'racist ideas . . . are the ideological product of material decline' (Miles and Phizacklea, 1981, 98). Immediate and easily identifiable targets, blacks served to draw the fire of working-class wrath away from 'the underlying dynamics of capitalist production' (Miles and Phizacklea, 1981, 93). It is these irrational but highly plausible fears which have been mercilessly articulated by the racist right in its fictitious portrayal of immigration 'as a kind of urban invasion, undermining settled white communities, displacing their inhabitants, creating slums and generally destroying . . . a prosperous, contented, law-abiding world' (Jones, 1983/4, 10; see also Lawrence, 1982 and Miles, 1982 on racism as 'commonsense'). Crucially too, the non-response from the leaders of the Labour Movement left this neo-facist ideology virtually uncontested. Because of the vacuum left by the Labour Party and the Trade Union movement, there was initially

> no progressive anti-racist political/ideological framework which would have enabled the working class to 'make sense' of a black presence in Britain. Before the working class could fashion a response from within its collective traditions and experiences of poverty and hardship, its reformist leadership had structured such a response around 'colour' as a problem (Joshi and Carter, 1984, 55).

5.7.2 State Racism

A further factor in the genesis and spread of racism is the interplay between white labour and state policy, the way in which the latter has both shaped and been shaped by the former. On state activity as a stimulus to popular racism, Miles and Phizacklea (1984) highlight the way that 'race' was officially defined as a domestic British problem as early as the Attlee government of 1945–51, with the setting up of a Cabinet committee to investigate ways of restricting Commonwealth immigration and the publication of a Royal Commission on Population, which underlined the need to ensure that migrant labour (urgently needed for post-war recovery) be drawn from 'good stock'. All this was accompanied by parliamentary debate of a kind not designed to allay public fears. The state itself from the very outset 'constructed a problem' of immigration and race (Miles and Phizacklea, 1984, 20–44). Ironically it has done so out of a concern with the possible threat to social order posed by the presence of 'unassimilable' elements. This is one of those instances when the material interests of capital (industrialists eager for new sources of cheap labour) clash with the long-term political interests of the capitalist order itself (Sivanandan, 1978; Green, 1979).

Having taken the lead in fanning working-class tensions, the state was then obliged to respond to them. Increasingly restrictive immigration legislation (Sivananadan, 1976; 1978; 1981/2) and increasingly repressive policing of black residential areas (Hall, 1980; Gilroy, 1980; 1982; Lea and Young, 1982; Scraton, 1982) are two important instances of state policy bowing to populist pressures to limit black freedoms. In turn, of course, state policy also feeds back into popular perceptions, reinforcing racism by lending it legitimacy. There can be little doubt that the passage of explicitly colour-biased immigration laws has done much to make racial hostility appear respectable and acceptable, and to heighten the sense of insecurity in the black communities. Effectively the state has presided over a changed morality of race discrimination.

Table 5.3.

STATE POLICY-POPULAR RACISM: CUMULATIVE FEEDBACKS

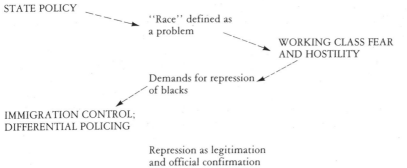

STATE POLICY

"Race" defined as
a problem

WORKING CLASS FEAR
AND HOSTILITY

Demands for repression
of blacks

IMMIGRATION CONTROL;
DIFFERENTIAL POLICING

Repression as legitimation
and official confirmation
of racism

5.7.3 The Black Response

A predictable result of these attacks is to drive both Asians and Afro-Caribbeans into defensive retreat, the ghetto being an obvious manifestation. *Avoidance* is still the key to black residential behaviour, a desire to maintain 'a certain distance from their white fellow workers and neighbours to ensure they were not directly abused' (Miles and Phizacklea, 1981, 96; see also Phillips, 1986 on the defensive residential behaviour of Bengalis in East London).

Naturally the style of avoidance has varied from one group to another. 'Gradually the West Indians began to set up their own clubs and churches and welfare associations. . . . The Indians and the Pakistanis on the other hand were mostly rural folk and found their life more readily in their temples and mosques and cultural associations' (Sivanandan, 1981/2, 113). In other words Asians brought with them a ready-made set of cultural institutions whereas Caribbeans were to a greater extent obliged to improvise ethnic lifestyles in order to cope with rejection by British society. Although something of an over-generalization, the observation by writers such as Hiro (1973) that Asians never intended to assimilate into British culture, whereas West Indians did but were not allowed to, is still a useful aid to understanding.

At this point we should stress that the ethnic community is more than simply a defensive enclosure. To borrow Boal's (1978) metaphor, it also has an 'attack' capability. In the first place this can mean that ethnic solidarity gives its members the collective strength and will to fight back against racism, a campaign which both Asians and West Indians have waged vigorously, both on the shop floor (Race Today Collective, 1979; Joshi and Carter, 1984) and in the place of residence (Sivanandan, 1981/2, Solomos *et al.*, 1982; Parmar, 1982). It can also mean a capacity to mobilize members of the group for advancement into the larger world. As we have noted above, the two classic forms of this as observed in the case of American minorities are economic advance through business owner-ship (especially by Jews and Orientals) and political advance via the mobilization of the ethnic vote (especially those of Irish and Italian extraction and, in the 1988 Presidential campaign, the Spanish speaking).

In Britain in the 1980s both these avenues are blocked. Advancement through political mobilization has been a non-option because in Britain there is no established set of institutions which can accommodate ethnic electoral politics. Party politics is rigidly structured around class interests and attempts to create ethnic political institutions from scratch have simply made matters worse.

> Government attempts during the late 1960s to create the institutions of ethnic 'community relations' meant the creation of institutions that were from the outset politically peripheralized. The notion of ethnic politics introduced from above in a political system already tightly organized along class lines served only to *reinforce* the political marginalization of ethnic minorities (Lea and Young, 1982, 17).

Lea and Young go on to argue that blacks suffer from a form of disenfranchisement by the absence of a legitimate force mobilized in their interests.

Advancement via entrepreneurialism and self-employment is an option which initially appeared to be particularly suited to Asians, whose solidary community structures and allegedly pronounced work ethic superficially suggest strong parallels with an earlier set of upwardly mobile immigrants, the Jews (Patterson, 1968; Rex and Tomlinson, 1979) but when the question came to be studied empirically, this hypothesis was widely rejected. Although there are isolated examples of successful black businesses in Britain, several recent studies conclude that such businesses are exceptional (Jones and McEvoy, 1986). Asian businesses, though numerous, are generally located near the bottom of the petty-capitalist stratum and geographically concentrated in the most unpromising locations. It would be stretching a point to portray this as social mobility. Even though Asian youth appears to achieve more than its Afro-Caribbean counterpart in educational attainment and entry into business and the professions (Rex, 1982a), this represents 'relatively less acute marginalization' (Lea and Young, 1982, 17).

5.7.3 *The Riots of the 1980s*

What we have described so far amounts to a set of pressures which by the 1980s brought forth an explosive reaction from the black (primarily Afro-Caribbean) communities in Brixton (South London), inner Liverpool (labelled Toxteth by the national press) and several other British cities. The inner-city 'riots' (or 'rebellion'?) of 1981 were the subject of an official enquiry (Scarman, 1981) and of voluminous academic debate (see, for example, Waller, 1981/2; Rex, 1982b; Cowell *et al.*, 1982; Centre for Contemporary Cultural Studies, 1982; on the Handsworth and Tottenham disturbances of 1985, see Hall, 1985 and Sivanandan, 1985).

In seeking explanations for the mass battles between black (and white) youth and the police, almost all the above writers draw attention to the structural conditions which we have emphasized in this chapter – the extreme economic marginalization ('an economically dispossessed black population' (Scarman, 1981, 6.27)) and political powerlessness of Britain's black population together with the blockage of potential escape routes and pathways to progress. Hemmed in by racism, blacks have largely been confined to those parts of the economy which have borne the brunt of recession/deindustrialization and to those geographical locations hit hardest by the decollectivization of consumption.

> As industry began to recede before the advance of technology, or simply died of the silicon age, it left once vital inner-city areas mired in poverty and decay and peopled

largely by a black underclass that had stemmed their decline for a while. But That-cherism has accelerated that decline . . . with the policies of a thousand cuts and the politics of the stick (Sivanandan, 1985, 13; see also Cross, 1982 on the 'manufacture of marginality').

Initially excluded from the 'affluent society', blacks are expected to endure the cutting edge of, for them, its collapse. The result is the emergence of inner-city communities 'excluded from the social, political, economic and cultural life of society . . . people begin to feel as if they are permanently out of sight of society at large, living behind God's back' (Hall, 1985, 12).

This then is alienation, a sense of estrangement and non-belonging. It is also a classic instance of a self-fulfilling prophecy: stigmatized as aliens from the moment of their first arrival, blacks eventually responded as aliens. It is noteworthy here that this outsider status is not entirely confined to the racially-excluded but is in some senses a generalized inner-city condition. Hence, during the riots of 1981, black youth was joined by similarly marginalized white youngsters seeking to vent their frustrations. Despite this apparent inter-racial solidarity, however, we should not lose sight of the primary distinction between these two sections, namely that one is excluded specifically on the basis of 'colour', ultimately more permanent than the (admittedly damaging) stigmata of low skills and poor education *per se* (Cross, 1982).

As we have underlined, both in this chapter and elsewhere, alienation can lead to apolitical responses such as apathy, withdrawal and even criminality. But in the case of some of Britain's blacks, conditions have been such as to encourage a more politicized response. One of these conditions has been a natural awareness of Third World liberation struggles (for example, in South Africa) and of the growth of black-power consciousness in the United States (Sivanandan, 1981/2). Another has been the direct attacks which these communities have suffered within their own living space. As Dunleavy (1977) asserts, there is nothing so likely to unite a community in active resistance than an external threat and there can be no doubt that racist attacks on people and property, together with frequent invasions of black neighbourhoods by neo-fascist marchers, have served to do precisely that.

On the subject of attacks by outside agencies, most profound in its effects has been the increasingly aggressive policing of black areas, a factor identified by Scarman (1981) and by practically every other serious analysis of the riots. Since the 1960s the traditional practice of 'consensus policing' – the maintenance of law and order with the consent, even the cooperation, of the general public – has in black neighbourhoods progressively given way to what Lea and Young (1982) call 'military policing'. This deterioration in police–public relations is documented in full by the above authors, who place a great deal of significance on the process of criminal labelling (see also Chapter 3). Its significance for British blacks, especially West Indian communities, is that they have come to be publicly stereotyped as prone to various forms of lawlessness, a stereotype which has given justification to the heavy policing tactics frequently employed in Brixton, Liverpool 8 and similar localities:

> High-profile strategies aimed less at particular identifiable offenders located with the aid of information obtained from the community, than against the community itself or, at least, the young. The distinction between offender and non-offender begins to blur and . . . any young person is as likely as any other to be caught in the net of stop and search (Lea and Young, 1982, 11).

To many commentators the riots can thus be seen as direct counter-attacks, triggered by particular instances of police heavy-handedness. While the aftermath of the riots of 1981 and 1985 has in some instances led to an attempt by senior officers to re-establish some kind of police–community relations, new powers given to the police in the 1986 Police and Criminal Evidence Act indicate that the velvet glove conceals an iron hand. On the ground 'military policing' persists, and to blacks it provides confirmation that they and their kin are not wanted: the state itself says so and sends its agents to convey the message.

5.8 Conclusion

In the view of the authors, the dominant ethos in British studies of racial segregation fails to accurately reflect the position of black minorities in contemporary society. Although geographers have borrowed heavily from positivist ecological analyses of the American city and have applied pluralist models largely derived from studies of ethnic rather than racial groups, there has been relatively little take-up of the more constraint-oriented interpretations of the position and status of racial groups, despite their growing significance in American research (Tabb and Sawyers, 1978, Wilson, 1979).

In particular, British geographers have failed to account adequately for the existence and extent of racial discrimination, despite the overwhelming body of evidence confirming both its presence and its intractability. Although it may lie beyond a geographer's conventional skills, there is also the need for an understanding of racial attitudes. The consciousness of blacks as alien and/or inferior helps their exploitation to be legitimized, to appear normal and acceptable (Gabriel and Ben-Tovim, 1978). The same false consciousness isolates them politically as a distinct class fraction who are seen to compete for marginal resources with working-class whites in times of recession.

Although British research has succeeded in outlining the pattern of black settlement in British cities and has afforded some interesting cultural detail, the need for a wider focus is paramount. The assimilationist view of black presence is moribund; there is little evidence of black–white integration from elsewhere in the world, and there is no reason for supposing the position in Britain will resolve any differently. If racial geography is not to appear even more remote from the real world, research needs to be presented with an increased awareness of racial attitudes, the allocative processes and the wider socio-economic structure of which they form part.

Recommended Reading

Research on the geography of ethnic and racial minorities in Britain is characterized by a wide range of empirical studies but relatively few overviews and/or theoretical insights. Although rather dated, the most succinct and balanced review is probably that by:

Boal, F.W. (1978), Ethnic Residential Segregation. In Herbert, D.T. and Johnston, R.J., eds., *Social Areas in Cities*, Chichester: Wiley, 57–95.

Other review chapters include:

Jones, P.N. (1979), Ethnic Areas in British Cities. In Herbert, D.T. and Smith, D.M., eds., *Social Problems and the City*, London: Oxford University Press, 158–85.

Peach, C. (1983), Ethnicity. In Pacione, M., ed., *Progress in Urban Geography*, London: Croom Helm, 103–27.

Both are, however, couched in the positivist tradition, and the chapter by Peach is explicitly concerned with ethnic rather than racial minorities.

The authors are aware of just three full length single-author books specifically on the geography of racial minorities. They are:

Lee, T.R. (1977), *Race and Residence*, Oxford: Clarendon Press.

Robinson, V. (1986), *Transients, Settlers and Refugees*, Oxford: Clarendon Press.

Smith, S.J. (1989), *The Politics of Race and Residence*. London: Polity.

We eagerly await the third of these.

There are four useful collections of (primarily) empirical material, namely:

Jackson, P. and Smith, S.J. (1981), eds., *Social Interaction and Ethnic Segregation*, London: Academic Press.

Peach, C., Robinson, V. and Smith, S.J. (1981), eds., *Ethnic Segregation in Cities*, London: Croom Helm.

Clarke, C., Ley, D. and Peach, C. (1984), eds., *Geography and Ethnic Pluralism*, London: Allen & Unwin.

Jackson, P. (1987), ed., *Race and Racism*, London: Allen & Unwin.

Of the above books, Jackson and Smith (1981) introduces the reader to a range of empirical research on British cities, mostly by doctoral students, while the Peach *et al.* (1981) collection includes articles by several important commentators on North America. Section C of Clarke *et al.* (1984) includes chapters on the Irish, West Indians and Asians in Britain but, as the title suggests, the material is set in the pluralist framework criticized in the above chapter. The most recent collection (Jackson, 1987) is based around a multi-disciplinary conference held in September 1985 which brought together researchers from a wide range of theoretical and disciplinary perspectives. This breadth of coverage makes the book a useful starting point for those interested in the interdisciplinary literature.

Those who wish to develop their understanding of the American research should consult the seminal papers and texts by Morrill (1965), Rose (1969), Taeuber and Taeuber (1965), Lieberson (1963; 1981) and Duncan and Lieberson (1959), all of which are listed below. For a more radical view on the black metropolis, Tabb and Sawyers (1978) offer an appropriate introduction to the literature.

In recent years there has also been a growing interest in the development of black business in Britain. The following edited collection provides an appropriate introduction:

Ward, R. and Jenkins, R. (1984), eds., *Ethnic Communities in Business*, Cambridge: Cambridge University Press.

6

The Urban Neighbourhood

In 1981, one of the present authors carried out a joint survey of residential attitudes in Finch House, a peripheral council estate on Merseyside (Liverpool Polytechnic, 1982). In geographical terms the survey area constitutes a true 'urban neighbourhood', being a small residential unit of approximately 600 dwellings bounded on all sides by major roads, open space or other ecological breaks and displaying considerable internal homogeneity – uniform dwelling type populated overwhelmingly by manual workers. Whether it also constitutes a 'community' is much more debatable. If we look upon community as a mutually interactive and supportive association of people, then this area would have conspicuously failed the test. Present here were all the hallmarks of the hard-to-let estate – rapid population turnover, high vacancy rates, destruction of unoccupied property, bitter residential dissatisfaction, petty crime, extensive vandalism, socially isolated elderly people afraid to venture forth after dark. All these are symptoms of social disintegration, the very opposite of community. Ostensibly this is a group of local residents alienated both from society at large and from one another, whose only form of collective action appears to be mutual self-destruction. Far from a positive resource, neighbours on this estate seem likely to constitute a hazard.

Yet, given the extreme stresses to which this population is subject, this would be a facile interpretation. Local social behaviour needs to be set in its context – in this case of crippling unemployment, high levels of welfare dependency, low housing quality, repeated municipal failure to supply adequate repairs and maintenance and an all-pervading sense of impotence. Local residents recognize that all these evils are externally generated and beyond the power of the individual to influence. Nevertheless, despite the low 'quality of life' here, many of the sentiments expressed about neighbours and neighbourhood were surprisingly positive. Only rarely was blame for the appalling physical conditions attributed to fellow residents. Many of the longer-term residents showed considerable commitment to the area and its people, declaring that, given the chance to participate in organized collective action to improve the estate, they would do so willingly.

The above case study is intended to set the scene for the present chapter by presenting a concrete example which relates to many of the issues to be raised. The urban neighbourhood is an essential component of social geography. As a small bounded space, uniform but distinct from within, the neighbourhood is the tile in the urban mosaic. Frequently it forms, or is thought to form, a basis for community development. But what is the link between local space and local community? Finch House is an apt example of the subtleties of this relationship. At a superficial level the problems of this estate suggest that community (social interaction, group solidarity, mutual bonds of loyalty) does not necessarily follow from neighbourhood proximity, a suggestion very much in tune with that school of thought which believes small-scale local institutions to be archaic relics with little part to play in large-scale modern society. Yet closer more sympathetic consideration would suggest that community bonds are remarkably tenacious even in the most adverse circumstances. Some degree of social disintegration may well be inevitable when a arbitrary mix of 'problem families' (often exceptionally disadvantaged people) is concentrated together in a confined and low-quality space by a combination of socio-economic circumstances and municipal house allocation procedures. What is apparently a solid working-class community turns out to be a collection of potentially conflicting sectional interests – young/old, waged/unwaged, established residents/recent arrivals, law-abiding/lawless, conventional families/single parents – the majority of whom are struggling to survive on inadequate resources. Nevertheless a vestigial community spirit still prevails even in the face of these colossal obstacles. When the slightest hope of progress is offered (in this case by interviewers whose promises were purely hypothetical) many people declare active support. This spirit lends credibility to those who see neighbourhood community as a force for positive change. This issue of community power and action will surface later in the present chapter (6.5). In the meantime, there are conceptual issues which must be clarified.

6.1 The Neighbourhood-Community Equation

The synonymous use of the terms neighbourhood and community is common practice in both academic and popular discourse. Much as 'land and people' and 'country and nation' trip off the tongue interchangeably, so too there is a widespread feeling that at the local level space and social network are so closely intertwined as to be indistinguishable for all practical purposes. Thus in an examination of 94 definitions of the term 'community' (assembled by Hillery, 1955), Bell and Newby (1971) find that no less than 70 of them use locality as one of their criteria. One immediate drawback here is that, for all the importance ascribed by so many authors to the neighbourhood factor, 'no one seems to be able to agree on exactly what it means or how it should be spatially specified' (Galster, 1986, 243). Neighbourhood has been used to denote spaces at almost every scale between the individual dwelling and the entire city (Hunter, 1979). Accordingly, much research has been directed to establishing some sort of principle whereby boundaries might be specified (Lee, 1968; Raban, 1975; Galster, 1986). We return to this at a later stage (6.3). For the moment it is sufficient to note the importance attached to neighbourhood and the widespread acceptance of a link between neighbourhood (local urban space bounded by the self-definitions and practices of its occupants) and community (networks of social interaction and bonding usually based on mutual interest). Often the correlation is portrayed as so strong as to amount to an equation: neighbourhood is community and vice-versa.

For geographers trained in the social relevance of space the logic of this appears unshakeable and might well be expressed in Johnston's (1974, 57) dictum that 'patterns of human behaviour are largely influenced by the environments in which they occur . . . [including the] . . . spatial environment, whose main variable is distance, which costs time, money and effort to cross.' If this principle is accepted, it provides an excellent starting point for studying the impact of spatial arrangements on community formation. Distance may be regarded as a constraint on interaction with other individuals, proximity as a positive resource facilitating social contact. *Ceteris paribus* we would expect a given person's social field to be concentrated in the immediate vicinity and to diminish with distance, a least effort distance-decay pattern which has been verified by a number of studies of marriage patterns and of other forms of social relationship (Ramsøy, 1966). Primary face-to-face social networks tend to be constructed within small-scale localized spaces inside the larger city.

On another related level this reasoning is further supported by the spatial-behavioural literature, particularly that which recognizes the human species as territorial by nature. One such commentator, Scherer (1972), describes men (sic) as like trees, needing roots, with groups of men creating a social attachment to a particular locality. Though obviously not divorced from the natural environment, urban dwellers are not immune to this 'sense of place' (Ley, 1983), drawing comfort, security and even personal identity from their own local patch of the built environment. But even more important than environmental bonds are the bonds forged with those who share that environment. It is in this sense that the neighbourhood–community equation assumes its full meaning – a local space containing friends and associates and supplying many immediate needs, functioning virtually as a microcosmic social system (Dennis, 1968).

Unhappily for those who like their models cut and dried, it must be conceded that this relationship between place-sharing and life-sharing is by no means universally operative, proximity being merely one of many possible preconditions for the development of human bonds. Neighbourhood offers a potential territorial base for community but it is neither sufficient nor necessary to ensure common goals, common action and common identity. This has long been recognized as a conceptual hazard. Lee (1968, 241) remarks on the 'elusiveness of neighbourhood' and continues

> If he [the social scientist] isolates it as a piece of territory, he often finds little or no correspondence with human behaviour; if he concentrates instead on social relationships, he finds that it does not synchronize with geography'.

More recently, Wellman and Leighton (1979, 366) have questioned the neighbourhood–community formula: 'when not found in the neighbourhood, community is assumed not to exist.' Both sets of authors see the persistence of the equation as due to the powerful influence of Park's (1926) theory of 'natural areas' in cities.

One author who has illuminated the non-deterministic nature of the space-community link is Webber (1963; 1964). As his oft-quoted seminal essays remind us, community frequently flourishes in the absence of a local residential base. In the modern 'urban non-place realm' (Webber, 1964, 79), personal mobility and ease of communication permit contacts to be forged outside the immediate locality, at the city-wide, national and global levels. This is precisely the sense in which we speak of a black community in Britain or even a worldwide Jewish community. These are but two of countless examples of communities cemented by common interests in the absence of common place, of 'community without propinquity' (Webber, 1963, 23).

On the other side of the coin, propinquity without community is an equally

Table 6.1 The neighbourhood continuum model

Types	Arbitrary	Physical	Homogeneous	Functional	Community
	Territory				
Common characteristics or dimensions		Environment			
			Social group		
				Functional interaction	
					Social interaction

Each stage in the continuum introduces a spatial or social dimension of the neighbourhood concept. As we move along the continuum each successive stage incorporates the dimensions of that preceding it.

Source: Blowers A. (1973), p. 56.

normal state. In the modern urban context, the mere fact that a collection of individuals lives in a single street is no guarantee of eye contact between them, much less of mutual sentiment and common action. Some additional basis for the association must be present over and above common residence. Indeed it may be only in rather rare and special circumstances that the fully-fledged urban neighbourhood community continues to exist at all. A striking feature of the modern city is the widespread occurrence of non-community neighbourhoods; sub-areas recognizable as distinctive bounded spaces, but possessing little or no social cohesion among their inhabitants. According to the *neighbourhood continuum model*, usefully expounded by Blowers (1973), the community neighbourhood is the last in a five-stage progression of neighbourhood types (Table 6.1). The four 'lesser' types – formal, physical, homogeneous and functional (in ascending order of complexity) – are all recognizable entities, whether or not community life is contained within. Formal neighbourhoods exist because local residents believe them to exist and identify with them as territorial reference points; physical and homogeneous neighbourhoods exist by virtue of their internal uniformity; functional neighbourhoods exist because the daily lives of many of their inhabitants revolve around a common central service point. As Table 6.1 suggests, the community neighbourhood incorporates an extra dimension, that of interaction. It is a socially interactive space inhabited by a close-knit network of households, most of whom are known to one another and who, to a high degree, participate in common social activities, exchange information (often through 'gossip networks' (Elias, 1978)), engage in mutual aid and support and are conscious of a common identity, a belonging together. The typical inhabitant numbers most of his/her friends, acquaintances and relatives amongst the neighbourhood population.

6.2 The Eclipse of Community?

How common is this locality-based social bonding in the modern city? Judging by the titles of such works as *The Eclipse of Community* (Stein, 1964), 'The Myth of Community Studies' (Stacey, 1969) and many others in a similar vein, the local urban community is a romantic vision which, if it ever existed at all, is now well down the path towards oblivion. Perhaps the best critique of the analytical usefulness of the term 'community' is Stacey's (1969) sharp corrective to woolly thinking. Aside from the sentimental meanings which customarily attach themselves to the term – community as vanished Garden of Eden – one of Stacey's objections is the sheer versatility with which social theorists have employed it, requiring it to perform a range of tasks beyond the scope of any self-respecting analytical concept. Among other usages, 'community' has been employed to denote:

- social relations in a defined geographic area
- a sense of belonging
- a mutual interest
- a sense of solidarity
- non-workplace relationships
- primary face-to-face relations

Indeed, as Stacey (1969, 15) notes, 'community also has been used to describe prisons and other more or less total institutions'. In other words this is a 'non-concept' (Hirst, 1980, 54), a verbal ragbag which can mean anything to anyone and therefore has very little descriptive, still less analytical, value. Required to do much work, it fails to do any work at all.

However, despite Stacey's objections, there is also a wealth of theoretical, empirical and polemical literature which asserts both the reality and the desirability of community, whether urban or rural, localized or diffuse. Pacione's observation that the idea of community has been subject to 'prolonged debate' (1984, 169) is something of an understatement. Ever since the founding fathers of social science, the dichotomy of 'community and society' has been the major framework within which sociologists have set their discussions of human associations and social change (Gusfield, 1975). The debate is thus a singularly long-running affair and one of great intellectual salience. It has also been heavily value-laden, with many of the early thinkers presenting the alleged decline of community as no less than the liberation of humankind, the eradication of old restraints – a view countered by writers such as Tonnies (1887), for whom community stood for a vanished golden age. Ever since Tonnies, community has symbolized 'the closer, warmer, more harmonious type of bonds between people' (Elias, 1978, xiii). Despite attempts by modern sociologists to strip the concept of its moral overtones, there remains 'confusion between what it [community] is (empirical description) and what the sociologist feels it should be (normative prescription)' (Bell and Newby, 1974, 2).

When the seeds of a belief of a decline in community were first planted well over a century ago, they were propagated by writers of such stature as to ensure a continuing influence on succeeding generations (see Gusfield, 1975). As we have noted above, for many of these pioneering social scientists the disappearance of old social bonds was a symptom of progress and modernization: a transition from feudalism to capitalism and thence to socialism (Marx); from religion to scientific rationalism (Comte); from mechanical solidarity to organic solidarity (Durkheim); from traditional to legal authority (Weber). The centrepiece of this progressive thinking was the 'autonomous individual' now about to be unlocked from the chains of place, tradition, religion, superstition and coercion represented by the old-style community.

Many of these strands were woven together in Tonnies's (1887) famous *Gemeinschaft* (community)/ *Gesellschaft* (mass society) dichotomy. *Gemein-*

schaft refers to forms of human association based on sentiment, loyalty, informality and close personal contact and is often considered to be embodied in the traditional village community. *Gesellschaft* refers to associations which are contractual in nature, rational, depersonalized and purged of emotion, entered into on the basis of calculation. Through industrialization, urbanization, the growth of state power, the rise of the large workplace and the spread of mass communication, society was moving away from gemeinschaft towards gesellschaft. In Tonnies's view this trend was to be deplored; membership of state, nation, firm, trade union or professional association was no substitute for community as a source of personal identity.

Subsequent contributions, many of them extremely weighty, have further elaborated the theme of modernization and community decline. Wirth, one of the most influential urbanists of the present century, identified urbanization itself as the decisive force, portraying the city as an 'enlarged sphere of independence' (Wirth, 1939, 497) and as the incarnation of the virtues of large-scale economic organization and personal mobility. The modern city thus represented the kind of benefits which can only be achieved through increasing scale: the economic benefits derived from specialized division of labour; the individual self-expression which comes from the breaking of traditional parochial ties. There is no place for the small community in this scheme of things. In essence, Webber's subsequent notion of the 'non-place urban realm' (1964, 79) – dispersed non-residential social networks replacing neighbourhood networks – is an extension of this reasoning. When the combined weight of these and other contributions is assessed there is every temptation to dismiss the neighbourhood community as an archaic relic, a survival which has little or no role in a modern city. (See Worthington, 1982 on the forces threatening the survival of neighbourhood community in contemporary Britain.)

Aside from the merits or flaws of this logic, geographers have their own special reason for caution in the face of any concept which appears to attribute determinant power to space. Interpreted literally, the neighbourhood community idea contains many of the classic pitfalls of the 'fetishism of space' in its implication that there exists some kind of causal connection between spatial proximity and social cohesion (Anderson, 1973: see also Hamnett's (1979) trenchant critique of spatial determinism). This warning is echoed by Wellman and Leighton (1979, 366):

> the identification of neighbourhood as a container for communal ties assumes the a priori organizing power of space. This is spatial determinism. Even if we grant that space–time costs encourage some relationships to be local, it does not necessarily follow that all communal ties are organized into solidary neighbourhood communities.

At most, spatial proximity should be regarded as a permissive condition, one which allows other more active processes of community formation to operate. In this as in any other context, the role of space is modificational rather than causal (Hamnett, 1979).

6.3 Neighbourhood as Empirical Reality

For all its authoritative weight, the eclipse-of-community school has never held unchallenged sway, theoretically or empirically. On the empirical front a host of authors on both sides of the Atlantic have at various times produced evidence purporting to prove the continuing survival and social relevance of local urban communities. In post-war Britain, such studies have ranged from the intuitive quasi-autobiographical essay of Hoggart (1957) through sociological studies of family and kinship fields (Bott, 1957; Young and Wilmott, 1957) to the spatial approach of geographers (Boal, 1969; Buttimer, 1972; Herbert, 1975; Pacione, 1984).

Somewhat surprisingly it is Hoggart's 'non-scientific' account of early post-war life in a working-class district of Leeds which provides some of the clearest testable hypotheses, notably about the requisite preconditions for neighbourhood social cohesion. If proximity itself is insufficient, what then is the basis of social solidarity? Many of the subsequent answers to this question are in effect unconscious confirmation of Hoggart's original insights. Indeed, his working-class locality appears as virtually incorporating every stage of the neighbourhood continuum, a community built upon territorial, social-class and functional foundations. The following community-forming elements may be inferred from Hoggart.

(a) *Territorial Identity* 'To the insider these are small worlds, each as homogeneous and well-defined as a village. Down below . . . the men stream up into their district, they know it as tribal areas' (Hoggart, 1957, 59–60).

Here a sense of mutual affinity is fostered by a common recognition of territorial bounds and of shared space, an observation which in a sense anticipates the spatial–behavioural findings on neighbourhood cognition (Ley, 1974; Herbert and Raine, 1976). This literature tends to argue that neighbourhood is a mental construct, a self-definition of one's local space: and that there is a remarkable degree of consensus among local inhabitants as to the bounds of that space. Seabrook (1984, 2) puts this into earthier terminology: 'The neighbourhoods were defined by the people who live there and have nothing to do with ward boundaries or parish limits or any other imposed administrative conveniences. People always know where their own neighbourhood ceases – at a main road, a canal, a row of shops, a park, a landmark.' Geographical analysis can provide confirmation of this – as a tendency at least – but it also suggests that the human mind recognizes a rather more complex system of spaces than Hoggart or Seabrook would suggest.

(b) *Class Consciousness* 'The world of "Them" is the world of the bosses, whether those bosses are private individuals, or as is increasingly the case today, public officials . . . "They" are "the people at the top", "the higher-ups", the people who give you your dole, call you up, tell you to go to war, fine you . . .' (Hoggart, 1957, 72–3).

It bears repeating that co-residence in a small space is of itself no guarantee of fraternal feelings and common interests. For people of Hoggart's neighbourhood, a consciousness of shared class membership was evidently the vital ingredient in the social cement binding them together. As Dennis (1968), Blowers (1973) and many others have indicated, social homogeneity is perhaps the greatest of all forces making for 'community spirit'. Gemeinshaft flourishes best in segregated 'social areas' where members share clearly defined attributes so that people can perceive their neighbours to be their social peers and similar in experience and outlook. Where neighbours can confidently assume that their material and social status are on a common level, then presumably there will be an absence of the kind of petty competitiveness which is the very antithesis of communalism.

Perhaps even more important is the sharing of common experiences and the common memories that these engender (Dennis, 1968). Hence, areas like mining settlements where the overwhelming majority share the same workplace are legendary for their internal cohesion. Furthermore this cohesion is at its most intense and durable where the memories are recollections of struggle and the experience is seen as one of oppression. There is no force more potent than the 'Common Enemy' for uniting people in self-defence. In Hoggart's neighbourhood residents obviously saw themselves as members of a group constantly subject to domination by authority in the shape of the employer and the bureaucrat. This was not seen as a particularly exploitative or coercive domination – after all, this was the expansive 1950s, not many years before prime minister Macmillan's famous 'you've never had it so good' speech. But authority, whether capital or state bureaucracy (interestingly no one seemed to bother to distinguish between the two) was to be viewed with constant suspicion: folk memories were long and after all it was 'They' who 'made you split the family in the Thirties to avoid a reduction in the Means Test allowance' (Hoggart, 1957, 73). 'They' were therefore never to be trusted. Rereading this with the benefit of three decades' hindsight, it seems that this attitude was well-grounded. Not long afterwards it was 'They' who came along to demolish neighbourhoods like this and disperse their populations to high-rise flats.

We should note briefly here that working-class identity was also underpinned by culture, an entire system of values, customs, speech and thought patterns, leisure activities, social institutions, even morality, which owed little to its middle-class counterpart and still less to 'High Culture'. This was a somewhat neglected topic at the time, though subsequently it won recognition as a serious area of high scholarship. We shall return to this issue later in relation to community/neighbourhood politics (6.5).

(c) *Continuity* 'unless he gets a council house, a working-class man is likely to live in his local area, perhaps even in the house he "got the keys for" the night before his wedding, all his life' (Hoggart 1957, 62).

Class consciousness as a social adhesive needs time to set, and hence Blowers (1973) notes continuity as a further precondition for neighbourhood community formation. The strength of many traditional working-class networks derived not least from their stable character over a lengthy period of time and all the classic studies of such areas (e.g. Young and Wilmott, 1957) stress the continuity of friendship and kinship bonds spanning several variations in the same vicinity: 'the family live near, have "always" lived near; each Christmas Day they all go to tea at Grandma's' (Hoggart, 1957, 62). In a sense social solidarity has grown organically in sharp contrast to such artificial implantations as New Towns and overspill estates.

By the same token the continuity factor implies that, given time, planted settlements of this type may themselves evolve into organic social networks. After all, the stable working class communities of today are nothing more than grown-up versions of the frontier settlements of yesterday, thrown up with chaotic rapidity by the Industrial Revolution and peopled by a disparate collection of migrants displaced from rural areas all over the British Isles and beyond. New settlements become old settlements, anonymity becomes familiarity, atomized individuals take on a collective identity. Building on this, we might suggest the operation of an evolutionary process of development, fully-fledged community status being reached through a series of intervening stages.

Were we to apply this logic to the overspill estate we described in the introduction to this chapter, Finch House, we might conclude that its problems stem in part from an interruption or even a reversal of this 'growing up' process. In many cases Britain's inter-war generation of council estates have enjoyed sufficient time for social networks to gel but, in Finch House, continuity has been ruptured. From the 1960s onwards, deaths, departures and their replacement by tenants often with severe socio-economic problems and with ingrained loyalties to other districts has led to considerable fragmentation. Such discontinuity and disintegration is by no means an abnormal occurrence in the council-housing sector. Indeed, it might be argued that by their very nature public housing allocation procedures pose considerable obstacles to community continuity, unless operated in a sensitive manner. Writing of parallel cases in the West Midlands, Seabrook (1984, 8) notes that, 'built at a time when it would have been felt improper to pamper the working class', this type of housing was only barely adequate at the outset and was often badly located, 'on poor sites where private builders were reluctant to speculate'. With the passage of time a combination of rising tenant expectations, the deaths of original tenants, mounting disrepair and the decay of the housing stock has reduced these neighbourhoods to hard-to-let status.

Lest we assume that discontinuity is a very recent problem, we should examine the historical evidence which suggests that it has been a constant menace to community formation throughout the entire modern period. 'Community' is a highly evocative symbol, one which inspires considerable idealism, an idealism which can spill over on occasion into romanticism. It is invariably assumed that community means continuity, which in turn denotes stability and harmony. Historically, the stereotype is one of a lost Golden Age of settled working-class communities undisturbed for generations until the post-war onslaught of urban renewal, suburbanization and mass culture. While this seems to fit the bill in Hoggart's case and while it is also unarguable that community was a highly valuable working-class resource, it is also true that the nineteenth and early twentieth centuries were times of great upheaval for much of the working-class. Stedman Jones's (1971, 1983) studies of Victorian London leave no doubt that working-class living space was invaded by commercial and industrial expansion to such an extent as to make continuity impossible in many areas.

> With the exception of a few outlying areas like Woolwich or Stratford, London working-class districts were shifting and unstable. The eviction of the poor from the central area continued and everywhere 'shooting the moon' (moving furniture from an apartment after dark before the landlord collected the rent) was a familiar feature of London working-class life. . . . The family as a working-class institution may have grown in importance but in London there was nothing very settled about the home' (Stedman Jones, 1983, 223).

Regional and local variations there certainly were, but throughout the last century any town or city experiencing rapid growth will also have undergone spatial upheaval with all its disruptive impacts on neighbourhood and community.

The above is intended as a caution against sentimentalized imagery and not as a denial of community *per se*. Despite the forces which inevitably inhibit non-market institutions in a market economy, there can be no doubt that from the very outset of industrial urbanism the territorially based social network played a vital role in the life of the city. To put it at its most basic, the neighbourhood community emerged because there was a need for it, a need which was most powerful among the poorest sections of the population, for whom some form of mutual aid was virtually a condition of survival under a brutal economic system. At the same time and more positively, the neighbourhood also supplied a vehicle for emotional needs such as fraternity, identity and sociability. Thus, as we shall see, community is not simply a matter of empirical verification but of theoretical justification.

Having evaluated these community-forming elements, we can conclude that Hoggart's home district came very close to the ideal-type community-neighbourhood as conceived in the abstract. Localism, class consciousness and continuity combined there to produce a version of the microcosmic social system (Dennis, 1968), a miniature replica of the various relationships and institutions of the larger world. Or to use Breton's (1964, 193) term, this was a place with a high degree of 'institutional completeness', supplying a remarkable range of everyday needs within its own bounds. Naturally enough, such self-contained microcosms can never be perfectly realized in practice. Never do we find an urban local area which supplies every inhabitant with work, for example, and even where workplaces are contained within, they are often controlled and owned from without. The same goes for 'social furniture' like pubs, schools and churches. Hence even in the traditional industrial working-class community, economic and social self-sufficiency stops well short of autonomy or self-determination. Yet this type of neighbourhood approaches about as far as it is possible to go along the road to institutional completeness. A sense of encapsulation in an enclosed world pervades Hoggart's entire book, a separation from the mainstream life of the city, made possible, even desirable, for the locals by the rich range of social opportunities within.

Can this model of neighbourhood community have any relevance today? It might be thought that, even ignoring for a moment the effects of urban renewal, individual expectations and personal mobility have increased far too much to permit such placebound lifestyles. Only in the case of groups with exceptional needs or constraints can the placebound community have direct relevance. Indeed the ethnic community is often cited as one of the few surviving instances, a kind of latter-day inheritor of the working-class neighbourhood. Yet it will become clear in the next section that any attempt to bury the urban neighbourhood community would be rather premature.

6.3.1 Geographical Confirmation of the Neighbourhood

Turning now to more recent work by British geographers, the most striking revelation is the degree to which key elements of the ideal-typical neighbourhood seems to have maintained their relevance even after decades of urban planning, slum clearance, housing municipalization and population relocation. One might have supposed these would have left no more than vestiges of the old inner-city districts. Whether consciously or not, British urban policy of the 1950s and 1960s seemed specifically designed to erect a Wirthian Utopia, with localism abolished and replaced by loyalty to a centralized polity and by a belief in a rational, scientifically ordered social universe (M.P. Smith, 1980). Apparently, however, many sections of the population (rather like the proles in Orwell's

Nineteen Eighty-Four) have persisted with their messy, sub-rational, sentimentalized lifestyles. Several geographical surveys have established that a great many urban citizens continue to use local space as a kind of psychological anchorage and to conduct much of their personal life within that space. Fairly representative of such studies is Pacione's exercise in Glasgow which, with true explorer's zeal, sets out to 'test for the existence of meaningful neighbourhood communities in the city' (1984, 169). After interviewing 760 respondents the author was able to conclude that there is a fair measure of local agreement on the perceived dividing lines between local areas (a finding which substantiates earlier work on Belfast (Boal, 1969), Swansea (Herbert, 1973) and Cardiff (Herbert and Raine, 1976)), and that these perceived areas provide the geographical base for a significant degree of community cohesion, as measured by ties of family friendship, participation in local activities and the use of local facilities.

Whether these findings are sufficient to justify the author's conclusion that the urban neighbourhood is alive and kicking – 'evidence from the present research strongly supports the continued existence and vitality of neighbourhood communities in the modern city' (Pacione, 1984, 180) – is perhaps open to some scepticism. Judging from the findings on friendship patterns, the 1980s community is a much looser network than its Hoggartian predecessor: in no area did the average resident find more than one-third of his/her friends in the same locality. Quite clearly the present-day Glaswegian has a much wider social space than would have been the case a generation earlier – proof positive, we might think, of greater mobility and choice. Network analysis (as advocated by Connell, 1973; see also Figure 6.1) is probably the only true measure of how close- (or loose-) knitted the neighbourhood is, but Pacione, in common with most other neighbourhood geographers, has chosen to ignore this rather cumbersome method. Whatever their definition of friendship, however, Pacione's respondents certainly demonstrate very positive attachments to place and to the institutions contained therein.

A further noteworthy feature here is that local settlement is by no means confined to manual workers and low-income groups. On the contrary, the most interactive area in Pacione's sample (judged by the number of local friendships) is one with a substantial proportion of people in 'middle class' occupations. In Cardiff, Herbert (1975) also found that more house-to-house visiting occurred in a middle-class rather than a working-class case-study area. Despite the obvious charge that this in itself is a class-biased measure – 'workers' are still more likely to meet in the pub than around the neighbour's cocktail cabinet – we must nevertheless recognize that middle-class districts are not necessarily devoid of community consciousness. As students of suburban life have argued, the commonly held view of middle and working-class lifestyles – the former privatized, status-conscious, suburban and nuclear-family based, the latter communal, cooperative and extended-kin based – is an exaggerated stereotype (Gans, 1962). In suburban locales, community of interest may develop on the basis of common family-cycle stage, for example, particularly where this leads to a feeling of shared need. Even relatively affluent households may be victims of the 'commuter widow' syndrome, where non-working wives with small children are isolated throughout the working week and thus driven into very close dependence upon local neighbourhood resources. Among these resources will be families in a similar predicament.

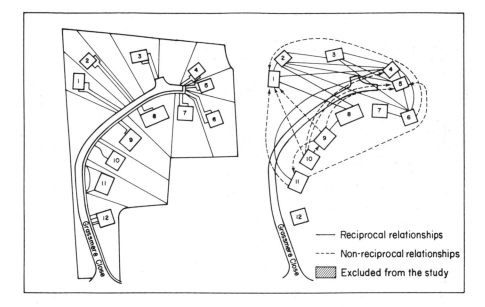

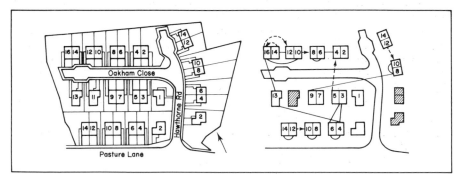

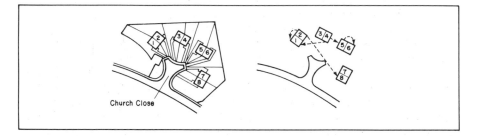

Figure 6.1 Sociograms of visiting relationships on three private estates:
   A Grasmere Estate: High patterns of activity. Young households with young
dependent children (nos. 2, 3, 4, 5 and 6) with a high level of interaction. Similar
relationship between another group of middle-aged housewives (nos. 1, 8, 9 and 10).
One elderly housewife (no, 12) was physically and socially isolated from the rest.
   B Huntley Estate: Average activity levels. Good relationships between next-door
neighbours and with some demographically similar households, e.g. retired people in
no. 4 and 6 Pasture Lane and no. 3 Oldham Close.
   C Church Close: Low level of interaction. Heterogeneous mix of households.
   Note: The names given are fictitious but refer to estates in the North Midlands.
   Source: Carey and Mapes 1972, 113–15.

6.4 Community and Neighbourhood as Theory

Community survival is not only empirically verifiable but also theoretically supportable, on the basis of the proposition that community (with or without propinquity) is functionally necessary to society. Gemeinschaft and gesellschaft are a union of opposites. On occasion, this coexistence has been presented as a virtual biological imperative, as in Scherer's (1972, xi) maxim, 'Communities still exist. If they did not, man himself could not survive'. In effect gemeinschaft and gesellschaft reflect two equally essential sides of the human personality – the emotional and the rational. Given that humans are eternally destined to feel as well as think, they are therefore destined to live in communities in some form or another.

Other writers have approached the question by returning afresh to the drawing board of the classical theorists, calling into question their implicit assumptions about modernization. For Gusfield (1975) the message conveyed by Marx, Weber and Wirth is of modernization as an irreversible historical progression from gemeinschaft to gesellschaft, a 'zero-sum game' in which community is totally eradicated and superseded by mass society. The change wrought by the industrial and democratic revolutions meant the rapid disappearance of one kind of human association and its replacement by rational ties (Gusfield, 1975), based on legal institutions, market relationships, mass electorates, political interest groups, and so on. Surviving pockets of community are simply the debris of yesteryear awaiting consignment to the planner's dustbin.

Gusfield's response is to construct what we might term a 'pluralistic' model of modern urban society, one in which community coexists with space or, more accurately, nests within it (both structurally and spatially). At the same time as citizens are increasingly drawn into contractual and bureaucratic associations, they also retain membership of informal face-to-face networks. Nowadays of course the nature of this membership is far less circumscribed than previously, even in the comparison with the fairly recent past described by Hoggart (1957), many of whose neighbours acted out every social role within the local milieu. Less placebound, the individual is also less exclusively tied to membership of a single all-embracing fraternity: 'In the highly pluralistic societies of the modern world . . . the same person is at once a Catholic, a student, a lawyer, a woman' (Gusfield, 1975, 42), labels which may actually conflict with one another in some circumstances but which may nevertheless attach themselves to a single person.

In certain respects this reasoning has been taken even further in the work of Kasarda and Janowitz (1974), whose use of a 'systemic' model of urban society treats community as a generic structure of mass society – i.e. a functionally necessary component. Presumably this means that it is necessary to have a counterweight to the anonymity and depersonalization of large-scale urbanism. While Wirth himself had assumed that personal identity and social order would come from loyalty to centralized institutions, Kasarda and Janowitz imply that this by itself is insufficient and that local 'urban villages' are necessary for social integration, perhaps to 'provide a local training ground for the development of larger loyalties to city and nation' (Bell and Newby, 1976, 193).

Although there is no lack of independent support for the pluralistic idea, it is by no means unqualified. Elias (1974) stresses that each individual operates in what amounts to a split-level world, dividing attention between the instrumentalist sphere of gesellschaft and the personal domain of gemeinschaft. In everyday terms this is no more than the distinction between public and private persona, but it is a point worth making nonetheless, since in traditional urban neighbourhoods where group life 'started from the home and worked outwards' (Hoggart, 1957, 68), such a distinction may not even have occurred to many of the residents. Yet, while upholding community survival, Elias is careful to point out that it is a much diminished survival. Whereas community and locality could once provide a very wide range of everyday needs – they were, to repeat Breton's (1964, 193) term, 'institutionally complete' – now they are reduced to catering solely for private emotional needs. While far from obliterating community, society has vitally encroached on it, assimilating a growing portion of each person's life and reducing community to the private domain. This reasoning may be linked to Gordon's (1964, 1978) concept of *structural assimilation*, which proposes that new immigrants to urban society initially base their life almost exclusively upon an ethnic community neighbourhood but subsequently undergo a process of differential assimilation. Ultimately they may become drawn into the various formal associations of the larger society but retain the ethnic community as a base for primary relationships (Table 6.2). Though specifically directed at minority groups, this reasoning might be taken as representative of the changing social relationships of the population en masse.

The pluralistic and structural assimilation models, with their conception of a multitude of relationships – both of the gesellschaft and gemeinschaft types – coexisting on a great many levels have their geographical equivalent in the literature of social space. Following Lee (1968), Buttimer (1972) has suggested we perceive and use a hierarchy of spaces, each level corresponding to a specific level and form of activity. These include 'micro-service' and 'macro-service' activity spaces, 'social participation' spaces and perceived home areas.

What of the role of neighbourhood in these complex hierarchical arrangements? Whereas all the above authors are united in a belief that community itself survives (albeit in a changed form), there is no such unanimity that community is bounded by neighbourhood. A definitive feature of the pluralistic society is increasing freedom of choice: the individual enters into a whole range of freely elected associations of which neighbourhood association is merely one of a whole range of competing alternatives. This is part of the logic underpinning the community-without-propinquity perspective. Whether based on kin, class or ethnicity, communal ties no longer rely on proximity. They can, as Gusfield (1975, 47) puts it, 'persist within the structure of class-oriented mobility, held together by occasional visits, 'phone calls and letters'. Evidently the keynote of this pluralistic city is mobility, liberation from the chains of place and the friction of distance, a viewpoint carried to extremes by writers such as Toffler (1970), with his image of modern people as the 'New Nomads'.

Yet such a vision of human command over space is one that distorts the realities of life for the great mass of humanity, even many dwellers in advanced urban nations. It ignores significant minorities whose residential and transport mobility is severely circumscribed for one reason or another: those too poor to afford a bus fare, much less a phone or car; housewives, large families, the old and the young, who enjoy far less mobility than active breadwinners and who are often forced into extreme dependence upon their local facilities; council tenants whose residential destination is often determined for them and frequently brings dissatisfaction; blacks and other ethnics whose residential access is often blocked by racism. In short, total human command of space is a fantasy of the sort that could

Table 6.2 Processes of assimilation

a. THE ASSIMILATION VARIABLES

SUBPROCESSOR OR CONDITION	TYPE OR STAGE OF ASSIMILATION	SPECIAL TERM
Change of cultural patterns to those of host society	Cultural or behavioural assimilation	Acculturation
Large-scale entrance into cliques, clubs, and institutions of host society, on primary group level	Structural assimilation	None
Large-scale intermarriage	Marital assimilation	Amalgamation
Development of sense of peoplehood based exclusively on host society	Identificational assimilation	None
Absence of prejudice	Attitude receptional assimilation	None
Absence of discrimination	Behaviour receptional assimilation	None
Absence of value and power conflict	Civic assimilation	None

b. *ASSIMILATION VARIABLES APPLIED TO SELECTED GROUPS IN THE UNITED STATES*

GROUP	TYPE OF ASSIMILATION						
	CULTURAL	STRUCTURAL	MARITAL	IDENTIFICATIONAL	ATTITUDE RECEPTIONAL	BEHAVIOUR RECEPTIONAL	CIVIC
Negroes	Variation by class	No	No	No	No	No	Yes
Jews	Substantially Yes	No	Substantially No	No	No	Partly	Mostly
Catholics (excluding Negro and Spanish-speaking)	Substantially Yes	Partly (variation by area	Partly	No	Partly	Mostly	Partly
Puerto Ricans	Mostly No	No	No	No	No	No	Partly

Source: Gordon 1978 p 169 and p 174.

only be concocted by a successful white male spiralist – a manager, a technocrat, an academic. It is a projection of self on to the rest of the human race.

6.5 Neighbourhood and Community as Ideology

We are now confronted by two sharply opposed images of neighbourhood community. Is it a sanctuary into which individuals can retreat at will from the stress-filled hurly-burly of the competitive society? Or is it a trap, a kind of social control mechanism for less successful citizens?

6.5.1 The Rehabilitation of Community

In the main it is the former image which is propagated by Tonnies and the more recent advocates of community life. According to such passionate community reconstructionalists as Nisbet (1970), the nineteenth-century faith in progress has been replaced in our own era by a fear of social breakdown. The 'autonomous individual' whose potentialities were supposed to be set free by modernization turns out to be a broken reed, unstable, inadequate and insecure when he is cut off from the channels of social membership and clear belief (Nisbet, 1970). In short, the dilution of community lies at the root of alienation. Insofar as modern urban dwellers suffer from this sense of non-identity, with its attendant passivity, powerlessness and sheer loneliness, its cause is to be found in the destruction of personal ties to place and people. New 'rational' loyalties to the larger society and polity as postulated by Wirth have evidently failed to compensate for the dwarfing of the citizen by large institutions; and equally by the physical gigantism of modern urban space, with its high-rise buildings, urban motorways, hypermarkets and all the other stock artifacts of the hi-tech consumer age. All this, coupled with an official passion for efficient, planned space use, has created an accurate environmental expression of impersonal sanitized formalism which is modern society at large. To advocates of community this is a world fit for androids, not warm-blooded carbon-based life-forms.

Clearly the reconstruction of urban living space has had a vital part to play in the disintegration of local social networks. It was perhaps inevitable that eventually there would be a popular backlash and that urban residents would attempt to reassert some sort of control over that space. Indeed the *urban social movements* discussed by Castells (1977; 1978; 1983; see also Chapter 2) are the living expression of this reassertion. Fairly predictably the state's response to grassroots protest has been to take it under its own wing and to assimilate it into the policy structure. Governments have been obliged to re-invent community for their own purposes and in Britain, the United States and several European countries community/neighbourhood development has become enshrined within the package of measures directed at the 'urban problem'. In Britain since the 1960s there has grown up a whole range of measures – from the decentralization of certain local authority services to support for community development programmes – which ostensibly operate to devolve a measure of power right down to the smallest local space (see 6.5.3). We say 'ostensibly' because in most instances the hand of the central state remains firmly on the tiller of power, crucially through the tight control of finances. Indeed, this control has become more overt in the 1980s as a result of public expenditure cuts.

On the surface this note of world-weary cynicism may seem out of place. Not only is community development more cost-effective than bureaucratic allocation (Worthington, 1982), it also has the virtue of infusing local residents with a sense of purpose, replacing apathy with zeal and passivity with active involvement. By granting the means to influence their own lives, it removes them from client status and welfare dependency (Ward, 1974). As a direct attack on alienation it even promises to make a wider social impact, for example in reducing crime and alleviating drug addiction and mental illness. Nor does community-as-policy appear to have any serious enemies: as an alternative both to the free market and to the centralized public sector, it promises a genuine 'third way', outside adversarial party politics, above class interests and appealing right across the political spectrum. Describing the broad consensus, Seabrook (1984) sees Conservatives as favouring community for its cost-effectiveness; cites the Liberal Party as a prime mover of community politics in the 1970s; and emphasizes the enthusiasm shown by many Labour local councils for the decentralization of services to neighbourhood offices. Community, then, is a slogan calculated to take the controversy out of urban policy, even to depoliticize it completely.

At this point caution should be counselled. Empirical evidence suggests that community-based policy is often conspicuously ineffective in satisfying the demands of local residents – for example the inner-city 'riots' often occurred in the very localities that had apparently received very heavy doses of the community medicine. At a deeper level there is a need to probe into the manner in which the small-scale local level of human life fits into the entire structure of the society at large. Only in this way can we make judgements about costs and benefits. In the ensuing section we argue that community is a *dependent* social formation (and neighbourhood a dependent space) rather analogous to Santos's (1979) 'lower circuit' of the urban economy. In this lower level outcomes are determined essentially by external forces. Like all dependent formations, its existence is useful, even essential, for capital accumulation and for the reproduction of capitalist social relations.

Attempts to rewrite history (or geography) from the popular viewpoint as opposed to that of the ruling class are frequently informed by the Marxian tradition. Yet, given Marx and Engels's own view of capitalist development, there is a distinct lack of Marxian vocabulary for community development or community politics (Mollenkopf, 1981) – at least in the sense that we perceive these issues in the modern city. For Marx and Engels community feeling and localism were incompatible with capital advance and, even as they wrote, were in the process of being swept away by the advance of modern production relations and modern values:

> In the conditions of the proletariat, those of old society at large are already virtually swamped. The proletarian is without property; his relation to his wife and children has no longer anything in common with the bourgeois family relations; modern industrial labour, modern subjection to capital . . . has stripped him of every trace of national character. Law, morality, religion are to him so many bourgeois prejudices (Marx and Engels, 1888, 68–9).

In itself, this destruction of traditional sentiments was no cause for celebration. But, at the same time, the dehumanization of relationships could be seen in a positive light as a basic contradiction of capitalism, one of the processes by which capitalism was signing its own death warrant: 'What the bourgeoisie . . . produces, above all, is its own gravediggers' (Marx and Engels, 1888, 71).

6.5.2 Capital versus Community

After a lengthy lapse, the capital / community antithesis has been taken up again by writers of our own time, often prompted by the post-1968 emergence of urban protest movements in much of the capitalist world (see Chapter 2: see also Castells, 1977; 1983). Whether standing on housing, transport or environmental issues, these urban social movements may be seen as community actively asserting itself in organized fashion in resistance to capital. Frequently, and of direct interest to us here, this resistance takes the form of *neighbourhood-based activism*, with residents of a given local area banding together against some form of external threat perceived as likely to disrupt or undermine the quality of local life. More often than not such threats come under the heading of 'redevelopment', implying the displacement of residential space to make way for offices, shops, road 'improvements' and car parking. Even where local residents escape house demolition and removal, they are unable to escape the 'attendant impacts on traffic patterns, neighbourhood aesthetics and the safety of children' (Cox and McCarthy, 1982, 197).

Redevelopment is essentially a capitalist threat. It consists of transforming the use of a parcel of land – in the above instance from residential to commercial, though in other cases it might involve a transfer from low-priced to high-priced housing – in order to yield greater profits from it. As Harvey (1978) argues, much of the long history of expansion and spatial change in the western city can be explained by reference to capital accumulation, the restructuring of urban space to yield higher returns to capital as a whole and to permit the expansion of those branches of capital directly engaged in the property market.

Having established this, we can now appreciate the basic antagonism between capital and community. It is the obstinate persistence of the latter that stands as a direct obstacle to the advancement of the former. Hence, as Marx and Engels noted, the process of accumulation invariably involves the destruction of community, since traditional community patterns impede the advance of accumulation (Mollenkopf, 1981). In a sense, of course, this is no more than a restatement of a widely observed social tension which has come to be expressed in such slogans as 'people before profits' or 'the costs of economic growth'. In the urban context, economic growth (accumulation) demands spatial change, which in turn impinges on people's communal space and consequently provokes political resistance. Community and neighbourhood must be restructured, sometimes obliterated altogether, before the next wave of accumulation can occur.

On the struggle between capital and labour within the neighbourhood, some of the most enlightening recent work is that by Cox (Cox, 1981; Cox and McCarthy, 1982). He begins by offering us an alternative definition of neighbourhood centred on class conflict, with capital and labour holding diametrically opposed views as to the proper use of neighbourhood space. For labour (i.e. local residents) that space is a set of resources – shops, schools, services, physical environment, friends, human contacts – absolutely essential to its own reproduction, both material and psychological. But for property owners, developers, finance houses, construction companies and eventual commercial land users it is a potential source of profit. Elsewhere this dualism has been conceptualized as local versus non-local land uses (Bunge and Bordessa, 1975; Hirst, 1980). Capital, in other words, values space as a commodity, whereas labour values it as a basis for home and community life. In pointing to this opposition between space-as-commodity and space-as-community, Cox (1981) is implicitly restating the classic Marxian dichotomy of use value / exchange value. In this context, use value is the non-monetary value which labour derives from occupying space, while exchange value is the monetary value which the capitalist market places upon that same space. Therefore development is the conversion of space from use value for labour to exchange value for capital.

However we conceptualize the process, it is usually one with negative effects for labour, one in which labour is forced to bear the costs of capital accumulation. Development acts to:

> Transform the spatial form of the city, substantially at labour's expense, and in a manner so spatially and temporally concentrated as to leave localized fractions of labour without a glimmer of doubt that development is impacting adversely upon them (Cox, 1981, 432).

Once again this reflects the basic capitalist contradiction between accumulation and reproduction (see Chapter 2). In order to accumulate, development capital must sweep aside the communal living space; but that living space, containing as it does housing and innumerable externalities necessary for workers' well-being, is the very means by which labour power is reproduced. And the reproduction of labour power is the *sine qua non* of continuing capital accumulation. From this it is evident that there are conflicts of interest within the capitalist class. The obliteration of neighbourhood and community is in the short-term interest of those fractions whose object is to realize the exchange value of land. This will be especially evident in inner-city areas abutting the Central Business District and in other accessible areas of 'potential' for commercial use.

In principle community and neighbourhood constitute formidable obstacles to capital at every possible level (Corrigan, 1975). At the material level capital accumulation may be hindered in various ways: by the locking up of valuable development land in the irrational and sub-optimal form of communal living space; by the persistence of neighbourhood mutual aid, informal non-market exchange directly opposed to the values and practices of the cash economy. Both Harvey (1978) and Cox (1981) note the importance of suburbanization as a means of converting people from 'neighbours' into atomized individuals fully integrated into mass-consumption lifestyles. Here the break-up of community opens up new opportunities for accumulation for the makers and sellers of consumer goods.

Politically too, the neighbourhood community can and very often has functioned as a building block for working-class resistance to capital (see Chapter 1). The 1984/5 miners' strike in Britain, notwithstanding its eventual outcome, was a very potent reminder of the effectiveness of communal and workplace bonds in supporting workplace struggles. Coal-mining settlements have long been regarded as archetypal models of workplace loyalty underpinned by local loyalty, the latter providing a source of material and political support. For all the media limelight directed at national leadership personalities, the wellspring of this strike was, as Samuel reminds us, local and communal:

> all the crucial initiatives came from below. This was the source of its peculiar energy. The real nerve centre was not the National Union of Mineworkers' headquarters . . . but the Miners' Welfare in the villages, curiously unvisited, or at any rate uncommented upon, by the industrial correspondents. . . . An even more spontaneous expression of local initiative . . . was the formation of Women's Action Groups, which transformed the strike from an industrial dispute into a communal act (Samuel, 1986, 14).

Without this material, political and moral support the strike would undoubtedly have collapsed much earlier.

The history of the labour movement in any advanced capitalist country would furnish countless examples of this supportive relationship. Quite simply, workplace solidarity is strongly linked to residential solidarity. Where workers share a common living space, an entire culture of resistance can arise in which there is massive overlap between social, political and recreational institutions. For Seabrook (1984, 4–5), the 'old industrial communities' were nothing less than political seedbeds 'out of which the great potential for change in the labour movement grew and which seemed to be on the edge of such spectacular success even as recently as 1945.' The seeds began germinating from the very beginning of the industrial city, as Corrigan (1975, 59) explains:

> Large numbers of workers and their families had begun living in close proximity to each other; each with the same troubles; each with the same interests. Not surprisingly they spoke to one another; shared their solution to problems; acted together; created a culture based on their material experience.

Happily for capital, however, all these coins have a reverse side. Materially, the traditional role of community within capitalism has been to provide a kind of hidden subsidy to capital. By providing one another on a mutual-aid basis with all manner of services now customarily supplied by the welfare state, working-class families and neighbours were actually bearing on their shoulders a portion of the cost of reproducing labour power. In the most poverty-striken areas, the existence of communal support effectively relieved capital of the need to pay a subsistence wage. Behind the romanticized images of the old-style working-class community lies an alternative reality – the exploitation of a human willingness to work for one another without monetary reward. Needless to say, these functions have been progressively usurped by the welfare state, but they still live on in such institutions as unpaid housewives' labour (Chapter 4), voluntary work and the charities. They also live on in many ethnic communities (see Chapter 5 on ethnicity as self-reproduction). Arguably such institutions have recently increased their contribution to self-reproduction, with the current drive to 'decollectivize' consumption leading to explicit attempts to resurrect community.

At the same time, also, the political interests of capital as a ruling class demand the retention of community and neighbourhood life or even, in certain circumstances, their expansion or revitalization. At the most basic level this is once again a question of social reproduction. The reproduction of labour power requires not only material consumption but also the satisfaction of emotional needs – what might be (pretentiously) called 'psychic' consumption. Family, neighbourhood and communal bonds are almost as essential as food and shelter in the reproduction of labour power. To eradicate them would destroy the means of emotional subsistence. The resulting alienation and social disintegration is the prime concern of such arch community advocates as Nisbet (1970), who sees the rebuilding of communal consciousness as a vital antidote to criminality, mental illness and many of the other scourges of our time. Yet it would be equally valid to invert this logic and assert that the ultimate beneficiary of community is capital rather than the population at large. Any institution which reproduces a contented apolitical labour forces clearly plays a major role in the maintenance of capitalist hegemony.

Viewing this problem historically, there are grounds for suggesting that the survival of community within mass urbanism has acted as a significant counter-revolutionary force, a channel through which much working-class energy has been siphoned off from political action. Without community, it is likely that the nineteenth-century rise of industrialism, with its transfer of population from a stable parochial society to rootless urbanism, would have created an unmanageable political crisis (Bell and Newby, 1976). Removed from feudal and aristocratic bondage and metamorphosed into a new urban proletariat, the masses posed a serious threat to the hegemony of capital.

Bell and Newby (1976) show a clear awareness of the link between geographical change and political change. Rural–urban migration and spatial concentration removes population from the strict controls of the old order – isolation in small self-contained villages, loyalty to locality, to landowner and to tightly defined community values – and reconstitutes it as a large dense mass, whose only unifying attribute is class, the common oppression of the propertyless. Old loyalties would therefore be replaced by a larger loyalty to a whole class, a revolutionary class-for-itself destined to overthrow capitalist rule. Somewhat paradoxically, this rather mechanistic view of humanity is not unreminiscent of that of Wirth (1938, 1939). Despite their opposed positions on the political spectrum, both Wirth and Marx are concerned with a transfer of loyalties from small to large institutions; in Wirth's (counter-revolutionary) case to the state, in Marx's (revolutionary) case to a class whose very purpose was to subvert that state.

In the event, working-class subversion has been directed not so much at the capitalist order as at the theories of Wirth and Marx. In its life outside work the new proletariat put down roots, and community life, while considerably less parochial than that of the old village, nevertheless reasserted itself along the lines of close-knit family/friendship networks, often contained within neighbourhood bounds. Far from emerging as a unified political force, the proletariat was once more separated from the political arena by community life and separated from itself by spatial fragmentation.

To elucidate this last point we turn once again to Hoggart's (1957) neighbourhood, a survivor of that generation of urban 'parishes' created by the Industrial Revolution. In Hoggart's view, subsequently echoed by Stedman Jones (1983) among others, working-class community embraces a complete culture, a system of values quite distinct from, and often diametrically opposed to, those of the dominant bourgeois culture. In this sense, working-class community life is a highly positive response to a highly negative structural position. Economically exploited and socially stigmatized they may have been, but manual workers refused to be culturally colonized. Far from existing as an impoverished dwarf version of national bourgeois culture, traditional working-class culture was self-defining, non-deferential, flourishing on its own terms (Stedman Jones 1983). Yet, at the same time as workers resisted cultural incorporation, so they resisted political involvement also. Instead of providing a means to a political end, the very strength of their culture provided a kind of end in itself. Deeply embedded in their values was a powerful puritanical ideology of 'putting up with things' (Hoggart, 1957, 91). 'When people feel they cannot do much . . . about their situation . . . they adopt attitudes towards that situation which allow them to have a liveable life under its shadow.' More recently this has been restated by Stedman Jones in his analysis of working-class culture in Victorian London: 'a working-class culture which showed itself staunchly impervious to middle-class attempts to guide it, but yet whose prevailing tone was not one of political combativity, but of an enclosed and defensive conservatism' (1983, 183). By the dawn of the Edwardian era institutions such as sport, the music-hall and the pub had assumed greater salience for the typical London worker than trade unionism or political debate. Clearly, in view of the arguments in Chapter I, allowance must be made for regional and local variations in class culture; but, nonetheless, some generalization is permissable. Whether 'stoical' like the Leeds workers or 'hedonistic' like the Cockneys, workers tended to treat home and community life as entirely separate from the more politicized world of work. Communal culture was defensive rather than offensive, a compensation for oppression in the workplace not a basis for challenging it.

This community-as-compensation effect is complemented by further constraints on class action. These include the neighbourhood itself, the fragmentation of the population into spatial compartments, each with its own parochial interests which, by definition, cut across class interests. Unlikely as it may seem, something very similar to the traditional rural localism noted by Bell and Newby (1976) has been re-erected in the decidedly non-feudal context of the modern city. This parochialism has been identified by Cox and McCarthy (1982) as a weakness inherent in neighbourhood political activism: because neighbourhood activism consists of co-residents defending their own 'turf' against commercial invasion, so the true class character of these conflicts is mystified. When, as so often happens, local activists are prepared to settle for displacing redevelopment to some other part of the city, the struggle becomes one between *places* rather than *classes*. Nothing could chime in more melodiously with the interests of capital and the state. Such particularism becomes even more marked when the state itself assumes responsibility for local 'community development':

> the selection of localities for community work often serves to isolate local groups from the broader class structure . . . interpreting the needs of local groups as not arising from their membership of the working class. Furthermore, locality-based work may promote sectarianism . . . if its 'parochial' achievements benefit active localized sections of the working class at the expense of other sections' (Hirst, 1980, 53; Townsend, 1979 has similar thoughts on the related subject of area deprivation policies).

Whatever the precise mechanisms of social control, the ruling class has rarely failed to appreciate the efficacy of community as political bromide. In Bell and Newby's (1976) view, capital and the state are even prepared to create artificial communities where necessary, as the history of model villages, garden cities and planned neighbourhood units will attest. This should be borne in mind during the ensuing discussion on community–state relationships.

6.5.3 Community and the State

It is now evident that if capital must remove community in the interests of accumulation then it must find some way of filling the resultant vacuum. If not, the system becomes unstable. This is precisely the kind of dilemma which the capitalist state has grown up to mediate, and over the past half century or more the various arms of the state have been exceedingly active in reshaping the geography of community. Here, as in other spheres of activity, the state has absorbed into itself the tensions of capital's own crisis, defraying many of the material, political and social costs which individual capital would otherwise incur in their drive to accumulate. In practical terms, capitalist urban redevelopment benefits from various kinds of state subsidization: for example public expenditure on transport and infrastructure essential to the conversion of urban land: and state responsibility for those residents displaced from the communal living space, a more hidden form of subsidy. Labour reproduction must go on but capital need not bear the immediate costs, preferring to delegate this responsibility to public housing programmes and state-supported owner-occupied housing (Chapter 2).

Over the past two decades there has been a sea-change in the state's declared stance on the community question, in Britain at least. In its earlier phase, the approach of policy-makers and executors was generally a negative one, rarely acknowledging that such institutions as neighbourhoods existed; or simply assuming, in line with mainstream social theory, that such pre-capitalist forms had no place in a modernizing society. Reaching its apogee in the first two post-war decades, this anti-communal ideology justified the truly enormous scale of city-centre redevelopment and inner-city clearance. Mention was occasionally made of the need to relocate people in 'balanced' neighbourhoods, and provision was actually made for this in the New Towns legislation. By and large, however, the concept of community was felt to be a dead duck.

Yet by 1969 the pendulum had swung so far that the very word 'community' – in the shape of the Community Development Projects – had become enshrined in a new area-based policy structure aimed at revitalizing the inner cities. The programme of comprehensive redevelopment, responsible for changing the face of British cities, had begun to meet more and more trenchant criticism from many quarters. Although much of this criticism was deterministic – suggesting that people were rendered inadequate by the simultaneous destruction of their communal ties and their part of the built environment – it became increasingly evident in social policy, featuring in state publications such as the Milner Holland Report on housing in London (Milner Holland Committee, 1965), the Plowden Report on primary education (Central Advisory Council for Education, 1967) and the Seebohm Report on the reform of the social services (Committee on Local Authority and Allied Personal Social Services, 1968). The idea was that 'pump-priming' activities could counteract the emergence of 'anti-social' attitudes, a notion developed from positive discrimination policies emerging in the United States in the early 1960s such as the Community Action Program, Operation Headstart and the Model Cities Program. Several British politicians and civil servants visited showpieces of the American urban programme and, with little appreciation of the policy's limitations and deficiencies, transported the ideas across the Atlantic. As we have seen, these ideas were already filtering through the policy arena when they were turned into action in 1968, the catalyst being Enoch Powell's notorious 'rivers of blood' speech, in which he predicted racial turmoil in the inner city and advocated the repatriation of Britain's black population. The response – 'a classic instance of Wilsonian decision-making: immediate, political, ultimately trivial and very reactive' (Lawless, 1979, 25) – was to set up a programme of Urban Aid, a limited central government fund allocated to local authorities and voluntary agencies for specific projects in selected inner cities. Despite, or perhaps because of, its restricted impact, from this seed of positive discrimination a whole bunch of policies began to bloom.

The adoption of the word 'community' came in the second, and perhaps most interesting, of these initiatives. In 1969 James Callaghan, then Home Secretary, announced the setting up of 12 experimental Community Development Projects lasting from 3–5 years. These projects were designed to improve the co-ordination and delivery of services and, more significantly, develop and tap community and individual self-help. Each project was to combine research and action, with detailed academic monitoring. The government's aims were clear, to seek pathological explanations of deprivation and impose 'acceptable' norms from above – indeed the *Objectives and Strategy* document issued to each team required them to collect information on personal care, family functioning, marriage rates and co-habitational stability! Initially the majority of the CDP teams did fulfil the state's requirements, setting up information and advice centres and liaising with the local authority, but they rapidly moved into advocacy – how to pressurize the service agency, how to assert one's rights. From this several of the more radical teams began to explain poverty and deprivation in terms of the economic and political structures within which individuals operate. From the state's point of view this exercise in community management went progressively further off the rails, increasing working-class political consciousness and developing a strong community commitment to make meaningful change – 'conflict theory in practice' (Lawless, 1979, 112). This in turn often led to disillusionment, a sense of futility and resignation among the local population, as an awareness of their place in the wider socio-economic order was reinforced and their desire for change was frustrated at every turn (see Higgins *et al.*, 1983, 12–46, for a full review).

While the scope and magnitude of the official community development effort has changed considerably since the 1960s, the CDPs still offer valuable lessons. On the face of it the CDPs seemed a positive attempt to release the human potential locked up in local communities by involving the residents and encouraging them to help themselves (Loney, 1983, 56). Yet almost from the outset the CDP approach was riddled with tension and ambiguities. The very designation of CDP areas, with its emphasis on 'deprived' neighbourhoods, itself carried the apparent admission that 'community' was only relevant as a kind of last resort for those with few resources and few alternatives. Allied to this was the initial guiding principle that the ills of these areas were largely internal in origin, self-generated by a 'culture of poverty' which was transmitted from generation to generation in a dreary cycle of low skills, low income, poor housing, inadequate child socialization, delinquency and educational underachievement. Herein lies the justification for attacking local conditions within the local area.

In retrospect it is hardly surprising that this approach attracted adverse criticism, though it is ironic that much of this emanated from within, from the CDPs own research workers (CDP, 1977; Holman, 1978; Loney, 1983). Quite simply the 'local pathology' explanation assumes a degree of autonomy which local neighbourhoods do not possess. The notion that local communities generate their own problems rests on the unrealistic assumption that urban neighbourhoods are enclosed spaces insulated from the wider world. As we have seen, urban communities of the poor and deprived do in some senses live encapsulated lives. Yet this cannot protect them from wider economic forces. At the critical level of determination, conditions within these urban local spaces are decided by processes operating at the urban, national and even global scales. The neighbourhood groups addressed by the CDPs were made up to a very large extent of sections of the population marginalized by various macro-level processes; the chronic unemployed excluded by rationalization and industrial relocation; the slum dwellers barred from decent housing by the managed scarcity of the capitalist market; blacks and immigrants specifically excluded by racism. On top of all this, these were communities depopulated and isolated in an environment ravaged by the processes of urban accumulation described in this chapter.

Immediately we might conclude that macro-level crises demand macro-level solutions and that community as a policy instrument is an irrelevance. Nonetheless the community bandwagon has continued to roll, gathering official credibility and a labyrinthine proliferation of programmes, projects and schemes in a variety of guises (see Table 6.3). Community had become bureaucratized, a visible contradiction in terms, since how can gemeinschaft be 'gesellified'? It is difficult to escape the conclusion that this bandwagon does little more than whitewash the consequences of economic decline and social inequality.

Yet there remains a glimmer of hope that, even within the present system, the balance between state and community (or between capital, state and labour in the communal living space) might be tipped less grotesquely in favour of the former. Although Blunkett and Green (1983, 1) see the current Conservative government in Britain as dedicated 'to destroy the working-class tradition of "collective" or "community" approaches to organizing social well-being – a charge to which past Labour governments are not immune – they argue that local government can, within constraints, adopt a different framework. This has been demonstrated in Sheffield City Council's own *Jobs Audit* (1985), which highlights how the city has attempted to exert some control over the means of production, distribution and exchange, theoretically 'putting the resources so created at the disposal of the community' (Blunkett and Green, 1983, 5). Although the paternalistic traditions of municipal socialism are far from extinct and individual officers are often resistant to change, many urban authorities have demonstrated a commitment to actually involve the community (Boddy and Fudge, 1984). The extent to which these laudable aims will be frustrated by bureaucratic and financial constraints remains to be fully explored (Raban, 1986). Nor, despite Blunkett and Green's (1983) comments, should we assume that the development of communal ties is attractive only to the political left. The financial savings made by devolving labour-intensive care and support on to the family, the community and the voluntary sector have not gone unnoticed amongst Conservative politicians and form an increasing element of Tory rhetoric.

Despite the *caveats* outlined above, we can perhaps conclude that the local state is, in some instances at least, beginning to respond more positively. A greater awareness of the needs and demands of neighbourhoods is being fostered, encouraging 'authority' to acknowledge that local residents are more than simply passive consumers of bureaucratically allocated rewards. Certain decision-making processes are being decentralized, involving consultation and even participation. But does this represent a radical shift in the balance of forces?

This question is best examined using one example of the dilemma of state–community interaction. Drawing from his experience as a Strathclyde community development manager, Worthington (1982) compares the cost of two delinquency assessment centres – astronomical at £8 millions capital outlay and £1½ millions per annum recurrent costs – with that of running local summer play schemes using volunteers supported by professional community workers. Not only was the latter method so cheap as to bear no comparison with the former, it also was seen to be effective in preventing crime. Worthington also deals with parallel initiatives in community health and education, reaching simi-

Table 6.3 Selected area-based initiatives, 1968 on

Initiative	Year Introduced	Comments
a. The Urban Experiments 1968–74		
i. MAJOR RESOURCE ALLOCATING PROJECTS		
Urban Aid The Urban Programme)	1968	Grants paid through Local Authorities in response to special social needs
Educational Priority Areas	1968	Additional resources for staff, preschooling facilities, community links in deprived areas
ii. IMPROVING THE CO-ORDINATION AND MANAGEMENT OF SERVICES		
Urban Guidelines	1972	Encourage Local Authorities to develop a 'total approach' to the problems of the urban environment
Area Management Trials	1972	Evaluate the potential of decentralised administration for alleviating deprivation and improving service delivery
Urban Development Unit (Home Office)	1973	Examine deprivation and co-ordinate action projects aimed at amelioration
Comprehensive Community Programme	1974	Co-ordinate different tiers and agencies of government
iii. INVESTIGATIVE PROJECTS		
Community Development Projects	1969	Neighbourhood-based experiments aimed at improving the coordination and delivery of services and facilitating individual and community self-help
Inner Area Studies	1972	Establish and evaluate specific projects; suggest ways in which deprived areas may benefit from new powers and resources
b. Local: Central Co-Operation and Partnership		
GEAR (Glasgow Eastern Area Renewal)	1976	Total approach to redevelopment involving regional and district councils, the Scottish Development Agency, Scottish Special Housing Association
Partnerships	1977	'Total approach' to 7 partnership areas, coordinated by central and local government. Progressively increasing economic bias
c. The Conservatives and the Inner Urban Economy, 1979 on		
Enterprise Zones	1980	Deregulation of selected areas to encourage enterprise (rate relief, relaxed planning controls etc.). Concept developed and extended in *Freeports*, areas in which the import/export and reprocessing of goods is allowed free of all taxation
Urban Development Corporations	1980	Based on New Town Development Corporations; wide powers to acquire, develop and sell/lease land and provide infrastructure. Partnership dominated by central government and private sector interests. Initially established in London Docklands and on Merseyside, recently extended to seven other locations.

Note: a. the progressive switch from community/social development initiatives to economic initiatives
b. the centralization of control away from local authorities (especially since 1979) and the developing role for the private sector
For further details of the initiatives listed above see Lawless 1979, 1986.

lar conclusions on their cost-effectiveness and efficiency. These findings seem to support Crenson's (1983) more general thesis that 'informal governance' at the neighbourhood level is now an essential supplement to the formal administration, without which the latter would be placed under intolerable strain.

Not unnaturally Worthington is enthusiastic about the Clydeside experience: 'In cost–benefit terms, the gains are enormous and the outlay tiny. If this is right, our community or society can nowadays ill afford to ignore the message' (1982, 149). In many ways this enthusiasm is justified. Common humanity demands that we applaud the selfless dedication of local workers, the apparent triumph of small over big, personal over impersonal, and the challenge to the rule of the expert. But we should pause before we hail urban neighbourhood development as an amalgam of Gandhi, Schumacher and Kropotkin. Let us consider the phrase 'cost–benefit'. For the state (and every other interest dedicated to maintaining the status quo), community development is a remarkably cheap (and effective) way of purchasing social order. It is cheap because, just as in the old-style neighbourhood, labour contributes to the cost of its own welfare – at least to the extent that schemes for community education, community health and self-policing are run by unwaged labour. Were these schemes costed on the basis of a proper living wage for all participants, then the discrepancy between institutional and communal provision would presumably be less spectacular.

For all their virtues then, it seems that projects of this type are consistent with the New Right principle of reducing collective consumption and returning to labour as much of the cost of its own reproduction as it will bear. As Saunders (1984) has implied, this involves the creation of a two-tier system. For the regularly employed majority the solution is privatization, the return of consumption to the market place. For the marginalized minority, sustained at declining levels by the waning welfare state, community provision is one of the alternatives. Here the true value of community and neighbourhood for the state probably lies in the ideological work they can do. As symbols of 'grassroots power', 'bottom-upwards development', 'small is beautiful', 'self-improvement' and other spiritually uplifting virtues they defy us to criticize them. To ask who ultimately benefits and at what cost to whom is to court charges of churlishness.

This has not stopped various writers (notably Stacey, 1969) from attempting to strip the term 'community' of its emotive content, nor others from portraying community development as a threat to working-class political integrity. While Corrigan (1975, 57) may appear extreme, even unkind, in arguing that 'the two major symbols of control' in capitalist society are 'the tank' and 'the community worker', there is yet a definite sense in which community development acts as a means of ideological control. It is tangible 'proof' that 'society cares'. In most respects pacification is less costly and more effective than repression and it is particularly important at the present time to win the consent of workers for the cuts in collective consumption which are being imposed upon them. With its emphasis on cost effectiveness, cheap (Training Agency) and voluntary labour, the community approach is in reality part of the cuts package. The rhetoric of grassroots care effectively disguises this, rendering loss as gain, decay as rebirth.

Recommended Reading

Much of the research on social interaction can be traced back to the seminal work of Elizabeth Bott, in particular *Family and Social Networks* (Tavistock 1957). We would add to this Richard Hoggart's realistic portrayal of life in post-war Leeds, *The Uses of Literacy* (Chatto & Windus 1957). More recently Bell and Newby's *Community Studies* (Allen & Unwin 1971) remains recommended reading, and the same authors contributed a brief chapter to Herbert and Johnston's edited collection *Social Areas in Cities Vol. 2* (Wiley 1976, 189–208). More substantial and more generalized, but still written with an empathy for the community studies tradition, is Lee and Newby's introductory textbook *The Problem of Sociology* (Hutchinson 1983). For a critical review of what she terms a 'non-concept', see Margaret Stacey's brief article in the *British Journal of Sociology* (1969, 134–46), 'The Myth of Community Studies', a good starting point for arguing the extent to which the term has been misused and abused. Richard Sennett takes this much further, claiming that the obsession with community leads us to devalue the macro-level forces which increasingly shape our society and distracts our attention from them. (*The Fall of Public Man*, Cambridge University Press 1977).

For geographers, a sound introduction can be found in Paul Knox's *Urban Social Geography* (Longman 1987, 55–96), while the behavioural/humanistic perspectives of David Ley (*A Social Geography of the City* (Harper & Row 1983)) may make stimulating reading. Andrew Blowers's Open University unit (*The Neighbourhood: Exploration of a Concept* (DT201, Unit 7, 1973) still seems a concise and well-organized basis for further study, while at the other end of the spectrum Michael Peter Smith's *The City and Social Theory* (Blackwell 1980) is decidedly challenging. On 'community as policy', selective use of Paul Lawless's two books, *Urban Deprivation and Government Initiative* (Faber & Faber 1979) and *The Evolution of Spatial Policy* (Pion 1986) is strongly recommended.

7
Contemporary Rural Society

7.1 Introduction: Nymphs and Shepherds

The critical probing of received wisdom has been an important aim of this book. Nowhere is this more necessary than in the study of the rural sphere, an area in which popular imagery is rife with stereotypes which do little service either to the country dweller or to the urbanite looking in on country life. In Britain, as in most other heavily urbanized countries, rural society has been subject to a process of cultural colonization in that the dominant images of rural life have been formulated by (middle-class) urbanites and projected on to the countryside. Just as in the process of Third World colonization, all cultural definitions are those of the colonizer. Popular attitudes have also been satisfied by official practice, as observed by Shaw (1979, 3) who notes that 'All of the early post-war planning initiatives were based on an urban viewpoint and even the planning of recreation, introduced by the National Parks Act 1949, was geared to the needs of city dwellers.' This theme of urban dominance: rural dependence, extending beyond the cultural to the economic and political realms, is one which constantly recurs throughout history and will constantly recur throughout this chapter.

Despite the pervasiveness of the dominant myths there has been no shortage in recent years of writers willing to tilt at them. One of the most persuasive of these has been Newby (1977; 1979), whose line of attack has received support from writers such as Pahl, 1968; Ambrose, 1974 and the contributors to Shaw, 1979. In a review of the English literary tradition on the subject, Newby (1977) identifies two interlocking images which together underpin the romanticized sentimental urban-biased definition of the countryside. The first of these is the 'Good Old Days' syndrome, the notion of a vanished Golden Age of yesteryear.

> One of the best examples can be seen in the works of Cobbett. In his critique of industrialism he used the opportunity to conjure up a rural arcadia peopled by merrie rustics and sturdy beef-eating yeomanry. The repressions and privations of old England were forgotten in a welter of nostalgia for the mythical lost paternalistic community' (Newby, 1977, 14; see also Williams, 1973).

The second related myth, in truth a corollary of the first, is that of the 'Good Life' or 'Rural Idyll', the feeling that despite change and modernization the countryside still retains sufficient of its ancient virtues to stand as a symbol of all that is best in the English character. '*Real* England has never been represented by the town but by the village, and the English countryside has been converted into a vast arcadian rural idyll' (Newby, 1977, 12). This attitude, even though it finds its apogee in English romantic literature and popular culture, has never been an exclusively English trait; it also finds a voice in the writings of many non-English classical social theorists, who have attributed many of the 'pathologies' of modern society to the decline of the rural community and its close-knit (allegedly) harmonious social bonds (see Tonnies, 1887 and the related discussion in Chapter 6). Implicitly we are given to understand that there is something unnatural about urban living, a feeling nowhere better expressed than in Rousseau's motto, 'Men are not meant to be crowded together in ant-hills but scattered over the earth to till it.'

The above sentiments show the way in which popular wisdom has historically been underpinned by much 'authoritative' wisdom, the latter reinforcing the former in its tendency firstly to reduce rural and urban to two simple categories each the antithesis of the other, and secondly to impute virtue to the one and vice to the other. At the same time, however, there have always been counter-images. In his own plea for an injection of realism into the rural question, Newby's (1977) central point is that rural class exploitation persists even now, centuries after feudalism, but goes largely unrecognized, obscured by a pink-tinged fog of pastoral imagery. Even though the farm labourer (or rural factory worker) may be grossly underpaid and overworked (Thomas and Winyard, 1979), this is offset by the very privilege of living in the countryside, a residential status coveted by thousands of frustrated suburbanites: 'metaphysical rewards have been deemed to be adequate compensation for his labour' (Newby, 1977, 15).

It would, however, be unfair to accuse the most recent generation of rural geographers and sociologists of lacking realism. Over the past three decades, rural social science has progressed away from (though not totally abandoned) its former preoccupation with 'traditional society' and with landscape, land resources and physical planning towards a new concern with social change in the countryside. This shifting focus is usefully summarized by Shaw (1979) and by Lewis (1983). One effect has been to force the question of *rural deprivation* on to the agenda (Shaw, 1979; Cooper, 1981; Knox and Cottam, 1981), recognizing at last that poverty, disadvantage, exclusion and indeed conflict over social resources are not solely confined to urban locations. The difference between urban and rural societies is chiefly one of visibility (Thomas and Winyard, 1979). Urban problems are conspicuous because they are spatially concentrated, because their victims tend occasionally to protest against them and because social deprivation is often visually reinforced by its hideous despoliated physical surroundings – decaying inner cities, badly planned and vandalized council estates. By contrast rural deprivation is concealed by population dispersal and by picturesque landscapes (where these have not been obliterated by 'prairie farming'). Penetrating beneath this veneer, a spate of recent studies have highlighted the serious (though regionally variable) nature of rural disadvantage, whether expressed as low incomes (Brown and Winyard, 1975; Thomas and Winyard, 1979; see also Newby, 1977; 1979 on farmworkers' wages), inadequate and distorted employment opportunities (Gilg, 1976; Packman, 1979; Cooper, 1981; Wright, 1983), housing stress (Rogers, 1976; 1983; Larkin, 1979; Newby, 1979), transport and accessibility problems (Moseley, 1979a; Coles, 1986; Hull, 1985), or simply an unsatisfactory general quality of life (Cooper, 1981; Knox and Cottam, 1981).

Despite this new awareness, the urge to seek 'metaphysical' advantages which might compensate for the material poverty of traditional rural life dies hard. Here the 'quality of life' approach adopted by recent researchers (for example

Knox and Cottam, 1981) poses serious problems, since it places considerable faith in the dictum that deprivation is ultimately definable only by the subject him/herself (see Lewis, 1983 for comments). Accordingly, we should make no judgement about social well-being without first taking account of 'views of the issue as seen through the eyes of the beholders' (Knox and Cottam, 1981, 45). For all its apparent sensitivity, such an argument would, at its most extreme, dismiss poverty – as measured by 'objective' yardsticks – as hardly mattering, providing its victim is unaware of it or expresses general satisfaction with life. Yet, since stoicism, endurance and lack of organized vocal protest are almost definitive characteristics of the rural poor, it is hardly surprising that empirical research finds ample evidence of 'satisfaction' with the quality of rural life. For example, an exhaustive study of two Scottish Highland villages found very high levels of *expressed* satisfaction with jobs, housing and social provision despite the fact that each of these communities contained inordinately large proportions of underprivileged households as measured by conventional indicators (Knox and Cottam, 1981).

While the above authors certainly do not seek to underplay the hardships suffered by their respondents, there is a clear danger that their findings might be used as ballast for the 'pastoral bliss' myth. This myth bears a striking resemblance to other pearls of reactionary wisdom such as 'the poor like living in slums' or 'blacks prefer ghettos'. Great care must be taken in assessing the self-judgements of people who lack any realistic alternatives to their present disadvantage and are therefore obliged to acquiesce in its inevitability. In such circumstances there may be a wilful refusal to admit one's dissatisfaction to an outsider – 'What people *believe* to be in their interests may not be what they *state* to be in their interests' (Rose *et al.*, 1979, 14) – or even to oneself.

Despite the recentness of the rural poverty debate, rural deprivation is as old as the communities themselves: 'rural people have always suffered hardships of some kind or another' (Lewis, 1983, 158). But as the following sections demonstrate, the nature and origins of this hardship have changed. For clarity we may express these changes crudely as a contrast between pre-modern and modern rural communities, dispensing for the moment with such subtleties as regional and local variations, the distinction between 'open' and 'closed' villages and the problem of historical overlap. Broadly speaking we may assert that in the pre-modern village poverty was largely internally generated, stemming from the exploitation of labourer and/or small tenant by large landowner. During and since the Industrial Revolution, however, rural affairs have become increasingly determined from outside as formerly self-contained communities have become progressively incorporated into a wider social and spatial system. Since that system is urban-dominated, we should therefore understand modern rural conditions as stemming largely from subordination to urban interests, not least those of the urbanites who have flocked to take up rural residence. This is not to say that all traces of the traditional land-owning hegemony have been obliterated; rather they now take a radically different form, with new alliances and new forms of conflict superimposed upon them. We now trace the development of these urban–rural spatial linkages and assess their consequences both practical and theoretical.

7.2 The 'Traditional' Rural Community

In geographical terms, the key feature of the European village in pre-industrial times was *spatial isolation*, a high degree of geographical separation from other settlements, entirely to be expected in a low-technology, low-mobility society (Ambrose, 1974). Economically, this could only mean a high level of self-sufficiency, each settlement functioning as a comparatively self-contained unit, largely reliant on its own resources and undertaking only a limited amount of localized agricultural exchange through the medium of a network of market towns.

The human relationships which evolve within this circumscribed geographical setting have provided grist for several generations of social researchers, anthropologists in particular (Arensberg and Kimball, 1949; Rees, 1950; Williams, 1964; Frankenberg, 1966; Jenkins, 1971; Cohen, 1982; see Newby (1986, 210–11) for a brief but telling critique). Much of the early work in this tradition was concerned with remote communities in which the old way of life had survived relatively unscathed and it did little to counter the idealized notion of the traditional community as an orderly harmonious organism. Such social harmony might be thought to derive from two major factors: *small population size*, with all community members mutually acquainted and commonly bound by ties of kinship, and *social homogeneity*, with most members sharing in the same basic work – agriculture and related services – and usually participating in shared cultural and social activities. Interdependence, common values, face-to-face contact, continuity, stability and tranquillity are the dominant motifs, with conflict largely confined to personal feuding.

Having said this, however, it would be a disservice to the community studies tradition to accuse its proponents of projecting a sentimentalized view of the countryside or of being unaware of more profound forms of conflict. Certainly Frankenburg devotes considerable space to the question of class conflict, while making the important point that in many rural areas such conflict takes place 'off-stage' (1966, 257). More recently this theme has been taken up again by Emmett:

> If class appears peripheral to the main business at hand in some useful community studies it is partly because the ruling class experienced by the majority as powerful, exploiting, ruling, is often distant and so invisible and faceless' (1982, 171).

Yet, while this tendency to direct attention at *intra*-class relations within confined localities is understandable (and valuable in its insights), it risks portraying the lives of ordinary village people as encapsulated, autonomous and unaffected by influences from above and outside. It is easy to exaggerate the extent to which the pre-industrial village community was unchanging and untouched from outside; and to forget the extent to which social order was a one-sided affair. There is nothing particularly virtuous about an imposed social harmony. As Newby (1979, 154) reminds us, 'The village inhabitants formed a community because they had to: they were imprisoned by constraints of various kinds, including poverty, so that reciprocal aid became a necessity.'

Reference to the more radical social historians whose gaze has encompassed centuries of inequality, repression, criminality, civil disorder and political protest in the English countryside enables us to place the entire question in a more satisfactory context (Thompson, 1963; Hobsbawm, 1971; Williams, 1973; Pearson, 1978; Charlesworth, 1983). The definitive characteristic of the traditional (feudal or quasi-feudal) rural order was a rigid social hierarchy, within which the surplus product of the masses was expropriated by an hereditary land-owning elite: Newby (1979) writes of the chronic poverty and occasionally cruel exploitation of the village labourer, a condition which persisted into the Victorian age and beyond. Linked to this was paternalism. Order was maintained not so much through coercion as through the acquiescence of the masses in their subordination, their acceptance of it as legitimate. As a reward for this deference and for services rendered, the lower orders received the protection of their lord. This form of social control was facilitated by the geography of settlement, the fragmentation of the population into small settlements isolated from one another and from external influences and containing the villager within a narrow set of horizons with no access to alternative standards on which to base expectations. This use of 'localism' as a means of social control is graphically described by Bell and Newby (1971).

History rarely comes neatly packaged and many of the institutions of the old order persisted long after the demise of feudalism and even after the onset of industrialization. Well into the nineteenth century, Ambrose's Sussex village was still a markedly hierarchical society, containing at least five distinguishable social strata, each demanding and receiving its standard ration of deference from those below (Ambrose, 1974). Added to this were subtle nuances within each stratum. For example, the consensus was that the publican was a lower life-form than other tradesmen. Newby (1977) describes a not dissimilar situation in nineteenth-century East Anglia and further suggests that certain relics of the *ancien régime* persist into the present, not least in the farmworker's acquiescence, accommodation and overt recognition of social 'superiors'.

To some extent the above observations are peculiar to lowland England. Given the rich regional variations of the British Isles, it would seem to be geographical heresy to build generalizations based on a single part of the country. In the remoter areas of upland Britain, with their dispersed settlement patterns, their livestock/grassland emphasis, their domination by small farms and, in some cases, their Celtic culture, the traditional social configuration might be expected to be quite distinct from that of the commercially-oriented arable lowland farming areas. It is noteworthy that the social distance between farmer and worker was much less in highland than in lowland Britain (Newby, 1979), especially where the labour 'lived in' on the farm and, as in Devon, was considered part of the family (Bouquet, 1982). On this basis, it is tempting to picture such rural areas as South-West England and Wales as virtually classless societies, as in the Welsh concept of *gwerin* or 'folk' (Williams, 1985), whose backbone was the small farmer working on a virtually equal basis with his employees.

Closer examination, however, refutes this. For all their 'sturdy independence', small farmers comprised an essentially subordinate class stratum (a 'lumpen bourgeoisie' in Wright Mills's (1956) apt term), often being tenants of large landowners and economically marginal. In nineteenth-century Devon, for example, the apparent social closeness of many farmers and workers masked a highly stratified society dominated by very large estates (Bouquet, 1982). Hoskins (1964, 242) speaks of the 'high importance of the squirearchy in Devon' and of 'the small farmers on their dirty isolated farms, hardly better off than the labourers who worked for the bigger men'. Under such conditions, as Wright Mills (1956) has stressed in another context, the tenant farmer has no option but to exploit himself, his family and his hired labour in order to secure survival. On the resultant condition of the Devon farm labourer, Hoskins (1964, 99) writes of:

> the degradation of a whole social class . . . able-bodied well-conducted men bringing up a family on a wage of eight shillings a week, sometimes only seven shillings, with an allowance of cider so sour that no one else would drink it, and no

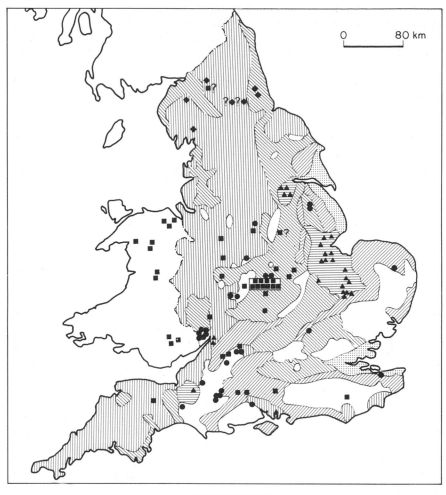

LAND PROTESTS

▲	Protests concerned with draining of fen or marsh
●	Protests concerned with forest or woodland
■	Protests concerned with enclosure or clearance
▫	Villages involved in the 1596 attempted rising
◆	Protests concerned with tenure disputes
P	Attacks on deer parks or emparkments
▨	Destruction of property titles (1640-49)
⊠	Camps (1549)

Farming regions 1500-1640

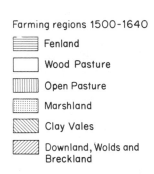

Fenland

Wood Pasture

Open Pasture

Marshland

Clay Vales

Downland, Wolds and Breckland

■? Approximate location, or protests known to be in that county but exact location unknown

Figure 7.1 Land protests in England and Wales, 1608–39.
Source: **Charlesworth, A. 1983**

privileges whatsoever in addition to their money wages. Wives were often employed as semi-slave labour as a condition of giving the husband work.

Essentially, then, in the matter of class oppression the difference between the British regions was one of style rather than content.

Even this picture of lop-sided predatory social relationships exaggerates the degree to which the pre-industrial village was a static orderly community functioning on the passivity of the masses. In Britain disruptive change had been proceeding from as early as the fifteenth century. For the rural labourer the commercialization of agriculture was often a traumatic process, which 'led the great landlords to evict tenant farmers and enclose common lands where peasants had previously grazed livestock . . . as peasants were "marginalized", that is, made marginal to the regular economy, they turned to crime' (Greenberg, 1981, 61; see also Thompson, 1963). They also turned in a more organized way to civil disobedience and riot. The frequency and scale of such violent resistance, as documented in Charlesworth, 1983, was such as to suggest that upheaval and open conflict were at least as typical of traditional rural life as harmony and tranquillity (Figure 7.1).

7.3 The Transformation of Rural Society

Whatever our view of social relationships in the pre-industrial village, there is little doubt that these have been subsequently erased or substantially modified by industrialization. As is now well documented, especially by Saville (1957), the primary effect of industrialization was to transform the economic base of rural life and in particular to undermine the role of agriculture as a source of employment and self-employment. This is a basic trend common to every advanced economy, although subject to great inter-regional and international variations in pace and timing, with Britain inevitably the first to feel its effects. In all these economies, the expansion of urban population created a burgeoning demand for agricultural produce but from the mid-nineteenth century onwards this was increasingly supplied by the application of capital and technology (for example, mechanization, selective breeding, the use of fertilizers) rather than by increased labour input. Indeed so readily was agricultural labour supplanted by capital that by the last quarter of the nineteenth century in Britain the absolute number of farm jobs had begun to decline and the rural population with it, with growing numbers of workers now migrating to the expanding urban-industrial sectors of the economy. Rural areas had lost their indigenous capacity to support their populations and had been transformed instead into vast cheap labour reservoirs for urban growth.

In England and Wales, rural depopulation emerged first as a *relative* loss in relation to urban areas and later as an *absolute* decline. The downward trend in agricultural employment was the primary cause of rural–urban migration. This process was, as Saville (1957) reminds us, reinforced by a linked decline in the 'secondary rural' population, the craftsmen, small manufacturers and service providers whose entire *raison d'être* was progressively undermined by the disappearance of the local agricultural workers who had formed their market (Stamp, 1949). We might perhaps see this as a case of cumulative causation – what Wallace and Drudy (1975) describe as the vicious circle of rural depopulation. On a broader spatial level rural depopulation is a logical outcome of several processes inherent in the growth of industrial capitalism (Lewis, 1979,

Table 7.1 A) Changes in the agricultural labour force 1841–1986: Great Britain and four selected regions.

	Labour force in agriculture: % of total + LQ									
	1841		1881		1921		1961		1986	
	%	LQ	%	LQ	%	LQ	%	LQ	%	LQ
South-East	19.4	0.87	9.7	0.77	5.2	0.71	2.1	0.56	0.9	0.60
East Anglia	41.4	1.85	33.3	2.66	28.0	3.83	14.3	3.86	4.5	3.00
South West	28.5	1.28	21.0	1.68	15.6	2.13	7.7	2.08	2.8	1.86
North West	9.0	0.40	4.6	0.37	2.7	0.36	1.6	0.43	0.7	0.46
GREAT BRITAIN	22.2	1.0	12.5	1.0	7.3	1.0	3.7	1.0	1.5	1.0

Changing numbers in Agriculture.

Number in 1841 ('000)	Percentage Change				Number in 1986 ('000)
	1841–81	1881–1921	1921–61	1961–86	
318	+ 3.7	− 23.3	− 37.1	− 57.2	68
128	+ 10.9	− 14.1	− 31.1	− 59.5	34
218	+ 2.7	− 25.0	− 28.9	− 61.1	44
77	+ 13.0	− 18.4	− 33.8	− 68.1	15
1526	+ 4.9	− 13.0	− 38.2	− 63.7	310

Source: Lee C. 1979: *British Regional Employment Statistics 1841–1971*. Cambridge: Cambridge University Press. Department of Employment 1987: *Employment Gazette*. London: Department of Employment.

Table 7.1 B) Date when agricultural employment overtaken by manufacturing employment: countries in Western Europe.

Before 1850	1880–1920	1920–1950	1950–1970	Since 1970
Great Britain	Germany Belgium Switzerland	Sweden Netherlands Austria	France Italy Norway Finland	Ireland Spain Portugal

Source: Mitchell B.R. 1975: *European Historical Statistics 1750–1970*. London: Macmillan.

21–7): the geographical centralization of all levels of the national economy; the growing dependency of the rural sector on the urban sector, its source of markets, technology, capital and decision-making; the increasing ease of movement and the consequent interdependence between urban and rural sectors.

Saville's (1957) work was one of several contemporaneous studies and reports which reflected a concern with rural depopulation and the decline of rural community which was supposedly a product of it. By this time; the question of agricultural and ancillary employment, though still central, had also to be seen alongside social and cultural deprivation. In a post-war society of rising expectations and greater social awareness, the words of the music hall song, 'How you gonna keep 'em down on the farm, now that they've seen Paree', came to have a figurative if not a literal truth. Young rural workers were no longer simply reacting to a contracting labour market: their migration was also an escape from housing market constraints (see 7.5.1); from the paucity and inaccessibility of social amenities; from a lack of personal mobility, exacerbated by the contraction of country transport services (St John Thomas, 1960); and even from a sense of inferiority, a recognition of the low status accorded to rural workers in the eyes of the outside world (Bracey, 1952; 1958; 1959; House, 1956; 1965; Wibberley, 1954; Sheppard, 1962).

Despite their optimistic prescriptions for rural rebirth, it is now clear that these studies were little more than funeral orations on the death of the old order. Decades of depopulation had reduced the British rural population to a residuum: the pre-industrial rural–urban population ratio of 4:1 had been completely reversed. The rural labour reservoir had been substantially drained, so much so that rural–urban migration would no longer play any significant role in the national space economy. If there was to be an economic and demographic resurgence in the English village, it would come from people and activities not normally regarded as rural: external forces having no direct links with the soil or with local territory (Ambrose, 1974).

At this point we enter an area of great confusion, hingeing on the apparently simple but actually tortuous question of what is rural and what is urban. Indeed the difficulties are so acute that 'even the most recent advanced urban geography textbooks do not attempt a precise definition of their field of study' (Lewis, 1979, 22). Although the entire question has been dismissed as virtually meaningless by some writers (notably Pahl, 1968), it has nonetheless triggered extensive debate and is central to any analysis of rural space. We may illustrate the practical difficulties by reference to post-war British population trends. It is now evident that the rural population reached its nadir in the 1930s. At the national level, each successive post-war census has recorded a definite shift from urban to rural districts, a gradual but decisive reversal of the time-honoured trend (Lewis, 1983, 150). This apparent rural repopulation has led to the coining of the term counter-urbanization (Hall *et al.*, 1973), a label which suggests a definite reaction against the growth of the city and a positive demographic and economic decentralization. Counter-urbanization is seen as a key spatial component of the post-industrial society, wherein constraints on industrial and residential location are eased and a concentrated space-economy is no longer necessary. (See Todd, 1983 for highly critical comments on the counter-urbanization hypothesis.)

Whatever the merits of the counter-urbanization thesis – and there is no doubt that decentralization is occurring (Cloke, 1985; see also Brown and

Wardell, 1981 and Fielding, 1982 for international comparisons) – we cannot conclude that any return to small-scale settlement must be accompanied by a return to the 'rural', whether defined by values, relationships, institutions or economic functions. On the contrary, repopulation may simply conceal a further erosion of traditional rural population and society. This has been evident for several decades. In an unpublished survey of 1950s population trends in 45 South Devon parishes, Jones (1968) found that a considerable proportion of population increase was generated by the urban sector, both national and local, and consisted of long-distance retirement migration, most notably from London and the West Midlands, and short-distance outflow of commuters from the local industrial and resort centres. In many parishes this was more than sufficient to outweigh the persistent outflow of ruralites displaced by agricultural mechanization, farm amalgamation and the erosion of traditional local activities such as quarrying and small manufacturing.

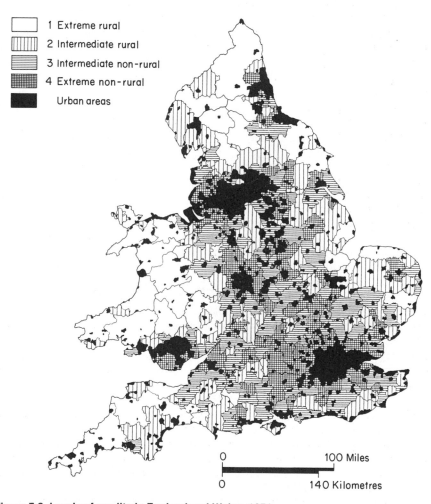

1 Extreme rural

2 Intermediate rural

3 Intermediate non-rural

4 Extreme non-rural

Urban areas

0   100 Miles

0   140 Kilometres

Figure 7.2 Levels of rurality in England and Wales, 1971.
Source: Cloke, P., 1977, 44.

The above findings have been substantially replicated by several authors, though the weight attached to the various components of population change differs according to locality (Figure 7.2). In the Home Counties for example, the dominant urban–rural outflow tends to be of commuters rather than retirees (Pahl, 1965b). This exodus of middle-class commuters from London is no recent event. It became significant very early in the present century with the spread of the rail network, accelerated rapidly in the 1930s and had swelled to major proportions by the 1950s as car ownership increased (Newby, 1979). By contrast in more distant upland regions such as North and West Wales second-home ownership bulks large (Coppock, 1977; Davies and O'Farrell, 1981). Irrespective of the source of repopulation, it would be more helpful to regard the process as a form of long-distance urbanization than as a rural resurrection *per se*. Increased affluence and mobility has allowed the suburb to detach itself from the parent city and leapfrog, sometimes to the far-flung rural periphery (Todd, 1983). In doing so, it usurps the space formerly occupied by those engaged in local land-based economic activity.

For the sake of accuracy we should note that rural repopulation has not stemmed exclusively from the outward migration of urbanites (Cloke, 1985, 17–20). Among other factors identified as contributing to the population turn-around in Britain and elsewhere are:

1. a spontaneous decentralization of manufacturing industry attracted by lower labour and land costs
2. official policies for industrial relocation, such as the initial siting of nuclear and hydro-electric power stations in regions
3. the activities of agencies such as the Highlands and Islands Development Board and the Welsh Development Agency which have had some success in attracting new employment to rural parts of their territories
4. the widespread growth of tourism with its considerable income-generating potential

These activities may perhaps be construed as more truly regenerative and less intrusive than urban–rural migration, in that their explicit effect is frequently the provision of job opportunities for the existing local population. Yet even here there are thorny problems of demarcation. Were we to define the rural population on strict functional grounds as comprising only those engaged in agriculture or the servicing of agriculturalists (Stamp's (1949) 'primary' and 'secondary' rural categories), then industrialization or any other form of non-agricultural development must necessarily be seen as a form of urbanization. Or perhaps we should distinguish between agricultural processing/servicing and other forms of manufacturing? Or between large and small-scale firms? Or between locally and externally owned businesses? So far as we are aware, the 'widespread and confused literature' (Lewis, 1983, 149) on rural communities does not yet provide accepted and watertight guidelines on these questions.

Judging from the emphasis in the rural literature, there are two associated problems seeking resolution here. One is the problem of selecting criteria which measure the definitive essence of rurality and the other is the issue of cut-off points – i.e. is there an imaginary line on the map at which such criteria cease to apply? Can urban and rural space be said to be bounded?

On the selection of criteria, an object lesson in clear-mindedness is provided by an unpublished Department of Environment paper (Department of

Environment, 1971, cited in Cloke, 1977), which uses three single variables for measuring rurality. The first is a spatial measure, *population density*, which reflects the commonsense conception of rural space as sparsely populated by small settlements. Physical definitions, however, are unsatisfying on their own. As Hall *et al.* (1973) point out, settlements are mere 'containers' of human activity and hence can be more fruitfully identified by their *functional* role (what they do) rather than by size and density (how they appear). Many settlements look like villages but perform characteristically urban functions. The functional role of the truly rural place is agricultural and by insisting on this as definitive we are able to distinguish the genuine rural community from the commuter satellite or metropolitan village, which may retain a rustic appearance but which derives most of its livelihood from a distant city centre. Though slightly pre-dating Hall *et al*, the Department of Environment paper implicitly adopts the logic of functional rurality by using *percentage employed in agriculture* and *percentage of commuters* as its second and third indicators.

When these functional criteria are directly applied to the settlement geography of Britain, they pose an unanswerable challenge to conventional myths about rural space. In particular they demonstrate the immense territorial extent of the urban commuting orbit and the degree to which former agricultural settlements have become assimilated within it. According to Figure 7.2, the great bulk of lowland England is now enmeshed within the overlapping orbits of the major conurbations and large towns. This map is taken from Cloke's (1977) exercise, in which his extreme rural category includes only the remotest regions of the outer periphery – the South-West peninsula, North and Mid Wales and the northern Pennines. Cloke's work, subsequently updated in Cloke and Edwards (1986), is important on several other counts, notably its attempt to construct and map an *index of rurality* for England and Wales. Recognizing that the Department of Environment's approach is perhaps over-simplified and that rural space is multi-dimensional, the author used 16 variables in a principal-components analysis. This enabled him to classify the administratively rural areas of England and Wales into four categories ranging from extreme rural through two intermediate classes to extreme non-rural. Although the selection of the 16 variables is somewhat questionable – for example a top-heavy age structure is included as an indicator of rurality, even though this could equally signify a resort town, a 1930s council estate or the location of a geriatric hospital – it must be said that this index of rurality achieves definite progress towards the precise measurement of rural attributes.

Cloke's method is also illuminating on the second vexed question, that of cut-off points. Although he chooses to group his data into four arbitrary classes, Cloke's component scores reveal no significant natural breaks or discontinuities and in this sense his findings are supportive of the *rural–urban continuum*. First propounded by Queen and Carpenter (1953), this well-known model rejects the urban versus rural stereotype and stresses instead the varying degrees of urbanism and rurality which characterize communities between the two extremes. In a society where cities no longer have walls and where urban encroachment has been blurring the rural–urban divide for centuries (Lewis, 1979), it makes no sense to think simply in binary categories. Thus at the level of literal truth the continuum model is a faithful representation of a society in which increasing numbers of people reside in the grey penumbric domains of 'Rurbia'. For the spatial analyst, the model is especially attractive, suggesting as it does a distance–decay effect, with urban influence progressively weakening towards the outer periphery. Although Cloke is careful not to claim too much for his findings, it is evident that they are broadly consistent with such an hypothesis, albeit in highly imperfect form.

7.3.1 The Cultural Dimension

One final aspect of rurality virtually ignored by the above contributors is culture. This is perhaps surprising in that cultural contrasts between urban and rural have been a major and long-lasting preoccupation of rural sociologists and anthropologists. How far may we speak of contrasting urban and rural ways of life? Seminal figures here have been Tonnies (1887) and Wirth (1938), the former of whom postulated a cultural antithesis between the large impersonal society (*gesellschaft*) with its formal agencies of allocation, socialization and control, and the small intimate community level of human experience (*gemeinschaft*), where individuals are integrated together through personal face-to-face contacts (Munters, 1982; see also Chapters 3 and 6). The vital point to grasp is that Tonnies's 'classic dichotomy . . . between *gemeinschaft* and *gesellschaft* was not in itself a typology of settlement form' (Saunders, 1981, 81). It was not his intention to let *gemeinschaft* stand for the village and *gesellschaft* for the city but rather to focus on society as a whole and in particular the manner in which industrialization was breaking down community bonds and substituting other less personal forms of association. It was Wirth's later intervention which was largely instrumental in projecting the country and the city as polar opposites, the former the repository of small-scale community values and the latter as the absence of these.

Even if Wirth's assertion is correct, we would once again expect urban–rural lifestyle contrasts to occur along a continuous gradient rather than across a cataclysmic divide. Several writers have acted upon this supposition, notably Frankenberg (1966), whose well-known anthology of community studies explicitly arranges its communities along a cultural continuum. This ranges from a Western Irish village, a representative of the simple rural face-to-face community, to Bethnal Green, a representative of urban complexity. Yet by the late 1960s, the continuum idea had come under heavy fire from a new generation of community sociologists (Mann, 1965; Pahl, 1965; 1968; 1975; see also Saunders, 1981). Among other criticisms, the continuum was thought to be unworkable because it contained too many misfits. What are we to make, for example, of those mining towns which contain large dense populations and are fully integrated into the urban economy, yet which exhibit all the traits of *gemeinschaft* – common work bonds, kin-centred social networks, primary social relations, and so on? (Dennis, Henriques and Slaughter, 1957). What are we to make of Mann's (1965) assertion that the inhabitants of Forest Row, a small Sussex commuter village, are more urbanized than the people of Huddersfield? More recently Baldwin and Bottoms (1976) emphasize the *gemeinschaft*-like qualities of Sheffield. Even more anomolous are the close-knit neighbourhood communities of immigrants which flourish *within* the largest city, people who are 'in the city but not of it'.

For critics of the rural–urban continuum, such cases serve to confirm that there are no specifically urban or rural ways of life and that the differences in lifestyle are just as great between one town and another as between urban and rural.

Class, age and ethnicity rather than settlement size are the crucial determinants of individual social behaviour and hence 'any attempt to tie particular patterns of social relationship to specific geographical milieux is a singularly fruitless exercise' (Pahl, 1968, 293; see also the discussion in Newby, 1980).

It is tempting to go even further than Pahl and to propose that virtually the whole British population is urbanized in the sense of sharing a common culture – 'a single national "way of life" shaped by a business culture, the mass media, advertising, state education and other quite general influences' (Dunleavy, 1982, 5) – transmitted from the leading urban centres. Indeed there can be few mature industrial societies where the 'folk culture' survives beyond occasional token vestiges. As Dunleavy explains, the process of industrialization has broken down the age-old urban–rural cleavage:

> an urban-rural dichotomy was appropriate as an analytic framework for analysing preindustrial societies . . . but as industrialization proceeded and societies entered an advanced stage, so the distinct pattern of life associated with farming production was in turn broken down and remoulded on the same lines as the rest of society (1982, 2; for a critique of the 'mass culture thesis' see Clarke, 1984).

Though the pace and timing of this transformation has varied markedly between nation-states, with Britain and the United States affected early and countries like Italy and France relatively late, no one can question the broad universality of the process. Further confirmation comes from Newby, who is clearly irritated by the entire debate:

> Who would wish now to argue the utility of 'urban' and 'rural'? . . . . Who too would wish to see a discussion of important socio-spatial issues become bogged down in definitional debates about 'urban' and 'rural'? There is now, surely, a general awareness that what constitutes 'rural' is wholly a matter of convenience (1986, 209).

We are inclined to the view that these critics have damaged their case by overstating it. To adopt Dunleavy's solution of rejecting 'any spatial reference for the "urban" and "rural" labels' (1982, 6) is to question the possibility that place and space can play any role in the shaping of society. In the light of the recent progress made by the 'society and space' school (see Chapter 1) such a proposition now seems dated. Dunleavy bases his claim that a distinctive rural lifestyle is now extinct on the drastic decline in agricultural employment. It is undeniable both that farm employment has shrivelled away to a tiny fraction of its former self and that agriculture was itself the source of rural culture, permeating socio-economic life in every village in the land. But to infer that its diminution automatically signals the demise of rurality is an oversimplification. Such an inference ignores the fact that despite the magnitude of its national decline agriculture remains highly significant locally, especially in the remoter parishes and regions. It also ignores the legacy of farming, the durability of social layers laid down in the distant past. Though shaped during former periods of agricultural dominance many of the characteristic social relations of rural areas continue to operate today, the persisting political influence of farmer and landowner being a case in point. Indeed the entire political culture of the rural working class – patterns of trade unionism (or lack of it), electoral behaviour, employer–employee relationships – is still so obviously different from urban and industrial areas.

A further aspect of the agricultural legacy, its spatial structure, has provided a major focus of interest for modern rural geographers. The inescapable and simple truth is that there remains a fundamental distinction between urban and rural spheres and that the distinction is physical (Lewis, 1979; 1983). Self-evidently, rural space is defined by relatively sparse population distributed in a scatter of small free-standing settlements whose original purpose was to house land workers. To invoke this 'commonsense' definition is not to retreat into crude physical determinism. On the contrary it is a reassertion of the active interplay between society and space as proclaimed by writers such as Massey (1984a; 1985) and Gregory and Urry (1985). This interplay is perhaps more clearly visible in the rural milieu than elsewhere, because the constraints on human action imposed by distance, inaccessibility and low population thresholds are considerable, even in a modern society supposedly liberated from the friction of space. While it is certainly true that the old rural order has been undermined by urban penetration and mass culture, it is also true that these modernizing processes have had to accommodate themselves to an inherited spatial system. The physical shape created by the old order materially affects the new, a perfect illustration of Massey's (1985, 4) dictum:

> Spatial distributions and geographical differentiation may be the result of social processes, but they also affect how these processes work. 'The spatial' is not just an outcome, it is also part of the explanation.

Massey's own case study of Cornwall (1984, 224–33) provides extended support for this thesis. In a superficial sense the characteristic settlement geography of this mainly rural area has outlived its usefulness. Cornwall's spatial structure was created in the past in response to a dominantly agricultural economy supplemented locally by mining, quarrying and fishing, all of which have declined drastically as sources of livelihood. Even so, the emergence of replacement activities (and inactivities) such as tourism, retirement and branch plant manufacturing can hardly be said to have created a new culture indistinguishable from that of more urbanized regions. We return to this theme in the final sections of this chapter.

7.4 Accessibility as a Specifically Rural Problem

For mainstream rural geographers and planners questions of accessibility, transport and mobility have similarly been seen as absolutely central to an understanding of rural affairs and this concern has been expressed in a rash of recent writing (Drudy and Drudy, 1979; Moseley, 1979a; 1979b; Nutley, 1980; 1983; Bannister, 1983; see also several of the contributions in Moseley, 1978 and Halsall and Turton, 1979). Less concerned than Massey to develop links between place, space and culture (a theme which remains to be more fully explored in the rural context), this literature has focused on the material effects of rural spatial structure, emphasizing its negative impact on the quality of life. Quite simply, inaccessibility – remoteness from essential goods, services and job opportunities, together with the money/time costs in overcoming this – constitutes the biggest single social hazard affecting the modern country dweller. Directly and indirectly the problem of inaccessibility is the greatest source of disutility and disadvantage to the greatest number of rural residents.

Before outlining the empirical findings on rural accessibility we might note that here, if anywhere, may lie one of the keys to the unresolved question of what differentiates rural from urban in modern mass society. Is it not simply relative accessibility which forms the critical divide? This is to argue that the definitive attribute of rural space is that its occupants experience a geographical constraint which is peculiar to this form of space, at least in its most disabling form.

Clearly such a proposition needs extensive justification. Given the technically complex measures devised to quantify rurality or the sophisticated arguments advanced to establish that rurality no longer exists in any meaningful way, it seems something of an anti-climax to attribute the whole answer to the elementary matter of population density. We can feel justified in this position only by recognizing that extensive land use and sparse settlement create a specifically 'rural' set of problems which are not faced by the inhabitants of large dense settlements.

> Transport availability and costs assume a greater importance in rural areas than in urban areas. Travel times and distances are usually greater and services and facilities less conveniently situated (Phillips and Williams, 1984, 130).

From recent research on rural shopping facilities (Moseley and Spencer, 1978), health provision (Heller, 1979) and educational provision (Watkins, 1979; Cooper, 1981), there emerges a consistent picture of precarious, often declining and sometimes downright inadequate standards of provision and in almost every case the authors note the additional difficulties of sheer physical access to such provision. Quite simply, a great many people live too far from the nearest service point and, in the widespread absence of frequent, rapid and affordable public transport, experience extreme inconvenience, even hardship, and in some cases are completely deterred from using such facilities. Naturally these hardships are most manifest in areas of Highland Britain (see Knox and Cottam, 1981 on the Scottish Highlands; Edwards, 1985 on Powys) and the more outlying parts of lowland Britain such as Norfolk (Owens, 1978). Even so, studies of areas such as the Cotswolds (Smith and Gant, 1981) and East Kent (Hull, 1985) suggest that the problem may be common to every area of small-scale scattered settlement. On Kent, Hull declares:

> The knowledge for many households that they are living in a relatively densely populated rural region, traversed by good communications and in close proximity to at least one urban area is of little comfort . . . in numerical terms the number of people likely to be suffering some degree of hardship is considerably greater than in the less densely populated and more remote parts of the British Isles (1985, 19).

In making the case for inaccessibility as a generically rural feature, Phillips and Williams (1984, 130) make a conceptual distinction between what they call 'physical inaccessibility' and 'social inaccessibility'. The second of these concepts refers to factors such as income, class, age, gender, race and educational qualifications, social constraints on access to resources. Elsewhere in this book, these conditions have been repeatedly highlighted as ubiquitous sources of social division and inequality, factors whose influence is fairly universal throughout society, respecting no rural–urban boundaries real or notional. Bannister's (1983) list of the disadvantaged sections of the rural population – low earners, the aged, the unemployed, children and the disabled – could be duplicated throughout the length and breadth of urban Britain. The distinction is that in rural areas these sections are also likely to be *transport-deprived* (see Wibberley's 1978 definition of the rural 'transport poor') and in this respect they are likely to be joined by other rural dwellers such as children, teenagers and many housewives from otherwise non-deprived households.

The crucial item is the car. Whereas in the urban milieu to be car-less is to be denied one of the present century's greatest consumer status symbols, we may nevertheless argue that urban carlessness is not necessarily disabling, given the spatial density of services, amenities and jobs on the one hand and of public transport nets on the other. By contrast to be without a car in a country village can mean genuine hardship, given the sparsity of rural facilities. In country areas the disadvantage of social inaccessibility is compounded by physical inaccessibility to a degree unthinkable for the majority of urban dwellers. A single example from the recent literature will suffice to illustrate the essential role of car ownership in the countryside. Hull's (1985) study of 16 Kent parishes finds that seven of these had no regular bus service, five were without a single shop, eleven lacked an NHS doctor and fourteen a chemist. There was no secondary school within the district, six parishes were without a primary school and almost a fifth of all children spent more than an hour each way in travel to school. Not surprisingly household car ownership amongst the population interviewed had risen (from 59 per cent (1973) to 70 per cent (1982)), but this still leaves a substantial group of immobile households dependent upon long inconvenient journeys by public transport. As noted previously this is ostensibly a highly central rural region, located within England's 'core'. Even here, however, the problems of geographical access are so acute as to effectively isolate a significant proportion of the population from full participation in the national space economy (for comparisons with Highland Britain, see Knox and Cottam, 1981; Edwards, 1985).

So far, then, we have advanced the proposition that, while urban and rural societies each contain a disadvantaged or marginalized minority, those in rural societies are additionally and decisively penalized by spatial constraints peculiar to their rural residence. Yet this pigeon-holing is valid only in a limited technical sense. Indeed most explanations of rural inaccessibility make the very mistake of seeing it largely in technical terms. Due to a deadly combination of small markets and large distances, rural space is uneconomic to service, whether for private outlets such as retailers or public providers such as the education system. The problem has risen to a head during the post-war era because consumer expectations have risen, depopulation has further reduced market potential, and consumer outlets have been subject to rationalization and concentration in a diminishing number of centres. Figure 7.3. traces the rationalization of the rail network in Norfolk and shows how rural residents have been squeezed not only contraction in the number of services but also in the means to get to them by public transport.

But how far is this a 'rural' problem as opposed to a problem of the whole society? It would appear quite legitimate to argue that declining service provision is dictated not by the nature of rural space but by the way that space is exploited. Like all other forms of space in capitalist society, it is exploited first and foremost for profit and only secondarily from other motives. Hence the pattern of social provision is a response to a particular set of class relations at a particular time in history. Arguably, if social resources were distributed according to need and not ability to pay then every member of society, rural and urban alike, would enjoy open access to life's necessities. Advanced industrialism certainly has the capacity to deliver this but the imperatives of capital accumulation ensure that it does not

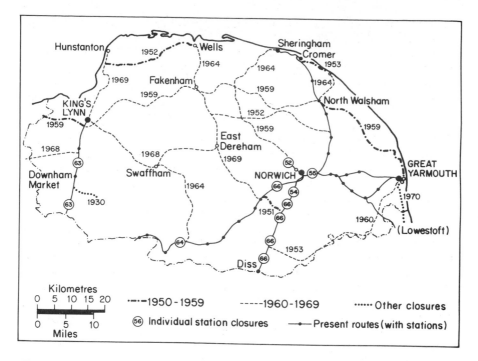

Figure 7.3 Post-war rail closures in Norfolk.
Source: Moseley 1979, 24.

do so. Given this capitalist dominance, then it is inevitable that rural space assumes enormous importance as a physical barrier but this does not reflect simply the inherent properties of rural as opposed to any other kind of space. In the last instance rural space, though a shaping influence, is not a determinant.

Having raised the issue of how rural space is used by producers, we now turn to its use by consumers, an equally germane item. Here we have a classic instance of Cox's (1981, 191) dictum that 'location is a social product rather than a thing in itself'. Although a truism, it bears repeating that accessibility is no absolute state but one which is highly variable from one section of the population to another depending on the ability to command the means of transport. Nowhere is this distinction more relevant than in the countryside. Indeed there are grounds for arguing that car ownership is the basis of a major consumption cleavage within modern rural society, since it is this which divides the mobile from the place-bound. It is only the former who can truly be said to enjoy choice through access to a comparatively wide range of jobs, services and other social resources. Meanwhile the latter are trapped in a dependence upon what the immediate vicinity may provide which, in the case of a small low-order settlement, is likely to be not much. Hence, in his model of modern rural housing strata, Ambrose (1974) actually incorporates car ownership as a defining variable – quite correctly in our view, since mobility will determine the utility which the house-holder is able to derive from the location of the home.

Because of the car's necessary role, rural ownership levels tend to exceed national norms. Yet this still leaves an important minority, the transport-poor. Obviously income is an important barrier here but, as the rural literature repeatedly stresses, a variety of non-class factors also act to exclude certain categories (Phillips and Williams, 1984). For example children and non-employed spouses in one-car families where the vehicle is necessary for the bread-winner's journey to work are victims of immobility for much of the time.

Superimposed upon this division between the 'privatized majority' of car owners and the 'marginalized minority' of the immobile is a further distinction, much highlighted in the rural literature, between incomers and localites. As previously noted, this process of repopulation by ex-urban dwellers is now the most visible feature of social change in the countryside and many writers have treated the resulting incomer/localite dualism as a conflict of interest (see, for example, Radford, 1970). By competing for limited resources, middle-class newcomers tend to heighten the deprivation suffered by many sections of the indigenous working class, a tendency inescapable in any process of 'gentri-fication' whatever the geographical milieu. Certainly the issue of transport and mobility contains the potential for such a clash of interests. Here we might suppose that the replacement of mainly low-income householders by more affluent car-owners would act to accelerate the erosion of public transport by reducing potential custom. Nor would it encourage the emergence of profit-seeking alternatives under the provisions of the 1986 Transport Act.

We shall return at greater length to incomer–localite relations in the next section. For the moment we do no more than add certain riders to the thesis that changing social composition undermines rural public transport:

1. By no means all newcomers can be numbered among the mobile. Among elderly retired in-migrants, for example, physical infirmity or the death of a car-driving spouse can create isolation of a particularly burdensome kind (Broschen and Himmighofen, 1983).
2. Over-emphasis on the role of newcomers can obscure the real root of inacces-sibility which, we repeat, is the under-provision of services and transport. Any search for ultimate causation would do well to focus on the general and rural transport policies of the central state and on the under-provision of education and welfare by rural local authorities, most of whom are dominated by property-owning interests dedicated to low-spending policies (Rose et al., 1979).

7.5 Conflict in the Modern Rural Community

Rural repopulation consists essentially of an urban colonization, the outward spread of a non-indigenous non-agricultural population originating from the towns and cities. Consequently many if not a majority of villages are now composed of two basic elements – a growing population of refugee urbanites grafted on to a much diminished rural rump. The essence of this dualism is captured by Lewis (1979, 155), who speaks of 'two communities within one place, each involving separate worlds within and beyond the locality', and by Newby (1979, 166), who depicts the indigenous working-class population as 'encapsulated', a 'community-within-a-community'.

For many students of the question, conflict of interest between these two elements is now one of the dominant motifs of modern rural life (Pahl, 1965; Ambrose, 1974; Newby, 1977). This is a view which, although valid in our judgement, nevertheless calls for some qualification. In the first place, any dualistic model of the modern rural community is technically a gross over-simplification, since neither newcomers nor indigenous ruralites constitute a group (and certainly not a class) in the strict sociological sense. In most cases the rural rump is likely to have maintained many of its characteristic internal divisions (landowner, farmer, farm worker, non-farm worker, small tradesperson).

This last point is important. Perhaps surprisingly in an advanced capitalist society the power of landed interests is highly persistent, at least in the rural areas themselves, where the long-standing lines of class division and conflict – landowner and large farmer on the one hand, landless worker and small tenant on the other – continue as a decisive influence. Naturally, of course, this power is no longer wielded exclusively within the confines of the enclosed village community. As is only to be expected in a modern democratic polity, much of it flows through channels of national and local state. At the national level, it may be argued that the post-war policies of the Ministry of Agriculture (responding to a substantial farming lobby both inside and outside Parliament) have had the effect of boosting the prosperity of the large at the expense of the small:

> there is a quite startling contrast between the undoubted success of this policy in terms of its stated goals (the expansion of home production, the increase in cost efficiency, the maintenance of a prosperous agriculture) and its mostly deleterious *social* effects on the countryside in either depopulating rural areas at an even faster rate than hitherto or polarizing them socially (Newby, 1979, 263).

Newby is arguing here that Central Government support for the expansion of cheap food production has benefited large owners of land and capital (including the new agribusinesses often owned or funded by City finance houses), while serving to marginalize small farmers (who are unable to compete) and to oust farm labour (who are replaced by machines). The effect on employment opportunities has been quite breathtaking, a 64 per cent fall in the national total of full-time farm workers in the United Kingdom between 1955 and 1977 (Packman, 1979). Though not by conscious design, agricultural policy has brought about a significantly regressive redistribution of real income (or more literally income-earning opportunities) in agricultural areas, a process vividly demonstrated by Wright's (1983) case study of Lincolnshire. As in so many other forms of rural deprivation, this problem may be understood as a consequence (in part at least) of urban dominance and rural dependency. For over a century, agricultural policy has been designed primarily to serve the needs of the urban majority rather than to protect the living standards of rural dwellers. The system has worked smoothly because these policy goals harmonize neatly with the interests of the richer, more powerful rural strata. It has been easy to overlook the fact that they run counter to the interests of lesser social strata.

Just as indigenous rural society is cut by internal divisions, so too is the incoming population. There is no necessary identity of interest between, for example, elderly retirees and commuting families with young children. Thus the situation appears as one in which a heterogeneous set of new groups is juxtaposed to an already stratified indigenous community. The resultant complexity has been recognized by Pahl (1965) in his designation of no less than eight rural social strata and by Ambrose (1974) who identifies seven strata defined by housing market position and personal mobility. Despite this, however, both authors insist that this stratification is secondary to the cardinal divide between indigenous and incoming residents. For all their internal diversity, each of these elements may be defined as distinct from and frequently opposed to one another in the sense that they are living in the countryside for different reasons and using its resources for different purposes. As Rogers (1976, 106) puts it, there is a 'dichotomy between those groups whose position in the countryside is a direct function of employment, of birth or of force of circumstances: and those who have entered the rural housing market in both a conscious and a well-equipped way'. Whereas for the localite, the rural locale represents first and foremost a livelihood, for most newcomers it is purely a place of residence and a refuge. As Newby (1977, 329) expresses it,

> The new immigrants possess a lifestyle in which the village is not the focus of their social activities, for not only do they not work locally, but their pattern of leisure activities – shopping, visiting friends and relatives, entertainment – often take them outside the immediate locality as well.

For these reasons the localite–newcomer split is a useful analytical device, at least at a preliminary level. But there still remains the question of how far and in what sense the interests of these two sections of the population conflict. Despite the different meanings they attach to rural life it is nevertheless possible to present localites and newcomers as harmoniously interdependent, each deriving benefits from the presence of the other. Intrusive though their presence may be, it is undeniable that in many instances urban newcomers have acted as the saviours of the rural physical fabric, restoring statutorily unfit and abandoned cottages which might otherwise have fallen into dereliction; as consumers whose spending power has thrown a lifeline to a village shop on the brink of extinction; as job creators, maintaining a demand for local builders, domestics and craftsmen. Lewis (1983, 150) makes the point that in the most extreme cases, where depopulation operates with no compensating inflow of urbanites, 'the whole fabric of the community collapses, since a small and ageing population find it impossible to sustain a viable set of activities'. When the impact of tourism is added to that of permanent settlers, there is a plausible case for outsiders as a dynamic leading sector in rural regeneration, a fountainhead from which benefits trickle down to the indigenous population.

This idealized vision of functional interdependence does not, however, accord with most of the recent evidence produced by the rural deprivation school. Any model of newcomer–localite interdependence in the countryside must perforce underplay two basic characteristics of the relationship between the parties: first, they are in direct competition for certain social resources, in indirect competition for others; second, that there is a wide gulf in income and purchasing power. On the second point, the literature leaves us in no doubt that the newcomer–localite confrontation is one between rich and, leaving aside large farmers and land-owners, poor (Pahl, 1965; Radford, 1970; Newby, 1979; Cooper, 1981). Consequently competition for scarce resources is inevitably an unequal contest, which will be dictated by the incomers to the disadvantage or even the exclusion of low-income rural workers. Inescapably, the entire balance of supply and demand will be distorted by a large and growing ex-urban element, entering markets which, even before their arrival, had traditionally failed to deliver the goods to a significant proportion of the rural population. We now expand this argument by reference to the key resource, housing.

7.5.1 Housing Provision

Perhaps more than any other area of rural resource provision, housing is a sphere where popular imagery (especially that beloved of estate agents – 'picturesque', 'quaint', 'secluded') belies reality as experienced by many local residents. It is also a sphere which has been consistently neglected in post-war Britain, where the housing crisis has almost invariably been presented as an urban crisis. Recent students of rural deprivation have berated academics and planners, central and local government alike for this myopia (Ambrose, 1974; Rogers, 1976; 1983). The errors of this 'urbocentric' bias are sharply exposed by Larkin (1979), who shows that the rural poor often suffer greater hardships than their urban counterparts; that council housing tends to fall even shorter of housing needs than in urban areas; and that caravans (Larkin's 'true ghettos of the rural poor') make up 10 per cent of 'permanent' dwellings in some rural districts. Some of his worst horror stories emanate from such exquisite picture-postcard locations as Dorset, South Oxfordshire and the Isle of Wight. Similarly Ambrose (1974, 200) directs a number of scathing criticisms at the system of rural housing allocation. The situation is summarized by Rogers (1983, 109) as follows:

> Relative to urban housing, housing in rural areas is generally older and in poorer condition. The twin factors of the age of the housing stock coupled with the low incomes of rural dwellers . . . mean that even in advanced societies rural dwellers have fewer housing amenities.

Initially one is tempted to attribute housing scarcity entirely to the urbanite invasion of the village, with its inflationary pressures and its unequal competition (Radford, 1970). Yet, while clearly important, this is far from the whole story. We would be wrong to assume that rural housing was a thatch and honeysuckle idyll prior to urban immigration. Historically the British countryside has been plagued by inadequate housing for well over a century, as various official enquiries from mid-Victorian days onwards will testify. Moreover the problem has stemmed from short supply rather than inflated demand. In turn this is a reflection of the low wages and political powerlessness of the rural working class, together with its dependence on rented (often tied) accommodation. This tied accommodation was mainly provided by landowners, farmers and other employers, who often found it in their interests to discourage labour mobility by restricting the supply of housing (see Newby, 1979 for an historical perspective on rural housing conflict).

To a great extent this class subordination persists today, though in a thoroughly modern form (Newby *et al.*, 1978; Barlow, 1986). As in urban areas, housing in the countryside has increasingly become a form of collective consumption, state intervention being used for much of the post-war period to try to plug the gaping holes in the free market. But, as Larkin (1979) establishes, rural council house provision generally falls well short of urban levels. This is mainly because landed property interests – this time in the guise of democratically elected local councillors – continue to control the instruments of housing provision. A case study of Suffolk (Rose *et al.*, 1979) revealed that in 1973 the chairman, vice-chairman and leader of the County Council were all farmers, as were the chairmen of all the key committees. This pattern was substantially replicated at district level within the county. Unamazingly, all these authorities were committed to policies of low spending, accurately reflecting the interests of these major ratepayers and property owners. Council housing, along with health and educational provision, is obviously a major sufferer here. Thus a principal urban–rural contrast is that working-class interests are substantially under-represented in the latter in comparison with the former. Unlike the major urban areas, where Labour Party and Trade Union organization has wrought significant reforms in the housing market, the rural working class lacks any effective political leverage. Because of physical isolation and dispersal it is in effect detached from the mainstream of working-class political life and in this respect if in no other it bears a strong resemblance to the disenfranchised minorities of the inner city – voiceless and disregarded. Its income levels, job opportunities and housing standards are all completely consistent with minority status.

Evidently then the advent of urban newcomers has not caused a rural housing crisis, but it has certainly aggravated it. Though in the first instance ostensibly no more than a replacement population filling in the 'holes' left by depopulation, the newcomers – whether commuters, retirees or second-home owners – have further eroded the already limited supply of accommodation open to low-income locals. In its effects, this rural colonization is not dissimilar to the process of gentrification observed in the urban setting. In the first place the influx of affluent home buyers pushes the price of owner-occupied property beyond the reach of all but the better-off local workers. Secondly, it makes inroads into the private rented sector by encouraging the conversion of such properties into owner-occupation or high-rent luxury accommodation (Newby, 1977, 231; Cooper, 1981). In addition to all this, newcomers tend to be opposed to the council building which could compensate for the shrinkage of the private rented sector. As property-owners, their natural interest is to side electorally with the existing vested interests and with low public spending on housing provision.

Perhaps the most notably regressive consequence of all this has been to drive low-income workers, and especially farm workers, into an even greater dependence on tied accommodation, a paternalistic housing relationship completely anachronistic in a late twentieth-century setting. Quite justifiably this is an highly emotive issue: 'there is probably no topic in farming which raises passions more than the tied cottage' (Newby, 1977, 182). Because the tied cottage is part of a farm worker's conditions of employment, the tenancy automatically lapses on termination of employment, giving the employer considerable leverage over his employee. Quite apart from reinforcing the traditional 'lord and master' power of the owner, this can also jeopardize the farm worker's lifelong residence in the village; as we have noted above, there is precious little accommodation to rent and none to buy in his price range.

7.6 The Political Economy of Rural Space: An Outline Agenda

To say that rural space and society have failed to excite the interest of modern Marxists would be a considerable understatement. This deficiency has been lamented by Bradley (1981, 581): 'scarcely any marxist writing has appeared outlining a political economy of rural society under advanced capitalism . . . the future direction for a marxian political economy of rural space is curiously unclear.' Although a number of recent commentators have begun to address themselves to this (Bradley and Lowe, 1984; Phillips and Williams, 1984), the gap remains substantially unplugged. We offer here a tentative outline which might perhaps encourage further work in this direction.

7.6.1 Rural Space and Capital Accumulation

Undoubtedly the modern Marxist's neglect of rural matters stems from the feeling that it is urban space which is at the 'sharp end' of capitalist development. By contrast rural space appears sheltered from the continual crises of accumulation and reproduction, a passive spectator on the sidelines of the class struggle arena. More 'urbocentric' even than bourgeois scholars, Marxists have still less excuse, for their methodology should tell them that rural and urban space are dialectically linked, contrasting expressions of a single unifying process. Certainly urban agglomeration has been absolutely central to the development of capitalist industrialization but throughout the entire modern period the existence of external non-urban space (both domestic and colonial) has been fundamental to the reproduction of urban space itself (Harvey, 1982, 417–19). As a general rule we might argue that the role of this external space has been that of a safety valve (or 'spatial fix' in Harvey's words), enabling capital to relieve the contradictions of accumulation–reproduction imposed by urbanism.

Naturally the precise role of the rural has changed as modern capitalism has evolved. For example, its earlier function as a migrant-labour reservoir, vital for the nineteenth-century expansion of British industry and the suppression of urban wage levels, is now only a shadow of its former self. Domestic rural–urban labour migration had shrunk to a trickle by the post-war period and had been supplanted by overseas migration from the New Commonwealth and elsewhere. Urban-based industry was still drawing on external rural space but the locus had shifted from Britain to overseas areas.

Even so, domestic rural areas still continue to function as labour reserves. The difference now is that their role is not so much to supply migrants for urban industry as to provide *in situ* labour for firms diffusing outwards from the cities. This is part of the broad trend discussed in Chapter 1 (see also Fothergill and Gudgin, 1979; Hudson, 1983; Massey, 1984a), whereby industrial capitals decentralize many of their branch operations to regional labour reserves. In the case of rural labour reserves, such capitals can take advantage of labour whose wages have historically been low and whose organizing and bargaining power is weak.

If we seek a single explanation for the vulnerability of rural workers, the answer lies in the chronic labour surplus created by agricultural modernization, a surplus which has not been fully eradicated even after decades of rural depopulation. All but the most drastically depopulated areas are replete with surplus labour and, while we may point to many instances of localized short-term shortages of specific types of workers, the general rule is that job opportunities have consistently fallen well below the number of potential job seekers. This endemic labour surplus takes various forms, though probably its most fundamental expression is a buyer's market for labour power, which has enabled farmers and other employers to maintain low wages and paternalistic employment relations. Aside from this, the surplus is manifested most obviously in persistent high unemployment in many localities (Packman, 1979; Cooper, 1981; Phillips and Williams 1984) and in all the classic symptoms of underemployment or concealed unemployment. Part-time and seasonal work abound and low activity rates are a hallmark of rural regions, not just in Britain but throughout the Western world. Female inactivity is especially pronounced, a combination of few job opportunities, poor pay and long journeys to work acting as an almost insurmountable deterrent, above all for married women.

In addition there is a fairly substantial 'hidden' surplus locked up in what might be described as a 'relic pre-capitalist economy'. In introducing this term, we do not wish to suggest that rural space is pre-capitalist space. On the contrary, its leading industry, agriculture, might be considered to be in the vanguard of modern capitalist advance, a factor stressed by most recent rural geography texts. Phillips and Williams, for example, leave little doubt that, in its motives, methods and social relations, the industry is as strongly capitalistic as any: 'with the decline of subsistence farming in Britain, agriculture does not lie outside the production process of capitalism but, rather, becomes a distinctive part of this process' (1984, 24; see also Newby, 1977; 1979).

At the same time, however, agriculture – if not the entire rural economy – exhibits a strongly dualistic streak. There still remain a great many activities which are essentially subsistence in nature, such as the small family farm, which in capitalist market terms is usually regarded as submarginal and whose purpose is to reproduce its occupants rather than to accumulate capital. It is not, in common parlance, 'run on business lines'. Strictly speaking it is better to regard such undertakings as 'petty capital of the simple reproduction type' rather than as 'pre-capitalist'. But whatever name we attach to these enterprises, their relevance to the present argument is that they represent a substantial but disguised labour reservoir. Some of this labour power is officially registered as employed or self-employed, even though its productive contribution may be peripheral and its monetary return marginal. Some of it is completely unrecorded and might be considered part of the 'black economy'. As Massey remarks in the case of Cornwall, female labour is again prominent:

> While many married women in Cornwall may not have been confined to the home and to domestic labour, and may even have appeared in official statistics as 'economically active', they have nonetheless rarely been working in capitalist wage relations and even more rarely within any labour process likely to provide the basis for solidarity and workplace organization (Massey, 1984a, 225–6).

Thus surplus labour not only remains numerous in absolute terms, it can also be assumed to be non-combative. In itself this must provide an active motive (though not the only one: see Massey, 1978) for mobile industries and services to relocate in green pastures. From this it becomes clear that the dispersal of urban-based production into the countryside is a key element in the post-1960 spatial restructuring of the British economy and the growth of a new spatial division of labour (see Urry, 1984 and Chapter 1). Moreover, despite the arguments already rehearsed in this chapter that rural space and society are no longer a distinct entity, the contrary suggestion here is that they do indeed occupy a distinctive niche in the new spatial division of labour. A clue to this is found in an early paper by Massey:

> Mass production and assembly work requiring only semi-skilled labour power . . . is increasingly located in areas where semi-skilled workers are not only available (since they are everywhere) but where wages are low and where there is no tradition . . . of militancy . . . [among] workers with no previous experience of capitalist relations of production (Massey, 1978, 117).

Though Massey does not hint at 'rural' in this argument, it is nevertheless rural areas and rural people who typically fit the bill. According to Massey, decentralization of low-level production is directed at localities characterized by the very forms of social relations and political culture which rural society is supposed to embody. This is by no means to assert that non-militancy, deference and an absence of the labour movement tradition are exclusively rural properties – working-class conservatism is ubiquitous. Nevertheless these qualities do tend to be commonplace in extensive areas of low population density, low-order settlements, a dispersed, fragmented, isolated labour force and a dominance, until recently, by agriculture as the major source of livelihood: areas, moreover, where agricultural property relations still play some part in determining local life chances. While there can be no rigid pigeon-holing, there appears nevertheless to be a particular use for those parts of the national space which have hitherto been less intensively colonized by the process of capital accumulation.

It seems likely that the bulk of British rural space is destined to remain comparatively empty for the foreseeable future. Areas of market towns and villages are characterized by Cooke (1986, 246) as 'diffusely industrialized labour markets', implying a pattern of small and medium-sized production units, normally spatially separated rather than clustered. It is this kind of space-economy and the industrial relations which go with it which offer the locational attractions for incoming firms. Clearly if a large industrial estate, office complex or overspill town were inserted into such a pattern, then the local labour market would be transformed, with emphatic effects on labour demand, wages, workplace and social relations, ultimately the entire local class structure. In general, however, capital's conquest of these areas is a gradual one, seemingly designed to incorporate them into a new spatial division of labour with minimal disturbance to existing arrangements. Except in localized instances, the scale and pace of growth has largely been sustainable on the basis of existing labour reserves, without pushing up labour demand to the point where wages must rise significantly. Further discussion on the impacts of exogenous manufacturing and service firms can be found in Thomas and Winyard (1979), Perry (1982), Massey (1984a, especially 224–33) and Phillips and Williams (1984). Even so, logic would point to a 'soft' colonization of rural space. To use Massey's geological metaphor, there are advantages for capital in preserving the existing social layer in rural localities even while gently depositing a new one. In this respect, rural areas stand in direct contrast to old urban-industrial areas, where recent under-development has had the effect of decomposing the old working-class layer. In the former, the traditional layer is functional for new capital accumulation, in the latter it is a barrier.

7.6.2 The Consumption of Rural Space

In the present chapter we have been very much preoccupied with the way rural resources are consumed and with conflict over resources, notably between indigenous and in-migrant populations. These are far from parish-pump squabbles. They reflect another macro-level change in capitalist development, this time in the sphere of reproduction. To summarize this change in a nutshell, we might say that rural resources have become 'nationalized'. By contrast with pre-industrial times when its role was mainly to reproduce labour power for agri-culture and other purely local activities, rural space now has a part to play in the process of reproduction at the national level. Bradley (1981, 585) highlights this when he defines rural space as the 'spatial dimension of collective consumption'. Somewhat cryptic in itself, this phrase can be translated as the exploitation of rural areas and their working class to underwrite the reproduction of the urban labour force. A clear instance of this is the cheap food policies pursued by a succession of British governments over a lengthy period, another state intervention in collective consumption, this time engineered through an alliance with rural capital i.e. hefty subsidies to large commercial farmers (Newby, 1979). This takes place at the expense of local petty capital (small uncompetitive farmers) and even more of rural labour, whose low money wages and poor social provision act in effect as an additional food subsidy.

More recently this urban consumption of rural space has expanded and diversified. In addition to subsidized food provision, rural Britain is also required to boost urban consumption standards by providing amenity and leisure space and, as we have seen, an extended dormitory zone for the more privileged middle-class strata of the urban workforce. As in other sectors of consumption, the state has played a direct role in the process: for example, the rural planning machinery set up in Britain in the early post-war years has placed very heavy emphasis on the leisure and environmental uses of rural land, even at times to the detriment of local labour and capital. Thus National Parks, Areas of Outstanding Natural Beauty and Green Belts have all emerged as loci of conflict. The interests of conservation frequently are directly opposed to those of farmers (for example land reclamation) and local workers (for example industrialization for job creation). These conflicts of interest are considered in more detail by Shaw (1979) and Phillips and Williams (1984).

7.7 Conclusion: The Ideology of 'Happy Poverty'

Finally we must ask how this exploitation of rural labour comes to be justified. Elsewhere in this book we have seen how ideological mechanisms such as racism (Chapter 5) and law and order crusades (Chapter 3) serve both to legitimize the position of marginal groups and to morally isolate them from the rest of the population. Following Newby's analysis of agricultural labour (Newby, 1977; 1979; Newby et al., 1978), we might identify two forms of ideological control which operate to render the poverty of farm workers (and many other sections of rural labour) morally unacceptable and politically unassailable.

1. *Workplace Relations* Through close and paternalistic contact in the workplace, farmers are able to impose their own view of social reality on their employees. Thus, despite the glaring visibility of the gap between them, farm workers identify overwhelmingly with farmers rather than with other members of the working class. Though this 'imposition of ruling-class ideas and values . . . is never more than patchy' (Newby et al., 1978, 279), farm workers' tacit acceptance nevertheless provides farmers with powerful justification for high profits won from underpaid labour.

2. *External Relations* Whether acquiescent or disaffected, rural labour is further handicapped by its invisibility in the eyes of the outside world, and its lack of voice to speak to urban members of the working class. Rural deprivation is thus either ignored or accepted as a 'normal' state of affairs. This cultural and political isolation is a product not simply of geographical distance but also of ignorance, an ignorance fostered by the caricature image which British popular culture has always projected on to the countryside and its inhabitants. Several components of this image have a direct bearing on the question at issue: the 'rural life as bucolic paradise' image, which suggests that living in the countryside is sufficient reward in itself so that poor wages, housing and

services do not constitute a genuine grievance; the 'ignorance-is-bliss' component, which implies that rural workers are culturally backward and know no better (Newby, 1977, 11); the ideology of the rural community, which suggests that the intimacy and richness of village life is sufficient compensation for material hardship (Newby et al., 1978). This condescending viewpoint is maintained even by urbanites who come to live in the countryside, for whom the local population is required to provide menial services and to function as quaint rustic stage props (Newby, 1977).

In making these observations we have travelled full circle back to our point of entry – the nymphs and shepherds romanticism which refuses to admit that rural society is in any way problematic. Over the past decade or so the work of many of the contributors reviewed here has done much to deglamourize the analysis of rural affairs. It remains to be seen, however, how far such realism can dent the complacency of those who have the power to decide the fate of rural areas and people.

7.8 Recommended Reading

There is a long tradition of geographical research on the physical, economic and social character of rural areas and this has been reflected in a spate of recent text books. Particularly recommended as an introduction to some of the issues highlighted above is David Phillips and Alan Williams's *Rural Britain: A Social Geography* (Blackwell 1984), an issue-based analysis of the character of contemporary rural society. More traditionally, Michael Pacione's *Rural Geography* (Harper & Row 1984) presents a solid and substantial overview of rural settlement, population, agriculture, resources and life quality in 20 descriptive chapters. The same author has put together *Progress in Rural Geography* (Croom Helm 1983). Like many other loosely edited collections there is occasional repetition and some unevenness in the quality and level of the individual contributions, but the book does contain some valuable articles. In particular Gareth Lewis's study of rural communities has been used quite extensively in the foregoing chapter; this builds on his earlier book, *Rural Communities: A Social Geography* (David & Charles 1979), which should be consulted for a more complete picture. Problems in defining rurality are clearly spelt out in the first chapter of Paul Cloke's *Key Settlements in Rural Areas* (Methuen 1979), while rural planning issues in the first three decades after the Second World War, not considered in any detail in our analysis, are carefully covered in Andrew Gilg's *Countryside Planning* (Methuen 1978). The same author's *Introduction to Rural Geography* (Arnold 1985) is also useful as a more broadly-based first text which accurately reflects the changing scope of contemporary rural studies. One issue of paramount importance to rural communities (and rural geographers) today is the decline in service provision and the consequent problems of accessibility, and no consideration of this issue should ignore the work of Malcolm Moseley (see, for example, *Accessibility: The Rural Challenge* (Methuen 1979)).

Finally, much of the most influential theoretical and empirical work reinterpreting rural life and highlighting the position of the rural poor has been generated by those whose disciplinary base is not explicitly geographical. In an essentially interdisciplinary field the work of Howard Newby is particularly important and his *The Deferential Worker* (Allen Lane 1977) and, especially, *Green and Pleasant Land* (Hutchinson 1979) make compelling reading; both are warmly recommended.

References

Abel-Smith, B. 1967: *Labour's Social Plans*. London: Fabian Society.
—— and Townsend, P. 1965: *The Poor and the Poorest*. London: Bell.
Abu-Lughod, J. 1969: Testing the Theory of Social Area Analysis: The Ecology of Cairo, Egypt. In *American Sociological Review* 34, 198–212.
—— and Foley, M. 1960: Consumer Strategies. In Foote N., Abu-Lughod, J., Foley, M. and Winnick, L. (editors), *Housing Choices and Constraints*. New York: McGraw Hill, 387–447.
Adams, J.S. 1969: Directional Bias in Intra-Urban Migration. In *Economic Geography* 45, 302–23.
Agnew, J. 1981: Home Ownership and the Capitalist Social Order. In Dear, M. and Scott, M. (editors), *Urbanisation and Urban Planning in Capitalist Society*. London: Methuen, 457–80.
Aldrich, H., Cater, J., Jones, T. and McEvoy, D. 1981: Business Development and Self-Segregation: Asian Enterprise in Three British Cities. In Peach, C., Robinson, V. and Smith, S. (editors), *Ethnic Segregation in Cities*. London: Croom Helm, 170–90.
—— 1983: From Periphery to Peripheral: The South Asian Petite Bourgeoisie in England. In Simpson, I. and Simpson, R. (editors), *Research in the Sociology of Work: Volume II*. Connecticut: JAI Press, 1–32.
——, Zimmer, P. and Jones, T. 1986: Small Business Still Speaks with the Same Voice: A Replication of 'The Voice of Small Business and the Politics of Survival'. In *Sociological Review* 34, 2, 335–56.
Allatt, P. 1984: Fear of Crime: The Effect of Improved Residential Security on a Difficult to Let Estate. In *The Howard Journal*, 23, 170–82.
Allen, S. 1982: Gender Inequality and Class Formation. In Giddens, A. and McKenzie, G. (editors), *Social Class and the Division of Labour*. Cambridge: Cambridge University Press.
—— and Wolkowitz, C. 1986: Homeworking and the Control of Women's Work. In Feminist Review (editors), *Waged Work*. London: Virago.
Alonso, W. 1960: A Theory of the Urban Land Market. In *Papers and Proceedings of the Regional Science Association* 6, 149–58.
Ambrose, P. 1974: *The Quiet Revolution: Social Change in a Sussex Village 1871–1971*. London: Chatto & Windus.
—— and Colenutt, R. 1975: *The Property Machine*. Harmondsworth: Penguin.
Amin, S. 1976: *Unequal Development*. Sussex: Harvester Press.
Anderson, J. 1973: Ideology in Geography: An Introduction. In *Antipode* 5, 1–6.
Anwar, M. 1979: *The Myth of Return: Pakistanis in Britain*. London: Heinemann.
Archbishop of Canterbury's Commission on Urban Priority Areas 1985: *Faith in the City*. London: Church House.
Arensberg, C.A. and Kimball, S.T. 1949: *Family and Community in Ireland*. Cambridge, Mass.: Harvard University Press.

Association of Metropolitan Authorities 1984a: *Inquiry into British Housing: The Association's Submission*. London: AMA.
—— 1984b: *Defects in Housing (Part II): Industrialised and System-Built Dwellings of the 1960s and 1970s*. London: AMA.
Baboolal, E. 1981: Black Residential Segregation in South London. In Jackson, P. and Smith, S. (editors), *Social Interaction and Ethnic Segregation*. London: Academic Press; 59–80.
Bagley, C. 1965: Juvenile Delinquency in Exeter: An Ecological and Comparative Study. In *Urban Studies* 2, 33–50.
Balchin, P. 1985: *Housing Policy: An Introduction*. London: Croom Helm.
Baldwin, J. 1979: Ecological and Areal Studies in Great Britain and the United States. In Morris, N. and Tonrey, M. (editors), *Crime and Justice: An Annual Review of Research*. Chicago: University of Chicago Press, 29–66.
—— and Bottoms, A. 1976: *The Urban Criminal*. London: Tavistock.
Bannister, D.J. 1983: Transport and Accessibility. In Pacione, M. (editor), *Progress in Rural Geography*. London: Croom Helm.
Banton, M. 1972: *Racial Minorities*. London: Fontana.
—— 1983: *Racial and Ethnic Competition*. Cambridge: Cambridge University Press.
Baran, P. and Sweezy, P. 1966: *Monopoly Capital*. New York: Monthly Review Press.
Barlow, J. 1986: Landowners, Property Ownership and the Rural Locality. In *International Journal of Urban and Regional Research* 10, 309–29.
Bassett, K. and Short, J.R. 1980: *Housing and Residential Structure: Alternative Approaches*. London: Macmillan.
Bechhofer, F., Elliott, B. and Rushworth, M. 1971: The Market Situation of Small Shopkeepers. In *Scottish Journal of Political Economy* 18, 161–80.
—— and Bland, R. 1974: The Petite Bourgeoisie in the Class Structure: The Case of the Small Shopkeeper. In Parkin, F. (editor), *The Social Analysis of Class Structure*. London: Tavistock.
Becker, H. 1963: *Outsiders: Studies in the Sociology of Deviance*. New York: Free Press.
Beechey, V. 1982: The Sexual Division of Labour and the Labour Process: A Critical Assessment of Braverman. In Wood, S. (editor), *The Degradation of Work*. London: Hutchinson, 54–73.
Bell, C. and Newby, H. 1971: *Community Studies: An Introduction to the Sociology of the Local Community*. London: Allen & Unwin.
—— 1976: Community, Communion, Class and Community Action: The Social Sources of the New Urban Politics. In Herbert, D.T. and Johnston, R.J. (editors), *Social Areas in Cities (Volume II): Spatial Perspectives on Problems and Policies*. London: Wiley, 189–207.
Benston, M. 1969: The Political Economy of Women's Liberation. In *Monthly Review* 21, 13–27.
Bentham, C. 1985: Trends in the Relationship between Earnings and Unemployment in the Counties of Great Britain. In *Area* 17, 4, 267–75.
Benwell Community Development Project 1978: *The Making of a Ruling Class: Two Centuries of Capital Development on Tyneside*. Newcastle: Benwell Community Development Project.
Berry, B.J.L. and Kasarda, J.D. 1977: *Contemporary Urban Sociology*. New York: Macmillan.
—— and Rees, P.H. 1969: The Factorial Ecology of Calcutta. In *American Journal of Sociology* 74, 445–91.
Berry, F. 1974: *Housing: The Great British Failure*. London: Charles Knight.
Bilton, T., Bonnett, K., Jones, P., Stanworth, M., Sheard, K. and Webster, A. 1981: *Introductory Sociology*. London: Macmillan.
Billig, M. 1978: Patterns of Racism: Interviews with National Front Members. In *Race and Class* XX, 2, 161–79.

Birch, D. 1979: *The Job Creation Process: MIT Program on Neighbourhood and Regional Change*. Cambridge, Mass.: MIT Press.

Blackaby, F. (editor) 1979: *De-Industrialisation*. London: Heinemann.

Blauner, R. 1972: *Racial Oppression in America*. New York: Harper & Row.

Blaut, J. 1975: Imperialism: The Marxist Theory and its Evolution. In *Antipode* 7, 2, 1–19.

Blowers, A. 1973: The Neighbourhood: Exploration of a Concept. In Open University DT201, Unit 7, *The City as a Social System*. Milton Keynes: Open University Press, 49–90.

Blunkett, D. and Green, G. 1983: *Building from the Bottom*. London: Fabian Society.

Boal, F. 1969: Territoriality on the Shankill–Falls Divide, Belfast. In *Irish Geography* 6, 30–50.

—— 1978: Ethnic Residential Segregation. In Herbert, D. and Johnston, R. (editors), *Social Areas in Cities*. Chichester: John Wiley, 57–95.

Boddy, M. 1976: The Structure of Mortgage Finance: Building Societies and the British Social Formation. In *Transactions, Institute of British Geographers* New Series 1, 1, 58–71.

—— 1980: *The Building Societies*. London: Macmillan.

—— 1981: The Property Sector in Late Capitalism: The Case of Britain. In Dear, M. and Scott, M. (editors), *Urbanisation and Urban Planning in Capitalist Societies*. London: Methuen, 267–86.

—— and Fudge, C. (editors) 1984: *Local Socialism?* London: Macmillan.

Boggs, C. 1976: *Gramsci's Marxism*. London: Pluto.

Boissevain, J. 1984: Small Entrepreneurs in Contemporary Europe. In Ward, R. and Jenkins, R. (editors), *Ethnic Communities in Business*. Cambridge: Cambridge University Press, 20–38.

Bonacich, E. and Modell, J. 1980: *The Economic Basis of Ethnic Solidarity: Small Businesses in the Japanese-American Community*. Berkeley: University of California Press.

Booth, C. 1902: *Life and Labour of the People of London*. London: Macmillan.

Bott, E. 1957: *Family and Social Networks: Roles, Norms and External Relations in Ordinary Urban Families*. London: Tavistock.

Bottoms, A. and Xanthos, P. 1981: Housing Policy and Crime in the British Public Sector. In Brantingham, P.J. and Brantingham, P.L. (editors), *Environmental Criminology*. London: Sage, 203–25.

Bouquet, M. 1982: Production and Reproduction of Family Farms in South West England. In *Sociologia Ruralis* 23, 261–75.

Bourne, L.S. 1981: *A Geography of Housing*. London: Edward Arnold.

Bowlby, J. 1953: *Child Care and the Growth of Love*. Harmondsworth: Pelican.

Bowlby, S., Foord, J. and McDowell, L. 1986: The Place of Gender in Locality Studies. In *Area* 18, 4, 327–31.

Box, S. 1983: *Power, Crime and Mystification*. London: Tavistock.

Bracey, H.E. 1952: *Social Provision in Rural Wiltshire*. London: Methuen.

—— 1958: Some Aspects of Rural Depopulation in the United Kingdom. In *Rural Sociology* 23, 385–91.

—— 1959: *English Rural Life: Village Activities, Organisations and Institutions*. London: Routledge & Kegan Paul.

Bradley, T. 1981: Capitalism and Countryside: Rural Society as Political Economy. In *International Journal of Urban and Regional Research* 5, 581–7.

—— and Lowe, P. (editors) 1984: *Locality and Rurality*. Norwich: Geobooks.

Braithwaite, J. 1981: The Myth of Social Class and Crime Reconsidered. In *American Sociological Review* 46, 36–51.

Brantingham, P.J. and Brantingham, P.L. 1981a: Introduction: The Dimensions of Crime. In Brantingham, P.J. and Brantingham, P.L. (editors), *Environmental Criminology*. London: Sage, 7–26.

—— 1981b: Notes on the Geometry of Crime. In Brantingham, P.J. and Brantingham, P.L. (editors), *Environmental Criminology*. London: Sage, 27–54.

Braverman, H. 1974: *Labour and Monopoly Capital: The Degradation of Work in the Twentieth Century*. New York: Monthly Review Press.

Breton, R. 1964: Institutional Completeness of Ethnic Communities and the Personal Relations of Immigrants. In *American Journal of Sociology* 70, 193–205.

Broschen, E. and Himmighofen, W. 1983: The Aged in the Countryside: Implications for Social Planning. In *Sociologia Ruralis* 23, 261–75.

Brown, C. 1984: *Black and White Britain*. London: Heinemann.

Brown, D.L. and Wardell, J.M. (editors) 1981: *New Directions in Urban-Rural Migration: The Population Turnaround in Rural America*. New York: Academic Press.

Brown, M. and Winyard, S. 1975: *Low Pay on the Farm*. London: Low Pay Unit.

Brown, R. 1978: Work. In Abrams, P. (editor), *Work, Urbanism and Inequality: UK Society Today*. London: Weidenfeld & Nicolson, 55–159.

Buck, T. 1979: Regional Class Differences: An International Study of Capitalism. In *International Journal of Urban and Regional Research* 3, 516–26.

Bull, P. 1983: Employment and Unemployment. In Pacione, M. (editor), *Progress in Urban Geography*. London: Croom Helm, 45–74.

Bunge, W. and Bordessa, R. 1975: *The Canadian Alternative: Survival, Expeditions and Urban Change*. Downsview, Ontario: Atkinson College, York University.

Burgess, E. 1925: *The City*. Chicago: University of Chicago Press.

Burney, E. 1967: *Housing on Trial*. Oxford: Oxford University Press.

Buttimer, A. 1972: Social Space and the Planning of Residential Areas. In *Environment and Behaviour* 4, 279–310.

Caradog Jones, D. 1934: *The Social Survey of Merseyside*. Liverpool: Liverpool University Press.

Carchedi, G. 1977: *On the Economic Identification of Social Classes*. London: Routledge & Kegan Paul.

Carey, L. and Mapes, R. 1972: *The Sociology of Planning: A Study of Social Activity on New Housing Estates*. London: Batsford.

Carney, J. 1980: Regions in Crisis: Accumulation, Regional Problems and Crisis Formation. In Carney, J., Hudson, R. and Lewis, J. (editors), *Regions in Crisis*. London: Croom Helm.

——, Hudson, R. and Lewis, J. (editors) 1980: *Regions in Crisis*. London: Croom Helm.

Carter, H. 1980: *The Study of Urban Geography*. London: Edward Arnold.

Cashmore, E. and Troyna, B. 1983: *Introduction to Race Relations*. London: Routledge & Kegan Paul.

Castells, M. 1977: *The Urban Question*. London: Edward Arnold.

—— 1978: *City, Class and Power*. London: Macmillan.

—— 1983: *The City and the Grassroots*. London: Edward Arnold.

Castles, S. 1984: *Here for Good: Western Europe's New Ethnic Minorities*. London: Pluto Press.

—— and Kosack, G. 1973: *Immigrant Workers and Class Structure in Western Europe*. London: Oxford University Press.

Cater, J. 1981: The Impact of Asian Estate Agents on Patterns of Ethnic Residence: A Case Study of Bradford. In Jackson, P. and Smith, S. (editors), *Social Interaction and Ethnic Segregation*. London: Academic Press, 163–83.

—— 1984: Acquiring Premises: A Case Study of Asians in Bradford. In Ward, R. and Jenkins, R. (editors), *Ethnic Communities in Business*. Cambridge: Cambridge University Press, 197–212.

—— and Jones, T. 1979: Ethnic Residential Space: The Case of Asians in Bradford. In *Tijdschrift voor Economische en Sociale Geografie* 70, 2, 86–97.

—— and Jones, T. 1987: Asian Ethnicity, Home Ownership and Social Reproduction. In Jackson, P. (editor), *Race and Racism*. London: Allen & Unwin.

Cell, J. 1982: *The Highest Stage of White Supremacy; Origins of Segregation in South Africa and the American South*. Cambridge: Cambridge University Press.

Central Advisory Council for Education 1967: *Children and their Primary Schools* (The Plowden Report). London: HMSO.

Central Statistical Office 1985a: *General Household Survey 1983*. London: HMSO.

—— 1985b: *Social Trends*. London: HMSO.

—— 1986a: *Annual Abstract of Statistics*. London: HMSO.

—— 1986b: *Regional Trends*. London: HMSO.

—— 1986c: *Social Trends*. London: HMSO.

—— 1987a: *General Household Survey 1985* London: HMSO.

—— 1987b: *Regional Trends*. London: HMSO.

—— 1987c: *Social Trends*. London: HMSO.

—— 1988a: *Regional Trends*. London: HMSO.

—— 1988b: *Social Trends*. London: HMSO.

Centre for Contemporary Cultural Studies 1982: *The Empire Strikes Back: Race and Racism in 70s Britain*. London: Hutchinson.

Chambers, G. and Thombs, J. 1984: *The British Crime Survey: Scotland*. Edinburgh: HMSO.

Charlesworth, A. (editor) 1983: *An Atlas of Rural Protest*. London: Croom Helm.

Chartered Institute of Public Finance and Accountancy 1988: *Education Statistics: Education Estimates 1987/8*. London: CIPFA.

Chisholm, M. 1974: Regional Policies for the 1970s. In *Geographical Journal* 140, 215–31.

City of Bradford Metropolitan District Council 1984: *Bradford in Figures*. Bradford: City of Bradford Metropolitan District Council.

Clark, C. 1967: *Population Growth and Land Use*. London: Macmillan.

——, Wilson, P. and Bradley, J. 1969: Industrial Location and Economic Potential in Western Europe. In *Regional Studies* 3, 2, 197–212.

Clarke, C., Ley, D. and Peach, C. (editors) 1984: *Geography and Ethnic Pluralism*. London: Allen & Unwin.

Clarke, J. 1984: 'There's no place like. . . .': Cultures of Difference. In Massey, D. and Allen, J. (editors), *Geography Matters: A Reader*. Cambridge: Cambridge University Press.

Clarke, R. 1984: Opportunity-Based Crime Rates. In *British Journal of Criminology* 24, 74–83.

Clinard, M. 1963: *The Sociology of Deviant Behaviour*. New York: Holt, Rinehardt & Winston.

Cloke, P. 1977: An Index of Rurality for England and Wales. In *Regional Studies* 11, 1, 31–46.

—— 1979: *Key Settlements in Rural Areas*. London: Methuen.

—— 1985: Counter-Urbanisation: A Rural Perspective. In *Geography* 70, 13–23.

—— and Edwards, G. 1986: Rurality in England and Wales 1981: A Replication of the 1971 Index. In *Regional Studies* 20, 289–306.

Coates, B., Johnston, R. and Knox, P. 1977: *Geography and Inequality*. Oxford: Oxford University Press.

Cockburn, C. 1977: *The Local State*. London: Pluto.

—— 1981: The Material of Male Power. In *Feminist Review* 9, Autumn, 41–58.

Cohen, A. 1974: Introduction. In Cohen A. (editor), *Urban Ethnicity*. London: Tavistock.

Cohen, A.P. (editor) 1982: *Belonging: Identity and Social Organisation in British Rural Cultures*. Manchester: Manchester University Press.

Cohen, P. 1979: Policing the Working Class City. In Fine, B., Kinsey, R., Lea, J., Picciotto, S. and Young, J. (editors), *Capitalism and the Rule of Law: From Deviancy Theory to Marxism*. London: Hutchinson.

Coleman, A. 1985: *Utopia on Trial*. London: Shipman.

Coles, O.B. 1986: Rural Transport Needs. In *Social Policy and Administration* 20, 58–73.

Collison, P. and Mogey, J. 1959: Residence and Social Class in Oxford. In *American Journal of Sociology* 54, 6, 599–605.

Commission on Local Authority and Allied Personal Social Services 1968: *The Seebohm Report*. London: HMSO.

Community Development Project 1977: *Gilding the Ghetto: The State and Poverty Experiments*. London: CDP Inter-Project Editorial Team.

Connell, J. 1973: Social Networks in Urban Society. In Clark, B.D. and Gleave, M.B. (editors), *Social Patterns in Cities*. London: Institute of British Geographers' Special Publication No. 5.

Cooke, P. 1982: Class Relations and Uneven Development in Wales. In Day, G. (editor), *Diversity and Decomposition in the Labour Market*. Aldershot: Gower.

—— 1985: Class Practices as Regional Markers: A Contribution to Labour Geography. In Gregory, D. and Urry, J. (editors), *Social Relations and Spatial Structures*. London: Macmillan, 213–41.

—— 1986: The Changing Urban and Regional System in the United Kingdom. In *Regional Studies* 20, 243–51.

Cooper, S. 1981: *Rural Poverty in the United Kingdom*. London: Policy Studies Institute.

Coppock, J.T. 1977: *Second Homes: Curse or Blessing?* London: Faber & Faber.

Cornfoot, T. 1982: The Economy of Merseyside 1945–82. In Gould, W. and Hodgkiss, A. (editors), *The Resources of Merseyside*. Liverpool: Liverpool University Press, 14–26.

Corrigan, P. 1979: The Local State: The Struggle for Democracy. In *Marxism Today*, July.

Cowell, D., Jones, T. and Young, J. (editors) 1982: *Policing the Riots*. London: Junction Books.

Cox, K.R. 1981: Capitalism and Conflict around the Communal Living Space. In Dear, M. and Scott, M.B. (editors), *Urbanisation and Urban Planning in Capitalist Society*. London: Methuen.

—— and McCarthy, J.J. 1982: Neighbourhood Activism as a Politics of Turf: A Critical Analysis. In Cox, K.R. and Johnston, R.J. (editors), *Conflict, Politics and the Urban Scene*. London: Longman, 196–219.

Cox, O.C. 1948: *Caste, Class and Race*. New York: Monthly Review Press.

Crompton, R. and Mann, M. 1986: Introduction. In Crompton, R. and Mann, M. (editors) *Gender and Stratification*. London: Polity Press.

Cross, M. 1982: The Manufacture of Marginality. In Cashmore, E. and Troyna, B. (editors), *Black Youth in Crisis*. London: Allen & Unwin, 35–52.

Crouch, C. (editor) 1979: *State and Economy in Contemporary Capitalism*. London: Croom Helm.

Cummings, S. (editor) 1980: *Self-Help in Urban America*. New York: Kennikat.

Dahrendorf, R. 1959: *Class and Class Conflict in Industrial Society*. London: Routledge.

Dahya, B. 1972: Pakistanis in England. In *New Community* 2, 25–33.

—— 1973: Pakistanis in Britain: Transients or Settlers? In *Race* 14, 241–77.

—— 1974: The Nature of Pakistani Ethnicity in Industrial Cities in Britain. In Cohen, A. (editor), *Urban Ethnicity* London: Tavistock, 77–118.

Dalla Costa 1972: *Women and the Subversion of the Community*. New York: Falling Wall Press.

Dalton, M. and Seaman, J.M. 1973: The Distribution of New Commonwealth Immigrants in the London Borough of Ealing 1961–66. In *Transactions, Institute of British Geographers* 58, 21–39.

Damette, F. 1980: The Regional Framework of Monopoly Exploitation: New Problems and Trends. In Carney, J., Hudson, R. and Lewis, J. (editors), *Regions in Crisis*. London: Croom Helm.

Dange, S. 1986: *Hindu Domestic Rituals*. Delhi: South Asia Books.

Daniel, W. 1968: *Racial Discrimination in England*. Harmondsworth: Penguin.

Darke, J. and Darke, R. 1979: *Who Needs Housing?* London: Macmillan.

Davidoff, L. 1986: The Role of Gender in the 'First Industrial Nation': Agriculture in

England 1780–1850. In Crompton, R. and Mann, M. (editors), *Gender and Stratification*. London: Polity Press.

Davidson, N. 1981: *Crime and Environment*. London: Croom Helm.

Davies, R.B. and O'Farrell, P.N. 1981: A Spatial and Temporal Analysis of Second Home Ownership in Wales. In *Geoforum* 12, 161–78.

Davies, W.K.D. 1978: Alternative Factorial Solutions and Urban Social Structure: A Data Analysis Exploration of Calgary in 1971. In *Canadian Geographer* 22, 273–97.

—— and Lewis, G.J. 1973: The Urban Dimensions of Leicester, England. In Clark, B.D. and Gleave, M.B. (editors), *Social Patterns in Cities*. London: Institute of British Geographers' Special Publication No. 5, 71–86.

Davison, R.B. 1962: *West Indian Migrants*. London: Oxford University Press.

Day, G. (editor) 1982: *Diversity and Decomposition in the Labour Market*. Farnborough: Gower.

Deakin, N. 1970: *Colour, Citizenship and British Society*. London: Panther.

Dear, M. and Scott, M. (editors) 1981: *Urbanisation and Urban Planning in Capitalist Society*. London: Methuen.

Dennis, N. 1968: The Popularity of the Neighbourhood Idea. In Pahl, R.E. (editor), *Readings in Urban Sociology*. London: Pergamon.

——, Henriques, F. and Slaughter, C. 1957: *Coal is our Life*. London: Eyre & Spottiswoode.

Dennis, R. 1978: The Decline of Manufacturing Employment in Greater London 1966–74. In *Urban Studies* 15, 63–73.

Department of Employment 1986a: *Employment Gazette*. London: Department of Employment.

—— 1986b: *Action for Jobs*. London: Department of Employment.

—— 1987: *1986 Labour Force Survey*. London: HMSO.

—— 1988: *Employment Gazette*. London: Department of Employment.

Department of Environment 1971: *The Nature of Rural Areas of England and Wales*. London: Department of the Environment Internal Working Paper.

—— 1977a: *Inner Area Studies; Liverpool, Birmingham and Lambeth. Summaries of Consultants' Final Reports*. London: HMSO.

—— 1977b: *Housing and Planning Technical Volume*, Volume III. London: HMSO.

De Vise, R. 1968: *Slum Medicine Chicago-Style: How the Medical Needs of the City's Negro Poor are Met*. Working Paper 3.IV.7., Chicago Regional Hospital Study. Chicago: Chicago RHS.

Dex, S. 1984: Evidence from the 1980 Women and Employment Survey. In *Employment Gazette*, December, 545–9.

Dicken, P. 1976: The Multi-Plant Enterprise and Geographic Space. In *Regional Studies* 10, 4, 401–12.

—— 1986: *Global Shift: Industrial Change in a Turbulent World*. London: Harper & Row.

Dickens, P. 1986: Review of Coleman's 'Utopia on Trial'. In *International Journal of Urban and Regional Research*, 10, 297–300.

——, Duncan, S., Goodwin, M. and Gray, F. 1985: *Housing, State and Localities*. London: Methuen.

Doherty, J. 1969: The Distribution and Concentration of Immigrants in London. In *Race Today* 1, 8, 227–31.

—— 1973: Race, Class and Residential Segregation. In *Antipode* 3, 45–51.

Drake, St C. and Cayton, H.R. 1962: *Black Metropolis: A Study of Negro Life in a Northern City*. New York: Harper & Row.

Drewnowski, J. 1974: *On Measuring and Planning the Quality of Life*. The Hague: Mouton.

Drudy, P.J. and Drudy, S.M. 1979: Population Mobility and Labour Supply in Rural Regions: North Norfolk and Galway. In *Regional Studies* 13, 91–9.

Duncan, O.D. and Duncan, B. 1955: A Methodological Analysis of Segregation Indexes.

In *American Sociological Review* 20, 210–17.

—— and Lieberson, S. 1959: Ethnic Segregation and Assimilation. In *American Journal of Sociology* 64, 364–74.

Duncan, S. 1977: *Housing Disadvantage and Residential Mobility: Immigrants and Institutions in a Northern Town*. Working Paper No. 5., Faculty of Urban and Regional Studies. Brighton: University of Sussex.

—— 1978: *Housing Reform, the Capitalist State and Social Democracy*. Working Paper No. 4, Centre for Urban and Regional Studies. Brighton: University of Sussex.

Dunleavy, P. 1977: Protest and Quiescence in Urban Politics: A Critique of some Pluralist and Structuralist Myths. In *International Journal of Urban and Regional Research* 1, 185–91.

—— 1980a: *Urban Political Analysis*. London: Macmillan.

—— 1981: *The Politics of Mass Housing in Britain 1945–75*. London: Oxford University Press.

—— 1982: Perspectives on Urban Studies. In Blowers, A., Brook, C., Dunleavy, P. and McDowell, L. (editors), *Urban Change and Conflict: An Interdisciplinary Reader*. London: Harper & Row/Open University.

Dunning, E. (with others) 1988; *The Roots of Football Hooliganism*. London: Routledge & Kegan Paul.

Durkheim, E. 1947: *Division of Labour in Society*. New York: Free Press of Glencoe.

Edwards, G.W. 1985; Rural Public Transport Alternatives in Central Powys. Unpublished paper presented to the Joint Rural Geography/Transport Geography symposium, Institute of British Geographers' Annual Conference, University of Leeds.

Elias, N. 1978: *What is Sociology?* London: Hutchinson.

Emmett, I. 1982: Fe godwn ni eto: Status and Change in a Welsh Industrial Town. In Cohen, A.P. (editor), *Belonging: Identity and Social Organisation in British Rural Cultures*. Manchester: Manchester University Press.

Engels, F. 1845: *The Conditions of the Working Class in England*. Manchester: no publisher cited.

—— 1984 reprint: *The Origins of the Family, Private Property and the State*. New York: International Publishers.

Equal Opportunities Commission 1987: *Women and Men in Britain: A Statistical Profile*. London: HMSO.

Erlich, I. 1973: Participation in Illegitimate Activities: A Theoretical and Empirical Investigation. In *Journal of Political Economy* 81, 521–65.

Evans, A.W. 1973: *The Economics of Residential Location*. London: Macmillan.

Evans, D. 1986: Geographical Analyses of Residential Burglary. In *The Geography of Crime*. Stoke on Trent: North Staffs. Polytechnic Occasional Papers in Geography 7.

—— and Oulds, G. 1984: Geographical Aspects of the Incidence of Residential Burglary in Newcastle under Lyme, UK. In *Tijdschrift voor Economische en Sociale Geografie* 75, 344–55.

Fielding, A.J. 1982: Counterurbanisation in Western Europe. In *Progress in Planning* 17, 1–52.

Fine, B., Kinsey, R., Lea, J., Picciotto, S. and Young, J. 1979: *Capitalism and the Rule of Law: From Deviancy Theory to Marxism*. London: Hutchinson.

Firn, J. 1975: External Control and Regional Development: The Case of Scotland. In *Environment and Planning A*, 7, 4, 393–414.

Foord, J. 1975: The Role of the Building Society Manager in the Urban Stratification System: Autonomy versus Constraint. In *Urban Studies* 12, 295–302.

—— and Gregson, N. 1986: Patriarchy: Towards a Reconceptualisation. In *Antipode* 18, 2, 186–211.

Foot, P. 1969: *The Rise of Enoch Powell*. Harmondsworth: Penguin.

Forrest, R. and Murie, A. 1986: Marginalisation and Subsidised Individualism: The Sale of Council Houses in the Restructuring of the British Welfare State. In *International*

Journal of Urban and Regional Research 10, 46–66.

Foster, J. 1974: *Class Struggle and the Industrial Revolution*. London: Methuen.

Fothergill, S. and Gudgin, G. 1982: *Unequal Growth: Urban and Regional Employment Change in the United Kingdom*. London: Heinemann.

Frank, A.G. 1969: *Capitalism and Underdevelopment in Latin America: Historical Studies of Chile and Brazil*. New York: Monthly Review Press.

Frankenberg, R. 1966: *Communities in Britain*. Harmondsworth: Penguin.

—— 1976: In the Production of Their Lives, Man (?) . . . Sex and Gender in British Community Studies. In Barker, D. and Allen, S. (editors), *Sexual Divisions and Society: Process and Change*. London: Tavistock, 25–51.

Friedrichs, D. 1980: Radical Criminology in the United States: An Interpretive Understanding. In Inciardi, J. (editor), *Radical Criminology: The Coming Crises*. New York: Sage.

Fryer, P. 1984: *Staying Power*. London: Pluto Press.

Gabriel, J. and Ben-Tovim, G. 1978: Marxism and the Concept of Racism. In *Economy and Society* 7, 2, 118–54.

Gaile, C. 1980: The Spread: Backwash Concept. In *Regional Studies* 14, 15–25.

Galbraith, J.K. 1977: *The Age of Uncertainty*. London: Andre Deutsch.

Galster, G.C. 1986: What is Neighbourhood? An Externality-Space Approach. In *International Journal of Urban and Regional Research* 10, 243–64.

Gamarnikov, E., Morgan, D., Purvis, J. and Taylorson, D. (editors) 1983: *Gender, Class and Work*. London: Heinemann.

Gans, H.J. 1962: *The Urban Villagers*. New York: Free Press of Glencoe.

—— 1979: Symbolic Ethnicity: The Future of Ethnic Groups and Cultures in America. In *Ethnic and Racial Studies* 2, 1–20.

Gershuny, J. 1978: *After Industrial Society? The Emerging Self-Service Economy*. London: Macmillan.

Gibb, A. 1983: *Glasgow: The Making of a City*. London: Croom Helm.

Giddens, A. and Mackenzie, G. (editors) 1982: *Social Class and the Division of Labour*. Cambridge: Cambridge University Press.

Gilg, A. 1976: Rural Employment. In Cherry, G. (editor), *Rural Planning Problems*. London: Leonard Hill.

Gill, O. 1977: *Luke Street*. London: Macmillan.

Gilroy, P. 1980: Managing the 'Underclass': A Further Note on the Sociology of Race Relations in Britain. In *Race and Class* 22, 47–62.

—— 1982: Police and Thieves. In Centre for Contemporary Cultural Studies, *The Empire Strikes Back: Race and Racism in 70s Britain*. London: Hutchinson.

—— 1987: *There Ain't No Black in the Union Jack*. London: Hutchinson.

Ginatempo, N. 1985: Social Reproduction and the Structure of Marginal Areas in Southern Italy: Some Remarks on the Role of the Family in the Present Crisis. In *International Journal of Urban and Regional Research* 9, 1, 99–112.

Glass, R. 1960: *Newcomers*. London: Allen & Unwin.

Glazer, N. and Moynihan, D.P. 1963: *Beyond the Melting Pot*. Cambridge, Mass.: MIT Press.

—— (editors) 1975: *Ethnicity: Theory and Experience*. Cambridge, Mass.: Harvard University Press.

Gordon, M.M. 1964: *Assimilation in American Life*. New York: Oxford University Press.

—— 1978: *Human Nature, Class, and Ethnicity*. New York: Oxford University Press.

Gould, W.T.S. and Hodgkiss, A.G. (editors) 1982: *The Resources of Merseyside*. Liverpool: Liverpool University Press.

Gray, F. 1975: Non-Explanation in Urban Geography. In *Area* 7, 228–35.

—— 1976: Selection and Allocation in Council Housing. In *Transactions, Institute of British Geographers* NS 1, 34–46.

—— 1977: The Management of Local Authority Housing. In Conference of Socialist Economists' Political Economy of Housing Workshop, *Housing and Class in Britain*.

London: Conference of Socialist Economists.

Green, A.D. 1979: On the Political Economy of Black Labour and the Racial Restructuring of the Working Class in England. In Centre for Contemporary Cultural Studies Occasional Paper, University of Birmingham. Birmingham: Centre for Contemporary Cultural Studies.

Greve Report 1986: *Homelessness in London*. London: GLC.

Greenberg, D.F. (editor) 1981: *Crime and Capitalism: Readings in Marxist Criminology*. Palo Alto, California: Mayfield.

Gregory, D. and Urry, J. (editors) 1985: *Social Relations and Spatial Structures*. London: Macmillan.

Guelke, L. 1978: Geography and Logical Positivism. In Herbert, D. and Johnston, R., (editors), *Geography and the Urban Environment* Volume I. London: Wiley, 35–61.

Guest, A.M. and Weed, J.A. 1976: Ethnic Residential Segregation: Patterns of Change. In *American Journal of Sociology* 81, 1088–111.

Gusfield, J.R. 1975: *Community: A Critical Response*. Oxford: Blackwell.

Gutzmore, C. 1983: Capital, 'Black Youth' and Crime. *Race and Class* 25, 13–30.

Habermas, J. 1973: *Legitimation Crisis*. Boston, Mass.: Beacon.

Hall, C. 1982: The Home Turned Upside Down? The Working Class Family in Cotton Textiles 1780–1850. In Whitelegg, E., Arnot, M., Bartels, E., Beechey, V., Birke, L., Himmelweit, S., Leonard, D., Ruehl, S. and Speakman, M. (editors), *The Changing Experience of Women*. Oxford: Basil Blackwell/Open University, 17–29.

Hall, P. 1984: Enterprises of Great Pith and Moment? In *Town and Country Planning*, November, 296–7.

——, Thomas, R., Gracey, H. and Drewett, R. 1973: *The Containment of Urban England*. London: Allen & Unwin.

Hall, S. 1980: *Drifting into a Law and Order Society*. London: Cohen Trust.

Hall, S. 1985: Cold Comfort Farm. In *New Socialist* 32, November, 10–12.

Hall, S., Critcher, C., Jefferson, T., Clarke, J. and Roberts, B. 1978: *Policing the Crisis: Mugging, the State and Law and Order*. London: Macmillan.

Halsall, D.A. and Turton, B.J. (editors) 1979: *Rural Transport Problems in Britain*. Keele: Department of Geography University of Keele for the Institute of British Geographers' Transport Geography Study Group.

Hamnett, C. 1973: Improvement Grants as an Indicator of Gentrification in Inner London. In *Area* 4, 252–61.

—— 1979: Area-Based Explanations: A Critical Appraisal. In Herbert, D.T. and Smith, D.M. (editors), *Social Problems and the City*. London: Oxford University Press, 244–60.

—— 1984: Housing the Two Nations. In *Urban Studies* 43, 389–405.

Hargreaves, D. 1967: *Social Relations in a Secondary School*. London: Routledge & Kegan Paul.

Harloe, M. 1981: The Recommodification of Housing. In Harloe, M. and Lebas, E. (editors), *City, Class and Capital: New Developments in the Political Economy of Cities and Regions*. London: Edward Arnold, 17–50.

—— and Paris, C. 1984: The Decollectivisation of Consumption: Housing and Local Government Finance in England and Wales 1979–81. In Szelyeni, I. (editor), *Cities in Recession*. London: Sage.

Harries, K.D. 1974: *The Geography of Crime and Justice*. New York: McGraw-Hill.

—— 1976: Cities and Crime: A Geographic Model. In *Criminology* 14, 369–86.

Harrison, B. 1982: The Politics and Economics of the Urban Enterprise Zone Proposal: A Critique. In *International Journal of Urban and Regional Research* 6, 3, 416–21.

Hartmann, H. 1981: The Unhappy Marriage of Marxism and Feminism: Towards a More Progressive Union. In Sargent, L. (editor), *Women and Revolution*. London: Pluto Press.

Hartshorne, R. 1938: Racial Maps of the United States. In *Geographical Review* 28, 276–88.

Harvey, D. 1973: *Social Justice and the City*. London: Edward Arnold.
—— 1975: The Geography of Capitalist Accumulation: A Reconstruction of the Marxian Theory. In *Antipode* 7, 2, 9–21.
—— 1977: Government Policies, Financial Institutions and Neighbourhood Change in United States Cities. In Harloe, M. (editor), *Captive Cities*. London: Wiley, 123–39.
—— 1978: Labor, Capital and Class Struggle around the Built Environment in Advanced Capitalist Societies. In Cox, K.R. (editor), *Urbanisation and Conflict in Market Societies*. New York: Methuen, 9–37.
—— 1981: The Spatial Fix: Hegel, von Thünen and Marx. In *Antipode* 13, 3, 1–12.
—— 1982: *The Limits to Capital*. Oxford: Basil Blackwell.
Hawley, A.H. 1981: *Urban Society*. New York: Wiley.
Hay, A. 1985: Scientific Method in Geography. In Johnston, R. (editor), *The Future of Geography*. London: Methuen, 129–42.
Hayford, A. 1974: The Geography of Women: An Historical Introduction. In *Antipode* 6, 2, 1–19.
Heller, T. 1979: Rural Health and Health Services. In Shaw, J.M. (editor), *Rural Deprivation and Planning*. Norwich: Geo Abstracts.
Herbert, D.T. 1972: *Urban Geography: A Social Perspective*. Newton Abbot: David & Charles.
—— 1973: Residential Mobility and Preference: A Study of Swansea. In Clark, B.D. and Gleave, M.B. (editors), *Social Patterns in Cities*. London: Institute of British Geographers' Special Publication No. 5, 103–21.
—— 1975: Urban Neighbourhoods and Socio-Geographic Research. In Phillips, A.D. and Turton, B.J. (editors), *Environment, Man and Economic Change*. London: Longman, 459–78.
—— 1976a: Urban Education: Problems and Policies. In Herbert, D.T. and Johnston, R.J. (editors), *Social Areas in Cities*. London: Wiley, 123–58.
—— 1976b: The Study of Delinquency Areas: A Social Geographical Approach. In *Transactions, Institute of British Geographers* New Series 1, 472–92.
—— 1977a: An Areal and Ecological Analysis of Delinquency Residence: Cardiff 1966 and 1971. In *Tijdschrift voor Economische en Sociale Geografie* 68, 83–99.
—— 1977b: Crime, Delinquency and the Urban Environment. In *Progress in Human Geography* 1, 208–39.
—— 1979: Urban Crime: A Geographical Perspective. In Herbert, D.T. and Smith, D.M. (editors), *Social Problems and the City*. Oxford: Oxford University Press, 117–38.
—— 1982: *The Geography of Urban Crime*. London: Longman.
—— 1983: Crime and Delinquency. In Pacione, M. (editor) *Progress in Urban Geography*. London: Croom Helm.
—— 1986: Geographers and Crime: An Overview. In Evans, D. (editor), *The Geography of Crime*. Stoke-on-Trent: North Staffs. Polytechnic Occasional Papers in Geography No. 7.
—— and de Silva, S. 1974: Social Dimensions of a Non-Western City: A Factorial Ecology of Colombo. In *Cambria* 1, 139–58.
—— and Smith, D.M. (editors) 1979: *Social Problems and the City*. Oxford: Oxford University Press.
—— and Raine, J.W. 1976: Defining Communities Within Urban Areas. In *Town Planning Review* 47, 325–38.
Higgins, J., Deakin, N., Edwards, J. and Wicks, M. 1983: *Government and Urban Poverty*. Oxford: Blackwell.
Hillery, G.A. 1955: Definitions of Community: Areas of Agreement. In *Rural Sociology* 20, 111–23.
Hiro, D. 1973: *Black British, White British*. London: Eyre & Spottiswoode.
Hirst, M. 1980: The Geographical Basis of Community Work. In *Community Development Journal* 15, 53–9.
Hirst, P. 1972: Marx and Engels on Law, Crime and Morality. In *Economy and Society*.

Hobsbawn, E. 1969: *Bandits*. New York: Delacorte.
—— 1971: *Primitive Rebels*. Manchester: Manchester University Press.
Hoggart, R. 1957: *The Uses of Literacy: Aspects of Working Class Life*. London: Chatto Windus.
Holman, R. 1978: *Poverty*. London: Martin Robertson.
Holtermann, S. 1975: Areas of Urban Deprivation in Great Britain: An Analysis of 1971 Census Data. In *Social Trends* 6, 33–47.
Hood, R. and Sparks, R. 1970: *Key Issues in Criminology*. London: Weidenfeld & Nicolson.
Hope, E. *et al* 1976: Homeworkers in North London. In Barker, D. and Allen, S. (editors), *Dependence and Exploitation in Work and Marriage*. London: Longman.
Hoskins, W.G. 1964: *A New Survey of England: Devon*. London: Collins.
House, J.W. 1956: *Northumbrian Tweedside: The Rural Problem*. Newcastle: University of Newcastle.
—— 1965: *Rural North-East England 1951-61*. Newcastle: University of Newcastle.
Hough, M. and Mayhew, P. 1983: *The British Crime Survey: The First Report*. London: HMSO.
—— 1985: *Taking Account of Crime: Key Findings from the British Crime Survey*. London: HMSO.
Hoyt, H. 1939: *The Structure and Growth of Residential Neighbourhoods in American Cities*. Washington, DC: Federal Housing Administration.
Hudson, R. 1978: Spatial Policy in Britain: Regional or Urban? In *Area* 10, 2, 121–2.
—— 1983: Regional Labour Reserves and Industrialisation in the EEC. In *Area* 15, 3, 223–30.
—— and Williams, A. 1986: *The United Kingdom*. London: Harper & Row.
Hull, A. 1985: Changing Patterns of Accessibility and Mobility in Sixteen Parishes in East Kent. In Cloke P. (editor), *Rural Accessibility and Mobility: Papers from the Joint Symposium of the Rural Geography and the Transport Geography Study Groups*. Lampeter: St David's University College Centre for Rural Transport Studies, 19–41.
Hunter, A. 1979: The Urban Neighbourhood: Its Analytical and Social Context. In *Urban Affairs Quarterly* 15, 267–88.
Husain, M.S. 1975: The Increase and Distribution of New Commonwealth Immigrants in Greater Nottingham. In *East Midlands Geographer* 6, 105–29.
Inciardi, J. (editor) 1980: *Radical Criminology: The Coming Crises*. New York: Sage.
Isard, W. 1956: *Location and Space Economy*. New York; Wiley.
Jackson, J.A. 1963: *The Irish in Britain*. London: Routledge.
Jackson, P. 1981: Paradoxes of Puerto Rican Segregation in New York. In Peach, C., Robinson, V. and Smith, S.J. (editors), *Ethnic Segregation in Cities*. London: Croom Helm, 109–26.
—— 1987: The Idea of 'Race' and the Geography of Racism. In Jackson P. (editor), *Race and Racism: Essays in Social Geography*. London: Allen & Unwin, 3–22.
—— and Smith, S.J. (editors) 1981: *Social Interaction and Ethnic Segregation*. London: Academic Press.
Jenkins, D. 1971: *The Agricultural Community in South West Wales at the Turn of the Turn of the Twentieth Century*. Cardiff: University of Wales Press.
Johnston, R.J. 1971: *Urban Residential Patterns: An Introductory Review*. London: Bell.
—— 1973: Residential Differentiation in Major New Zealand Urban Areas: A Comparative Factorial Ecology. In Clark, B.D. and Gleave, M.B. (editors), *Social Patterns in Cities*. London: Institute of British Geographers' Special Publication No. 5, 143–68.
—— 1974: Social Distance, Proximity and Social Contact: Eleven Cul-de-Sacs in Christchurch, New Zealand. In *Geografisker Annaler* 56, 57–67.
—— 1983: *Geography and Geographers*. London: Edward Arnold.
—— and Brack, E.V. 1983: Appointment and Promotion in the Academic Labour Market: A Preliminary Study of British University Departments of Geography

1933–82. In *Transactions, Institute of British Geographers* 8, 100–11.

Jones, E. 1980: Social Geography. In Brown, E. (editor), *Geography Yesterday and Tomorrow*. Oxford: Oxford University Press, 251–62.

—— and Eyles, J. 1977: *An Introduction to Social Geography*. Oxford: Oxford University Press.

Jones, G. (editor) 1980: *New Approaches to the Study of Central: Local Government Relationships*. Farnborough: Gower.

Jones, H. 1958: Approaches to an Ecological Study. In *British Journal of Delinquency* 8, 277–93.

Jones, P. 1967: The Segregation of Immigrant Communities in the City of Birmingham 1961. In *University of Hull Occasional Papers in Geography*, No. 7. Hull: University of Hull Department of Geography.

—— 1970: Some Aspects of the Changing Distribution of Coloured Immigrants in Birmingham 1961–66. In *Transactions, Institute of British Geographers* 50, 199–219.

—— 1976: Some Aspects of the Changing Distribution of Colored Immigrants in Birmingham, England. In *Annals, Association of American Geographers* 66, 89–102.

—— 1978: The Distribution and Diffusion of the Coloured Population in England and Wales 1961–71. In *Transactions, Institute of British Geographers* New Series 3, 4, 513–33.

—— 1979: Ethnic Areas in British Cities. In Herbert, D.T. and Smith, D.M. (editors), *Social Problems and the City*. London: Oxford University Press, 158–85.

Jones, T.P. 1968: Rural Depopulation in South Devon. Unpublished MA thesis, University of London.

—— 1983/4: Residential Segregation and Ethnic Autonomy. In *New Community* 13, 1, 10–22.

—— and McEvoy, D. 1978: Race and Space in Cloud-Cuckoo Land. In *Area* 10, 3, 162–6.

—— 1986: Ethnic Enterprise: The Popular Image. In Curran, J., Stanworth, J. and Watkins, D. (editors), *The Survival of the Small Firm: The Economics of Survival and Entrepreneurship*. Aldershot: Gower.

Joshi, H.E. 1984: Women's Participation in Paid Work: Further Analysis of the Women and Employment Survey. In *Department of Employment Research Papers*, No. 45. London: Department of Employment.

Joshi, S. and Carter, R. 1984: The Role of Labour in the Creation of a Racist Britain. In *Race and Class* 25, 53–70.

Kantrowitz, N. 1969: *Negro and Puerto Rican Populations of New York City in the Twentieth Century*. New York: American Geographical Society.

—— 1981: Ethnic Segregation: Social Reality and Academic Myth. In Peach, C., Robinson, V. and Smith, S. (editors), *Ethnic Segregation in Cities*. London: Croom Helm, 43–57.

Karn, V. 1977/8: The Financing of Owner Occupation and its Impact on Ethnic Minorities. In *New Community* 6, 49–64.

Kasarda, J.D. and Janowitz, M. 1974: Community Attachment in Mass Society. In *American Sociological Review* 39, 328–39.

Kearsley, G. and Srivastava, S.R. 1974: The Spatial Evolution of Glasgow's Asian Community. In *Scottish Geographical Magazine* 90, 110–24.

Keeble, D. 1976: *Industrial Location and Planning in the United Kingdom*. London: Methuen.

—— 1977: Spatial Policy in Britain; Regional or Urban? In *Area* 9, 1, 3–8.

—— 1984: Industrial Location and Regional Development. In Short, J. and Kirby, A. (editors), *The Human Geography of Contemporary Britain*. London: Macmillan, 40–51.

——, Owens, P. and Thompson, C. 1982: *Centrality, Peripherality and EEC Regional Development*. Cambridge: Cambridge University Press.

Kemeny, J. 1980: Home Ownership and Privatisation. In *International Journal of Urban and Regional Research* 4, 372–88.

King, R. 1982: Southern Europe: Dependency or Development? In *Geography* 67, 221–34.

Kinsey, R. 1984: *Merseyside Crime Survey: First Report*. Liverpool: Merseyside County Council.

Kirby, A. 1982: *The Politics of Location*. London: Methuen.

——, Knox, P.L. and Pinch, S. (editors) 1984: *Public Service Provision and Urban Development*. London: Croom Helm.

Kirby, D.A. 1979, Slum Housing and Residential Renewal. London: Longman.

Knox, P.L. 1975: *Social Well-Being: A Spatial Perspective*. Oxford: Oxford University Press.

—— 1987: *Urban Social Geography: An Introduction*. London: Longman (2nd edition).

—— and Cottam, M.B. 1981: A Welfare Approach to Rural Geography: Contrasting Perspectives on the Quality of Highland Life. In *Transactions, Institute of British Geographers* New Series 6, 433–50.

Lacey, C. 1970: *Hightown Grammar: The School as a Social System*. Manchester: Manchester University Press.

Lambert, C. 1976: Building Societies, Surveyors and the Older Areas of Birmingham. Working Paper No. 38, Centre for Urban and Regional Studies. Birmingham: University of Birmingham.

Land, H. 1976: Women: Supporters or Supported? In Barker, D. and Allen, S. (editors), *Sexual Divisions in Society*. London: Tavistock.

Lander, B. 1954: *Towards an Understanding of Juvenile Delinquency*. New York: Columbia University Press.

Lapple, D. and Van Hoogerstrand, P. 1980: Remarks on the Spatial Structure of Capitalist Development: The Case of the Netherlands. In Carney, J., Hudson, R. and Lewis, J. (editors), *Regions in Crisis*. London: Croom Helm, 117–66.

Larkin, A. 1979: Rural Housing and Housing Needs. In Shaw, J.M. (editor), *Rural Deprivation and Planning*. Norwich: Geo Books, 71–80.

Laski, H. 1968: *Reflections on the Revolution of our Time*. London: Frank Cass.

Law, C.M. 1981: *British Regional Development since World War I*. London: Methuen.

Lawless, P. 1979: *Urban Deprivation and Government Initiative*. London: Faber & Faber.

—— 1981: *Britain's Inner Cities: Problems and Policies*. London: Harper & Row.

—— 1986: *The Evolution of Spatial Policy*. London: Pion.

Lawrence, E. 1982: Just Plain Common Sense: The 'Roots' of Racism. In Centre for Contemporary Cultural Studies, *The Empire Strikes Back: Race and Racism in 70s Britain*. London: Hutchinson.

Lawton, R. 1982: From the Port of Liverpool to the Conurbation of Merseyside. In Gould, W.T.S. and Hodgkiss, A. (editors), *The Resources of Merseyside*. Liverpool: Liverpool University Press.

—— and Cunningham, C. (editors) 1969: *Merseyside: Social and Economic Studies*. London: Longman.

Lea, J. and Young, J. 1982: The Riots in Britain 1981: Urban Violence and Political Marginalisation. In Cowell, D., Jones, T. and Young, J. (editors), *Policing the Riots*. London: Junction Books.

—— 1984: *What is to be Done about Law and Order?* Harmondsworth: Penguin.

Lee, C. 1979: *British Regional Employment Statistics 1841-1971*. Cambridge: Cambridge University Press.

Lee, D. and Newby, H. 1983: *The Problem of Sociology*. London: Hutchinson.

Lee, T.R. 1968: Urban Neighbourhood as a Socio-Spatial Scheme. In *Human Relations* 21, 241–67.

—— 1973: Immigrants in London: Trends in Distribution and Concentration 1961–71. In *New Community* 2, 2, 145–58.

—— 1977: *Race and Residence: The Concentration and Dispersal of Immigrants in London*. Oxford: Clarendon Press.

Lemert, E.M. 1967: *Human Deviance, Social Problems and Social Control*. New York: Prentice-Hall.

Lenin, V.I. 1964: *The Development of Capitalism in Russia*. Moscow: Progress Publishers.

Lewis, G.J. 1979: *Rural Communities: A Social Geography*. Newton Abbot: David & Charles.

—— 1983: Rural Communities. In Pacione, M. (editor), *Progress in Rural Geography*. London: Croom Helm.

Lewis, O. 1967: *The Children of Sanchez*. New York: Random House.

Ley, D. 1974: *The Black Inner City as Frontier Outpost: Images and Behavior of a Philadelphia Neighbourhood*. Washington, DC: Association of American Geographers' Monograph Series No. 7.

—— 1983: *A Social Geography of the City*. New York: Harper & Row.

Lieberson, S. 1963: *Ethnic Patterns in American Cities*. New York: Free Press of Glencoe.

—— 1981: An Asymmetrical Approach to Segregation. In Peach, C., Robinson, V. and Smith, S. (editors), *Ethnic Segregation in Cities*. London: Croom Helm, 61–82.

Liepitz, A. 1980: The Structuralisation of Space, the Problem of Land, and Spatial Policy. In Carney, J., Hudson, R. and Lewis, J. (editors), *Regions in Crisis*. London: Croom Helm, 60–75.

Light, I. 1972: *Ethnic Enterprise in America: Business and Welfare Among Chinese, Japanese and Blacks*. Berkeley: University of California Press.

Little, J. 1986: Feminist Perspectives in Rural Geography: An Introduction. In *Journal of Rural Studies* 2, 1, 1–8.

Liverpool Polytechnic 1982: *The Finch House Study*. Liverpool: Liverpool Polytechnic for Knowsley Metropolitan Borough Council and the National Association for the Care and Rehabilitation of Offenders.

Lloyd, G. 1984: Policies in Search of an Opportunity. In *Town and Country Planning*, November, 298–300.

Lloyd, P. 1969: The Impact of Development Area Policies on Merseyside 1949–1967. In Lawton, R. and Cunningham, C. (editors), *Merseyside: Social and Economic Studies*. London: Longman, 374–410.

—— 1979: The Components of Industrial Change for Merseyside Inner Area. In *Urban Studies* 16, 45–60.

—— and Dicken, P. 1981: *Modern Western Society*. London: Harper & Row.

—— and Shutt, J. 1983: Recession and Restructuring in the North West Region. North West Industry Research Unit Working Paper No. 13. Manchester: University of Manchester Department of Geography.

Lockwood, D. 1982: Fatalism: Durkheim's Hidden Theory of Order. In Giddens, A. and MacKenzie, G. (editors), *Social Class and the Division of Labour*. Cambridge: Cambridge University Press, 101–18.

Loney, M. 1983: *Community Against Government: The British Community Development Project 1968–78*. London: Heinemann.

Lowman, J. 1982: Crime, Criminal Justice Policy and the Urban Environment. In Herbert, D.T. and Johnston, R.J. (editors), *Geography and the Urban Environment (Vol. 5)*. London: Wiley, 307–41.

Lown, J. 1983: Not so much a Factory, More a Form of Patriarchy: Gender and Class During Industrialisation. In Gamarnikov, E., Morgan, D., Purvis, J. and Taylorson, D. (editors), *Gender, Class and Work*. London: Heinemann.

McCracken, K. 1975: Household Awareness Spaces and Intra-Urban Migration Search Behaviour. In *Professional Geographer* 27, 166–70.

McCrone, G. 1969: *Regional Policy in Britain*. London: Allen & Unwin.

McDowell, L. 1983: Towards an Understanding of the Gender Division of Urban Space. In *Environment and Planning D*, 1, 59–72.

—— 1986: Beyond Patriarchy: A Class-Based Explanation of Women's Subordination. In *Antipode* 18, 3, 311–21.

—— and Massey, D. 1984: A Woman's Place? In Massey, D. and Allen, J. (editors) *Geography Matters!* Cambridge: Cambridge University Press/Open University Press, 128–47.

McKay, D. and Cox, A. 1979: *The Politics of Urban Change*. London: Croom Helm.

Mackay, R. 1976: The Impact of the Regional Employment Premium. In Whiting, A. (editor), *The Economics of Industrial Subsidies*. London: HMSO.

MacKenzie, G. 1982: Class Boundaries and the Labour Process. In Giddens, A. and MacKenzie, G. (editors), *Social Class and the Division of Labour*. Cambridge: Cambridge University Press, 63–86.

Mack, J. and Lansley, S. 1985: *Poor Britain*. London: Allen & Unwin.

Malpass, P. (editor) 1986: *Housing in Crisis*. London: Croom Helm.

Mandel, E. 1975: *Late Capitalism*. London: New Left Books.

Manion, T. 1982: Access to Local Authority Housing: A Case Study of Liverpool. Unpublished Ph.D. thesis, University of Lancaster.

Mann, P. 1965: *An Approach to Urban Sociology*. London: Routledge & Kegan Paul.

Marable, M. 1984: *Race, Reform and Rebellion: The Second Reconstruction of Black America, 1945–82*. New York: Macmillan.

Marglin, S. 1976: *Value and Price in the Labour Surplus Economy*. London: Oxford University Press.

Martin, J. and Roberts, C. 1984: *Women and Employment*. London: HMSO.

Martin, R. 1985: Women and Redundancy: Some Case Studies in Manufacturing Industries. In *Employment Gazette*, February, 59–63.

Martin, R. and Rowthorn, B. (editors) 1986: *The Geography of De-Industrialisation*. London: Macmillan.

Martinelli, P. 1985: Public Policy and Industrial Development in Southern Italy: The Anatomy of a Dependent Region. In *International Journal of Urban and Regional Research* 9, 1, 47–81.

Marx, K. 1973: *Grundrisse*. Harmondsworth: Penguin.

—— and Engels, F. 1848: *The Manifesto of the Communist Party*. Harmondsworth: Penguin (1969 reprint; see also 1888 edition).

Mason, C. 1981: Manufacturing Decentralisation: Some Evidence from Greater Manchester. In *Environment and Planning A* 13, 869–84.

Massey, D. 1978: Regionalism: Some Current Issues. In *Capital and Class* 6, 106–25.

—— 1979: In What Sense a Regional Problem? In *Regional Studies* 13, 233–43.

—— 1982: Enterprise Zones: A Political Issue. In *International Journal of Urban and Regional Research* 6, 3, 429–34.

—— 1984a: *Spatial Divisions of Labour: Social Structures and the Geography of Production*. London: Macmillan.

—— 1984b: Foreword. In Women and Geography Study Group of the Institute of British Geographers, *Geography and Gender*. London: Hutchinson, 11–13.

—— 1985: New Directions in Space. In Gregory, D. and Urry, J. (editors), *Social Relations and Spatial Structures*. London: Macmillan, 9–19.

—— and Allen, J. (editors) 1984: *Geography Matters!* Cambridge: Cambridge University Press/Open University Press.

—— and Batey, P. (editors) 1977: *Alternative Frameworks for Analysis*. London: Pion.

—— and Meegan, R. 1978: Industrial Restructuring versus the Cities. In *Urban Studies* 15, 273–88.

—— 1982: *The Anatomy of Job Loss: The How, Where and Why of Employment Decline*. London: Methuen.

—— (editors) 1985: *Politics and Method: Contrasting Studies in Industrial Geography*. London: Methuen.

Matrix, 1984: *Making Space: Women and the Man-Made Environment*. London: Pluto Press.

Mawby, R. 1977: Defensible Space: A Theoretical and Empirical Appraisal. In *Urban Studies* 14, 169–79.

—— 1979: *Policing the City*. Farnborough: Gower.

—— 1981: Police Practices and Crime Rates: A Case Study of a British City. In Brantingham, P.J. and Brantingham, P.L. (editors), *Environmental Criminology*. London: Sage.

—— 1986: The Geography of Crime and the Criminal Justice System: Gatekeepers as Commuters. In Evans, D. (editor), *The Geography of Crime*. Stoke-on-Trent: North Staffs. Polytechnic Occasional Paper in Geography No. 7.

Maxfield, M.G. 1984: *Fear of Crime in England and Wales*. London: HMSO.

Mayhew, H. 1862: *London Labour and the London Poor*. London: Griffin Bohn.

Mayhew, P. 1981: Crime in Public View: Surveillance and Crime Prevention. In Brantingham, P.J. and Brantingham, P.L. (editors), *Environmental Criminology*. London: Sage.

—— and Smith, L. 1985: Crime in England and Wales and Scotland: A British Crime Survey Comparison. In *British Journal of Criminology* 25, 148–59.

Meier, A. and Rudwick, E.M. 1966: *From Plantation to Ghetto*. New York: Hill & Wang.

Melling, J. (editor) 1980: *Housing, Social Policy and the State*. London: Croom Helm.

Merrett, S. 1979: *State Housing in Britain*. London: Routledge & Kegan Paul.

—— 1982: *Owner-Occupation in Britain*. London: Routledge & Kegan Paul (with F. Gray).

Merseyside County Council 1986: *Policing: An Agenda for Merseyside*. Liverpool: Merseyside County Council.

Merseyside Socialist Research Group 1980: *Merseyside in Crisis*. Manchester: Free Press.

Middleton, C. 1983: Patriarchal Exploitation and the Rise of English Capitalism. In Gamarnikov, E., Morgan, D., Purvis, J. and Taylorson, D. (editors), *Gender, Class and Work*. London: Heinemann.

Miles, R. 1982: *Racism and Migrant Labour*. London: Routledge & Kegan Paul.

—— and Phizacklea, A. 1981: Racism and Capitalist Decline. In Harloe, M. (editor), *New Perspectives in Urban Change and Conflict*. London: Heinemann, 80–100.

—— 1984: *White Man's Country: Racism in British Politics*. London: Pluto.

Millett, K. 1969: *Sexual Politics*. New York: Abacus.

Milner Holland Committee 1965: *Report of the Committee on Housing in Greater London*. London: HMSO.

Mitchell, B.R. 1975: *European Historical Statistics 1750–1970*. London: Macmillan.

Mollenkopf, J. 1981: Community and Accumulation. In Dear, M. and Scott, M. (editors), *Urbanisation and Urban Planning in Capitalist Society*. London: Methuen.

Moore, B. and Rhodes, J. 1973: Evaluating the Effects of British Regional Economic Policy. In *Economic Journal* 83, 87–110.

—— and Tyler, P. 1977: The Impact of Regional Policy in the 1970s. In *CES Review* 1, 67–77.

Morgan, K.O. 1982: *Rebirth of A Nation: Wales 1880–1980*. Oxford: Clarendon Press/University of Wales Press.

Morgan, K. and Sayer, A. 1985: A 'Modern' Industry in a 'Mature' Region: The Redrafting of Management: Labour Relations. In *International Journal of Urban and Regional Research* 9, 3, 383–404.

Morrill, R.L. 1965: The Negro Ghetto: Problems and Alternatives. In *Geographical Review* 55, 339–61.

—— and Donaldson, O.F. 1972: Geographical Perspectives on the History of Black America. In *Economic Geography* 48, 1, 1–23.

Morris, R.N. 1968: *Urban Sociology*. London: Allen & Unwin.

Morton, P. 1971: 'A Woman's Work is Never Done', or the Production, Maintenance and Reproduction of Labour Power. In Albach, E. (editor), *From Feminism to Liberation*. Cambridge, Mass.: Schenkman.

Moseley, M.J. (editor) 1978: *Social Issues in Rural Norfolk*. Norwich: University of East Anglia Centre for East Anglian Studies.

—— 1979a: *Accessibility: The Rural Challenge*. London: Methuen.
—— 1979b: Rural Mobility and Accessibility. In Shaw, J.M. (editor), *Rural Deprivation and Planning*. Norwich: Geo Abstracts.
—— and Spencer, M.B. 1978: Access to Shops: The Situation in Rural Norfolk. In .Moseley, M.J. (editor), *Social Issues in Rural Norfolk*. Norwich: University of East Anglia Centre for East Anglian Studies.
Muir, R..1970 (reprint): *A History of Liverpool*. London: Williams & Norgate.
Munters, Q.J. 1982: Ferdinand Tonnies and Contemporary Rural Sociology. In *Sociologia Ruralis* 22, 305–10.
Murdie, R.A. 1969: The Factor Ecology of Metropolitan Toronto 1951–61. Chicago: University of Chicago Department of Geography Research Paper No. 116.
Murgatroyd, L., Savage, M., Shapiro, D., Urry, J., Walby, S., Warde, A. and Lawson, M.J. 1985: *Localities, Class and Gender*. London: Pion.
—— and Urry, J. 1985: The Class and Gender Restructuring of the Lancaster Economy 1950–80. In Murgatroyd, L., Savage, M., Shapiro, D., Urry, J., Walby, S., Warde, A. and Lawson, M.J., *Localities, Class and Gender*. London: Pion, 30–53.
Muth, R.F. 1969: *Cities and Housing*. Chicago: University of Chicago Press.
Myrdal, G. 1944: *An American Dilemma*. New York: Harper.
—— 1957: *Economic Theory and Underdeveloped Regions*. New York: Methuen.
National Advisory Committee for Civil Disorders 1967: *Report of the Kerner Commission*. New York: Bantam.
Nationwide Building Society 1986: *Lending to Women - Nationwide*. London: Nationwide Building Society.
—— 1987: *House Prices in 1986*. London: Nationwide Building Society.
Nationwide Anglia Building Society 1988: *Lending to Women*. London: Nationwide Anglia Building Society.
Nettler, G. 1984: *Explaining Crime*. New York: McGraw-Hill.
Newby, H. 1977: *The Deferential Worker*. Harmondsworth: Penguin.
—— 1979: *Green and Pleasant Land? Social Change in Rural England*. London: Hutchinson.
—— 1980: Trend Report: Rural Sociology. In *Current Sociology* 28, 1–141.
—— 1986: Locality and Rurality: The Restructuring of Rural Social Relations. In *Regional Studies* 20, 209–15.
Newman, O. 1972: *Defensible Space*. New York: Macmillan.
Newman, W.M. 1973: *American Pluralism: A Study of Minority Groups and Social Theory*. New York: Harper & Row.
Nisbet, R.A. 1970: *The Quest for Community*. London: Oxford University Press.
Nutley, S.D. 1980: Accessibility, Mobility and Transport-Related Welfare: The Case of Rural Wales. In *Geoforum* 11, 335–52.
—— 1983: *Transport Policy Appraisal and Personal Mobility in Rural Wales*. Norwich: Geo Books.
Oakley, A. 1974: *The Sociology of Housework*. London: Martin Robertson.
—— 1981: *Subject Women*. London: Martin Robertson.
O'Dowd, L. and Rolston, B. 1985: Bringing Hong Kong to Belfast? The Case of an Enterprise Zone. In *International Journal of Urban and Regional Research* 9, 2, 218–32.
Ossofsky, W. 1963: *Harlem: The Making of a Ghetto*. New York: Harper & Row.
Owens, S.E. 1978: Changing Accessibility in Two North Norfolk Villages, Trunch and Southrepps, from the 1850s to the 1970s. In Moseley, M.J. (editor), *Social Issues in Rural Norfolk*. Norwich: University of East Anglia Centre for East Anglian Studies.
Pacione, M. 1984: Neighbourhood Communities in Glasgow. In *Scottish Geographical Magazine*, 169–81.
Packman, J. 1979: Rural Employment: Problems and Planning. In Shaw, J.M. (editor), *Rural Deprivation and Planning*. Norwich: Geo Abstracts.
Pahl, R.E. 1965a: Trends in Social Geography. In Chorley, R.J. and Haggett, P. (editors),

Frontiers in Geographical Teaching. London: Methuen, 81–100.

—— 1965b: *Urbs in Rure: Urban Influences on Rural Areas within the London Metropolitan Region.* London: London School of Economics Geography Papers No. 2.

—— 1968: The Rural: Urban Continuum. In Pahl, R.E. (editor), *Readings in Urban Sociology.* London: Pergamon.

—— 1969: Urban Social Theory and Research. In *Environment and Planning A* 1, 143–53.

—— 1970: *Patterns of Urban Life.* London: Longman.

—— 1975: *Whose City?* Harmondsworth: Penguin.

—— 1979: Socio-Political Factors in Resource Allocation. In Herbert, D.T. and Smith, D.M. (editors), *Social Problems and the City.* London: Oxford University Press.

—— 1980: Employment, Work, and the Domestic Division of Labour. In *International Journal of Urban and Regional Research* 4, 1, 1–20.

—— 1985: The Restructuring of Capital, the Local Political Economy and Household Work Strategies. In Gregory, D. and Urry, J. (editors), *Social Relations and Spatial Structures.* London: Macmillan, 242–64.

Park, R.E. 1926: The Urban Community as a Spatial Pattern and a Moral Order. In Burgess, E. (editor), *The Urban Community.* Chicago: University of Chicago Press, 3–18.

—— 1952: Human Communities: The City and Human Ecology. In Hughes, E. (editor), *The Collected Writings of Robert Park.* New York: Free Press of Glencoe.

——, Burgess, E.W. and McKenzie, R.D. 1925: *The City.* Chicago: University of Chicago Press.

Parkin, F. 1979: *Marxism and Class Theory: A Bourgeois Critique.* London: Tavistock.

Parmar, P. 1982: Gender, Race and Class: Asian Women in Resistance. In Centre for Contemporary Cultural Studies, *The Empire Strikes Back: Race and Racism in 70s Britain.* London: Hutchinson.

Patterson, S. 1963: *Dark Strangers.* London: Tavistock.

—— 1968: *Immigrants in Industry.* London: Oxford University Press.

Peach, C. 1968: *West Indian Migration to Britain: A Social Geography.* London: Oxford University Press.

—— 1975a: Introduction: The Spatial Analysis of Ethnicity and Class. In Peach C. (editor), *Urban Social Segregation.* London: Longman, 1–17.

—— 1975b: Immigrants in the Inner City. In *Geographical Journal* 141, 3, 372–9.

—— 1983: Ethnicity. In Pacione, M. (editor), *Progress in Urban Geography.* London: Croom Helm, 103–27.

——, Robinson, V. and Smith, S. (editors) 1981: *Ethnic Segregation in Cities.* London: Croom Helm.

Pearson, G. 1978: Goths and Vandals: Crime in History. In *Contemporary Crises* 2, 119–39.

Peet, R. 1975: The Geography of Crime: A Political Critique. In *Professional Geographer* 27, 277–80.

Perry, R. 1982: The Role of the Small Manufacturing Business in Cornwall's Economic Development. In Shaw, G. and Williams, A.M. (editors), *Economic Development and Policy in Cornwall.* Exeter: South-West Papers in Geography No. 2.

Phillips, D. and Williams, A. 1984: *Rural Britain: A Social Geography.* Oxford: Blackwell.

Phillips, D.A. 1986: What Price Equality? A Report on the Allocation of GLC Housing in Tower Hamlets. In *GLC Housing and Research Policy Report No. 9,* London: Greater London Council.

Phillips, P.D. 1980: Characteristics and Typology of the Journey to Crime. In Georges-Abeyie, D. and Harries, K.D. (editors), *Crime: A Spatial Perspective.* New York: Columbia.

Pickvance, C. 1977: Explaining State Intervention. Paper presented to the CES

Conference 'Urban Change and Conflict', University of York.
—— 1981: Policies as Chameleons: An Interpretation of Regional Policy and Office Policy in Britain. In Dear, M. and Scott, M. (editors), *Urbanisation and Urban Planning in Capitalist Society*. London: Methuen, 231–65.
Pinch, S. 1984: Inequalities in Pre-School Provision: A Geographical Perspective. In Kirby, A., Knox, P.L. and Pinch, S., *Public Service Provision and Urban Development*. London: Croom Helm, 231–82.
—— 1985: *Cities and Services*. London: Routledge & Kegan Paul.
Platt, T. 1981: 'Street' Crime: A View from the Left. In Platt, T. and Takagi, P. (editors), *Crime and Social Justice*. New York: Macmillan.
—— and Takagi, P. 1981: Intellectuals for Law and Order: A Critique of the New 'Realists'. In Platt, T. and Takagi, P. (editors), *Crime and Social Justice*. New York: Macmillan.
Poulantzas, N. 1975: *Classes in Contemporary Capitalism*. London: New Left Books.
Prais, S.J. 1976: *The Evolution of Giant Firms in Britain*. Cambridge: Cambridge University Press.
Preteceille, E. 1981: Collective Consumption, the State and the Crisis of Capitalist Society. In Harloe, M. and Lebas, E. (editors), *City, Class and Capital*. London: Edward Arnold.
Queen, S.A. and Carpenter, D.B. 1953: *The American City*. New York: McGraw-Hill.
Quinney, R. 1977: *Class, State and Crime*. London: Longman.
Raban, C. 1986: The Municipalities and the Future of the Welfare State. In Lawless, P. and Raban, C. (editors), *The Contemporary British City*. London: Harper & Row, 163–9.
Raban, J. 1975: *Soft City*. London: Fontana.
Race Today Collective 1979: New Perspectives on the Asian Struggle. In *Race Today* 11, November/December, 103–9.
Radford, E. 1970: *The New Villagers: Urban Pressure on Rural Areas in Worcestershire*. London: Frank Cass.
Ramsøy, N.R. 1966: Assortive Mating and the Structure of Cities. In *American Sociological Review* 31, 773–86.
Rattansi, A. 1982: Marx and the Abolition of the Division of Labour. In Giddens, A. and MacKenzie, G. (editors), *Social Class and the Division of Labour*. Cambridge: Cambridge University Press, 12–28.
Rees, A.D. 1950: *Life in a Welsh Countryside*. Cardiff: University of Wales Press.
Rees, P.H. 1970: The Factorial Ecology of Metropolitan Chicago, 1960. In Berry, B.J.L. and Horton, F.W. (editors), *Geographical Perspectives on Urban Systems*. Englewood Cliffs, NJ: Prentice-Hall.
Rex, J. 1968: The Sociology of the Zone of Transition. In Pahl, R. (editor), Readings in Urban Sociology. Oxford: Pergamon, 211–31.
—— 1970: *Race Relations in Sociological Theory*. London: Weidenfeld & Nicolson.
—— 1973: *Race, Colonialism and the City*. London: Routledge & Kegan Paul.
—— 1982a: West Indian and Asian Youth. In Cashmore, E. and Troyna, B. (editors), *Black Youth in Crisis*. London: Allen & Unwin.
—— 1982b: The 1981 Urban Riots in Britain. In *The International Journal of Urban and Regional Research* 6, 1, 88–113.
—— and Moore, R. 1967: *Race, Community and Conflict: A Study in Sparkbrook*. London: Oxford University Press.
—— and Tomlinson, S. (editors) 1979: *Colonial Immigrants in a British City*. London: Routledge & Kegan Paul.
Rich, D.C. 1980: Locational Disadvantage and the Regional Problem: Manufacturing Industry in Scotland 1961–71. In *Regional Studies* 14, 399–417.
Robinson, V. 1979: Contrasts Between Asian and White Housing Choice. In *New Community* 7, 195–201.
—— 1980: Asians and Council Housing. In *Urban Studies* 17, 323–31.

—— 1986: *Transients, Settlers and Refugees*. Oxford: Clarendon Press.

Robson, B.T. 1969: *Urban Analysis*. London: Cambridge University Press.

Rogers, A.W. 1976: Rural Housing. In Cherry, G.E. (editor), *Rural Planning Problems*. London: Leonard Hill.

—— 1983: Housing. In Pacione, M. (editor), *Progress in Rural Geography*. London: Croom Helm, 106–29.

Rose, D., Saunders, P., Newby, H. and Bell, C. 1979: The Economic and Political Basis of Rural Deprivation: A Case Study. In Shaw, J.M. (editor), *Rural Deprivation and Planning*. Norwich: Geo Books.

Rose, H.M. 1969: Social Processes in the City: Race and Urban Residential Choice. *Commission on College Geography Resource Paper No. 6*. Washington, DC: Association of American Geographers.

—— 1970: The Development of an Urban Sub-System: The Case of the Negro Ghetto. In *Annals, Association of American Geographers* 60, 1, 1–17.

—— 1972: The Spatial Development of Black Residential Sub-Systems. In *Economic Geography* 48, 43–66.

Rosentraub, M. and Taebel, D. 1980: Jewish Enterprise in Transition. In Cummings, S. (editor), *Self-Help in Urban America*. Port Washington: Kennikat.

Rowbotham, S. 1973: *Hidden From History: 300 Years of Women's Oppression and the Fight Against It*. London: Pluto Press.

Royal Commission on the Distribution of the Industrial Population 1940: *Report of the Royal Commission on the Distribution of the Industrial Population* (The Barlow Report). London: HMSO.

Runciman, W. 1966: *Relative Deprivation and Social Justice*. London: Routledge & Kegan Paul.

Runnymede Trust and the Radical Statistics Race Group 1980: *Britain's Black Population*. London: Heinemann.

Salaman, G. 1981: *Class and the Corporation*. London: Fontana.

—— 1982: Managing the Frontier of Control. In Giddens, A. and MacKenzie, G. (editors), *Social Class and the Division of Labour*. Cambridge: Cambridge University Press.

Santos, M.M. 1979: *The Shared Space*. London: Methuen.

Sargent, L. (editor) 1981: *The Unhappy Marriage of Marxism and Feminism*. London: Pluto Press.

Sampson, R.J. and Castellano, T.C. 1982: Economic Inequality and Personal Victimisation: An Areal Perspective. In *British Journal of Criminology* 22, 363–85.

Saunders, P. 1979: *Urban Politics*. London: Hutchinson.

—— 1981: *Social Theory and the Urban Question*. London: Hutchinson.

—— 1984: Beyond Housing Classes: The Sociological Significance of Private Property Rights in the Means of Consumption. In *International Journal of Urban and Regional Research* 8, 202–27.

Saville, J. 1957: *Rural Depopulation in England and Wales 1851–1951*. London: Routledge & Kegan Paul.

Scarman, Lord 1981: *The Brixton Disorders, 10–12 April 1981*. Command 8427. London: HMSO.

Scase, R. 1982: The Petty Bourgeoisie and Modern Capitalism: A Consideration of Recent Theories. In Giddens, A. and MacKenzie, G. (editors), *Social Class and the Division of Labour*. Cambridge: Cambridge University Press, 148–61.

—— and Goffee, R. 1980: *The Real World of the Small Businessman*. London: Croom Helm.

Scherer, J. 1972: *Contemporary Community*. London: Tavistock.

Schmid, C.F. 1960: Urban Crime Areas. In *American Sociological Review* 25, 527–54 and 655–78.

Scott, P. 1972: The Spatial Analysis of Crime and Delinquency. In *Australian Geographical Studies* 10, 1–18.

Scraton, P. 1982: Policing and Institutionalised Racism on Merseyside. In Cowell, D.,
    Jones, T. and Young, J. (editors), *Policing the Riots*. London: Junction Books.
Seabrook, J. 1982: *Unemployment*. London: Quartet.
—— 1984: *The Idea of Neighbourhood: What Local Politics Should be About*. London:
    Pluto Press.
Seager, J. and Olson, A. 1986: *Women in the World: An International Atlas*. London:
    Pluto.
Secchi B. 1977: Central and Peripheral Regions in the Process of Economic Development:
    The Italian case. In Massey, D. and Batey, P. (editors), *Alternative Frameworks for
    Analysis*. London: Pion, 36–51.
Seers, D. 1979: The Periphery of Europe. In Seers, D., Schaffer, B. and Kiljunen, M-L.
    (editors), *Underdeveloped Europe: Studies in Core: Periphery Relations*. Sussex:
    Harvester, 3–34.
——, Schaffer, B. and Kiljunen, M-L. (editors) 1979: *Underdeveloped Europe: Studies in
    Core: Periphery Relations*. Sussex: Harvester.
Sennett, R. 1977: *The Fall of Public Man*. Cambridge: Cambridge University Press.
Shaw, C.R. and McKay, H.D. 1942: *Juvenile Delinquency and Urban Areas*. Chicago:
    University of Chicago Press.
Shaw, J.M. (editor) 1979: *Rural Deprivation and Planning*. Norwich: Geo Books.
Sheffield City Council 1985: *Sheffield Jobs Audit*. London: Sheffield City Council
    Department of Employment and Economic Development.
Sheppard, J. 1962: Rural Population Changes since 1851: Three Sample Studies. In
    Sociological Review 10, 81–95.
Shevky, E. and Williams, M. 1949: *The Social Areas of Los Angeles*. Los Angeles:
    University of California Press.
Shin, Y., Jedlicka, D. and Lee, E. 1977: Homicide Among Blacks. In *Phylon* 38,
    398–407.
Short, J.R. 1982: *Housing in Britain*. London: Macmillan.
Showler, B. and Sinfield, A. (editors) 1982: *The Workless State*. London: Martin
    Robertson.
Simmie, J. 1985: The Spatial Division of Labour in London 1978–81. In *International
    Journal of Urban and Regional Research* 9, 557–69.
Sims, R. 1981: Spatial Separation Between Asian Religious Minorities: An Aid to
    Explanation or Obfuscation? In Jackson, P. and Smith, S. (editors), *Social
    Interaction and Ethnic Segregation*. London: Academic Press, 123–36.
Sivanandan, A. 1976: Race, Class and the State: The Black Experience in Britain. *Race
    and Class Pamphlet No. 1*. London: Institute of Race Relations.
—— 1978: From Immigration Control to 'Induced Repatriation'. In Sivanandan, A.
    (editor), *A Different Hunger*. London: Pluto.
—— 1981/2: From Resistance to Rebellion: Asian and Afro-Caribbean Struggles in
    Britain. In *Race and Class* 23, 111–52.
—— 1985: Britain's Gulags. In *New Socialist* 32, November, 13–15.
Skellington, R. 1980: Council House Allocation in a Multi-Racial Town. *Faculty of Social
    Sciences Occasional Paper No. 2*. Milton Keynes: Open University.
Smith, D. (editor) 1980: *People and a Proletariat: Essays in the History of Wales
    1780–1980*. London: Pluto Press.
Smith, D.J. 1976: *The Facts of Racial Disadvantage: A National Survey*. London: Political
    and Economic Planning.
Smith, D.M. 1974: Who Gets What Where and How: A Welfare Focus for Human
    Geography. In *Geography* 59, 289–97.
—— 1977: *Human Geography: A Welfare Approach*. London: Edward Arnold.
—— 1979: *Where the Grass is Greener: Living in an Unequal World*. Harmondsworth:
    Pelican.
Smith, J. and Gant, R. 1981: Transport Provision and Rural Change: A Case Study from
    the Cotswolds. In Whitelegg J. (editor), *The Spirit and Purpose of Transport*

Geography. Lancaster: University of Lancaster Department of Geography for the Transport Geography Study Group, Institute of British Geographers.

Smith, M.P. 1980: *The City and Social Theory*. Oxford: Basil Blackwell.

Smith, P. 1978: Domestic Labour and Marx's Theory of Value. In Wolpe, A. and Kuhn, A. (editors), *Feminism and Materialism*. London: Routledge & Kegan Paul.

Smith, S.J. 1982: Victimisation in the Inner City: A British Case Study. In *British Journal of Criminology* 22, 386–402.

—— 1986: Social and Spatial Aspects of the Fear of Crime. In Evans, D. (editor) *The Geography of Crime*. Stoke-on-Trent: North Staffs. Polytechnic Occasional Paper in Geography No. 7.

—— 1987: Fear of Crime: Beyond a Geography of Deviance. In *Progress in Human Geography* 11, 1, 1–23.

Solomos, J., Findlay, B., Jones, S. and Gilroy, P. 1982: The Organic Crisis of British Capitalism and Race: The Experience of the Seventies. In Centre for Contemporary Cultural Studies, *The Empire Strikes Back: Race and Racism in 70s Britain*. London: Hutchinson.

Spear, A. 1967: *Black Chicago: The Making of a Negro Ghetto 1890-1920*. Chicago: University of Chicago Press.

Spender, D. 1982: *Invisible Women: The Schooling Scandal*. London: Writers' and Readers' Publishing Group.

Stacey, M. 1969: The Myth of Community Studies. In *British Journal of Sociology* 20, 134–47.

—— 1981: The Division of Labour Revisited, or Overcoming the 'Two Adams'. In Abrams, P., Deem, R., Finch, J. and Rock, P. (editors), *Practice and Progress: British Sociology 1950-80*. London: Allen & Unwin.

—— 1986: Gender and Stratification: One Central Issue or Two? In Crompton, R. and Mann, M. (editors), *Gender and Stratification*. London: Polity Press.

Stamp, D. 1949: The Planning of Land Use. In *Advancement of Science* 6, 224–33.

Stanworth, M. 1983: *Gender and Schooling*. London: Hutchinson.

Stedman Jones, G. 1971: *Outcast London: A Study in the Relationship between Classes in Victorian Society*. Oxford: Clarendon Press.

—— 1983: *Languages of Class: Studies in English Working Class History 1832-1982*. Cambridge: Cambridge University Press.

Stein, M. 1964: *The Eclipse of Community*. New York: Harper & Row.

St John Thomas, D. 1960: *Rural Transport: A Report*. London: Routledge & Kegan Paul.

Tabb, W.K. 1975: *The Political Economy of the Black Ghetto*. New York: Norton.

—— and Sawyers, L. (editors) 1978: *Marxism and the Metropolis*. New York: Oxford University Press.

Taeuber, K.E. and Taeuber, A.F. 1965: *Negroes in Cities*. Chicago: Aldine.

Takagi, P. 1977: The Management of Police Killings. In *Crime and Social Justice* 8, 34–43.

Tawney, R.H. 1931: *Inequality*. London: Allen & Unwin.

Taylor, I., Walton, P. and Young, J. 1981: *The New Criminology: For a Social Theory of Deviance*. London: Routledge & Kegan Paul.

Taylor, P. and Hadfield, H. 1982: Housing and the State: A Case Study and Structuralist Interpretation. In Cox, K.R. and Johnston, R.J. (editors), *Conflict, Politics and the Urban Scene*. London: Longman, 241–63.

—— and Johnston, R.J. 1984: The Geography of the British State. In Short, J. and Kirby, A. (editors), *The Human Geography of Contemporary Britain*. London: Macmillan, 40–51.

Thomas, C. and Winyard, S. 1979: Rural Incomes. In Shaw, J.M. (editor), *Rural Deprivation and Planning*. Norwich: Geo Books.

Thompson, E.P. 1963: *The Making of the English Working Class*. London: Gollancz.

Timms, D.W.G. 1965: The Spatial Distribution of Social Deviants in Luton, England. In *Australian and New Zealand Journal of Sociology* 1, 38–52.

—— 1971: *The Urban Mosaic*. London: Cambridge University Press.

Titmuss, R.M. 1968: *Commitment to Welfare*. London: Allen & Unwin.

Tittle, C.R., Villemez, W.J. and Smith, D.A. 1978: The Myth of Social Class and Criminality: An Empirical Assessment of Empirical Evidence. In *American Sociological Review* 43, 643–56.

Tivers, J. 1985: *Women Attached: The Daily Lives of Women with Young Children*. London: Croom Helm.

Toby, J. 1980: The New Criminology is the Old Baloney. In Inciardi, J. (editor), *Radical Criminology: The Coming Crises*. New York: Sage.

Todd, D. 1983: Observations on the Relevance of the Industrial-Urban Hypothesis for Rural Development. In *Geoforum* 14, 45–54.

Toffler, A. 1970: *Future Shock*. London: Bodley Head Press.

Tonnies, F. 1887 (reprinted 1963): *Community and Association*. New York: Harper & Row.

Townroe, P. 1975: Branch Plants and Regional Development. In *Town Planning Review* 46, 1, 47–62.

Townsend, A. 1980: Unemployment and the Government's New 'Regional' Aid. In *Area* 12, 1, 47–62.

—— 1983: *The Impact of Recession*. London: Croom Helm.

——, Smith, C. and Johnson, M. 1978: Employees' Experiences of North-east England: Survey Evidence on Some Implications of British Regional Policy. In *Environment and Planning A* 10, 1345–62.

Townsend, P. 1979: *Poverty in the United Kingdom*. Harmondsworth: Penguin.

—— and Davidson, N. 1982: *Inequalities in Health*. Harmondsworth: Penguin.

——, Phillimore, P. and Beattie, A. 1988: *Health and Deprivation: Inequality and the North*. London: Croom Helm.

United States Law Enforcement Assistance Agency 1975: *Criminal Victimisation Surveys in the Nation's Five Largest Cities*. Washington, DC: United States Government Publishing Office.

Urry, J. 1984: Capitalist Restructuring, Recomposition and the Regions. In Bradley, T. and Lowe, P. (editors), *Locality and Rurality*. Norwich: Geo Books.

Van Amserfoort, H. 1978: Minority as a Sociological Concept. In *Ethnic and Racial Studies* 1, 2, 218–34.

Van Den Berghe, P.L. 1978: *Race and Racism*. New York: Wiley.

Van Den Haag, E. 1975: *Punishing Criminals*. New York: Basic Books.

Van Valey, T.L., Roof, W.C. and Wilcox, J.E. 1977: Trends in Residential Segregation 1960–70. In *American Journal of Sociology* 82, 826–44.

Vogel, L. 1981: Marxism and Feminism, the Unhappy Marriage: Trial Separation or Something Else? In Sargent, L. (editor), *The Unhappy Marriage of Marxism and Feminism*. London: Pluto Press.

—— 1983: *Marxism and the Oppression of Women*. London: Pluto.

Wabe, J.S. 1986: The Regional Impact of De-industrialisation in the European Community. In *Regional Studies* 20, 1, 27–36.

Walby, S. 1986: Gender, Class and Stratification: Towards a New Approach. In Crompton, R. and Mann, M. (editors), *Gender and Stratification*. London: Polity Press.

Walker, R. 1985: Class, Division of Labour and Employment in Space. In Gregory, D. and Urry, J. (editors), *Social Relations and Spatial Structures*. London: Macmillan, 164–89.

Wallace, D.B. and Drudy, P.J. 1975: *Social Problems of Rural Communities*. Newcastle: University of Newcastle upon Tyne Agricultural Adjustment Unit.

Waller, P.J. 1981/2: The Riots in Toxteth, Liverpool: A Survey. In *New Community* 9, 344–53.

Wallman, S. 1982: *Living in South London: Perspectives on Battersea 1971–81*. Aldershot: Gower.

Walvin, J. 1973: *Black and White: The Negro and English Society 1555-1945*. London: Allen Lane.

Wannop, U.A. 1986: Glasgow/Clydeside: A Century of Metropolitan Evolution. In Gordon, G. (editor), *Regional Cities in the United Kingdom 1890-1980*. London: Harper & Row, 83-100. .

Ward, C. 1974: *Tenants Take Over*. London: Architectural Press.

Ward, R. 1982: Race, Housing and Wealth. In *New Community* 10, 1, 3-15.

—— and Jenkins, R. (editors) 1984: *Ethnic Communities in Business*. Cambridge: Cambridge University Press.

—— and Reeves, F. 1980: West Indians in Business in Britain. Memorandum to the Home Affairs Committee (Race and Immigration sub-Committee), 15 December. London: House of Commons.

Wardle, A. 1985: Spatial Change, Policies, and the Division of Labour. In Gregory, D. and Urry, J. (editors), *Social Relations and Spatial Structures*. London: Macmillan, 190-212.

Warr, M. 1984: Fear of Victimisation: Why are Women and the Elderly More Afraid? In *Social Science Quarterly* 65, 691-702.

Watkins, R. 1979: Educational Disadvantage in Rural Areas. In Shaw, J.M. (editor), *Rural Deprivation and Planning*. Norwich: Geo Books.

Watts, H.D. 1981: *The Branch Plant Economy: A Study of External Control*. London: Longman.

Weaver, R. 1948: *The Negro Ghetto*. New York: no publisher cited.

Webber, M. 1963: Order in Diversity: Community Without Propinquity. In Wingo, L. (editor), *Cities and Space*. Baltimore: Johns Hopkins University Press.

—— 1964: The Urban Place and the Non-Place Urban Realm. In Webber, M. *et al* (editors), *Explorations into Urban Structure*. Philadelphia: University of Pennsylvania Press.

Weber, A. 1909: *Theory of the Location of Industries*. Cambridge, Mass.: Harvard University Press (1926 translation).

Weber, M. 1947: *The Theory of Social and Economic Organisation*. Glencoe, Illinois: Free Press.

Weclawowicz, G. 1979: The Structure of Socio-Economic Space in Warsaw 1931 and 1970: A Study in Factorial Ecology. In French, R.A. and Hamilton, F.E.I. (editors), *The Socialist City*. London: Wiley, 378-423.

Wellman, B. and Leighton, B. 1979: Networks, Neighbourhoods and Communities: Approaches to the Study of the Community Question. In *Urban Affairs Quarterly* 13, 363-90.

West, J. 1980: A Political Economy of the Family in Capitalism: Women, Reproduction and Wage Labour. In Nichols, T. (editor), *Capital and Labour: A Marxist Primer*. London: Fontana.

Westergaard, J. and Resler, H. 1975: *Class in a Capitalist Society: A Study of Contemporary Britain*. London: Heinemann.

Wibberley, G.P. 1954: Problem Rural Areas in Great Britain. In *Geographical Journal*, 120, 43-58.

—— 1978: Mobility and the Countryside. In Cresswell, R. (editor), *Rural Transport and Country Planning*. London: Leonard Hill.

Wiles, P. 1975: Criminal Statistics and Sociological Explanations of Crime. In Carson, W. and Wiles, P. (editors), *The Sociology of Crime and Delinquency in Britain: Volume One*. London: Martin Robertson.

Williams, G. 1980: Locating a Welsh Working Class: The Frontier Years. In Smith, D. (editor), *People and a Proletariat: Essays in the History of Wales 1780-1980*. London: Pluto Press, 16-46.

—— 1984: Women Workers in Contemporary Wales 1962-82. In *Welsh History Review* 11, 4.

—— 1985: *When Was Wales? A History of the Welsh*. Harmondsworth: Pelican.

Williams, N., Sewel, J. and Twine, F. 1986: Council House Allocation and Tenant Incomes. In *Area* 18, 131–40.

Williams, P. 1976: The Role of Institutions in the Inner London Housing Market: The Case of Islington. In *Transactions, Institute of British Geographers*, NS 1, 72–82.

—— 1978: Building Societies and the Inner City. In *Transactions, Institute of British Geographers*, NS 3, 1, 23–34.

Williams, R. 1973: *The Country and the City*. London: Chatto & Windus.

Williams, W.M. 1964: *A West Country Village: Ashworthy*. London: Routledge & Kegan Paul.

Wilson, F.O. 1979: *Residential Consumption, Economic Opportunity and Race*. New York: Academic Press.

Winchester, S.W.C. 1974: Immigrant Areas in Coventry in 1971. In *New Community* 4, 97–104.

Wirth, L. 1938: Urbanism as a Way of Life. In *American Journal of Sociology* 44, 1–24.

—— 1939: Localism, Regionalism and Centralisation. In *American Journal of Sociology*, 44.

—— 1945: The Problem of Minority Groups. In Linton, R. (editor), *The Science of Man in the World Crisis*. New York: Columbia University Press.

Wolpert, J. 1965: Behavioural Aspects of the Decision to Migrate. In *Papers and Proceedings of the Regional Science Association* 15, 159–69.

Worthington, A. 1982: Why Local Government Should Encourage Community Development. In *Community Development Journal* 17, 147–54.

Women and Geography Study Group of the Institute of British Geographers 1984: *Geography and Gender*. London: Hutchinson.

Wright, S. 1983: Pigeon-Holed Policies; Agriculture, Employment and Industrial Development in a Lincolnshire Case Study. In *Sociologia Ruralis* 23, 242–60.

Wright Mills, C. 1956: *White-Collar: The American Middle Class*. New York: Oxford University Press.

Wrigley, E.A. 1962: *Industrial Growth and Population Change: A Regional Study of the Coalfield Areas of New Europe in the Late Nineteenth Century*. Cambridge: Cambridge University Press.

Young, I. 1981: Beyond the Unhappy Marriage: A Critique of Dual Systems Theory. In Sargent, L. (editor), *The Unhappy Marriage of Marxism and Feminism*. London: Pluto Press.

Young, J. 1979: Left Idealism, Reformism and Beyond: From New Criminology to Marxism. In Fine, B., Kinsey, R., Lea, J., Picciotto, S. and Young, J. (editors), *Capitalism and the Rule of Law: From Deviancy Theory to Marxism*. London: Hutchinson.

Young, M. and Wilmott, P. 1957: *Family and Kinship in East London*. London: Routledge & Kegan Paul.

Zukin, S. (editor) 1985: *Industrial Policy: Business and Politics in the United States and France*. New York: Praeger.

Author Index

Subject Index